# GAMMA RADIATION

Edited by **Feriz Adrovic**

**Gamma Radiation**
Edited by Feriz Adrovic

**Published by InTech**
Janeza Trdine 9, 51000 Rijeka, Croatia



**Publishing Process Manager** Daria Nahtigal

**Technical Editor** Teodora Smiljanic

**Cover Designer** InTech Design Team

Additional hard copies can be obtained from orders@intechopen.com

Gamma Radiation, Edited by Feriz Adrovic
p. cm.
ISBN 978-953-51-0316-5

For EU product safety concerns:
IN TECH d.o.o., Prolaz Marije Krucifikse Kozulić 3, 51000 Rijeka, Croatia,
info@intechopen.com or visit our website at intechopen.com.

# Meet the editor

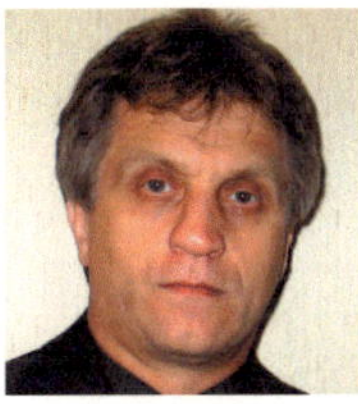

Born in 1955, Berane, Montenegro. Currently works as Associate Professor - Department of Physics, University of Tuzla, Bosnia and Herzegovina, and as a Full Professor - Faculty of Science, Kosovska Mitrovica, Kosovo. Research interests: Applied Nuclear physics, Radon, Gamma spectrometry, Radiation Protection. In very fruitful career he participated in research project of uranium prospection in Kosovo, as well as the development of new dosimeters for detecting radon. At the University of Tuzla he established a respectable scientific-research laboratory for detection, dosimetry and radiation protection. He achieved fruitful and intensive cooperation with renowned research institutions such as Institute of Nuclear Sciences Vinca, Polish Institute for Nuclear Research in Krakow, Institute of Nuclear Sciences in Dubna, Russia. Dr. Adrović attended the 17th World Conference on nuclear track detectors in Dubna 1994, IRPA 9 in Vienna 1996, Cairo International Conference on the sources, transport and disposal of hazardous materials 1998; then IRPA 11 in Madrid 2004, and IRPA 12 in Buenos Aires 2008.

# Contents

# Preface

Gamma rays were discovered by Villard more than a hundred years ago. Rutherford gave the name $\gamma$-rays to the electrically neutral radiation emitted by the naturally occurring radioactive materials.

Gamma rays are electromagnetic radiation, photons, which arises due to energy changes in the nucleus. It does not appear as an independent form of radiation, but it follows $\alpha$ and $\beta$ decay, in those cases when the newly created core of the offspring is found in the excited state. This state is very short, while the excited nucleus in its transition to a lower energy level or the primary energy state, mosty usually transmits $\gamma$- photons or transfers the energy to some of the atom electrons. The product nuclide in such an excited state either falls directly to the ground state or descends in steps to lower energy states through the dissipation of energy as gamma radiation.

So, it's the word about the deexcitation process that usually occurs immediately after excitation of the core to some of the higher energy levels. The energy of the emitted $\gamma$- photon is strictly determined by the difference of energy states of the nucleus at this transition. Therefore, the energy spectrum of $\gamma$- radiation is linear, i.e. we are talking about a discrete energy spectrum of the emitted $\gamma$- rays.

Gamma rays bombard our bodies constantly. They come from the naturally radioactive materials in rocks and the soil. We take some of these materials into our bodies from the air we breathe and the water we drink. Gamma rays can be produced in labs through the process of nuclear collision and also through the artificial radioactivity that accompanies these interactions. The high-energy nuclei needed for the collisions are accelerated by devices such as the cyclotron and synchrotron.

Gamma radiation is of interest to persons working in a wide spectrum of disciplines. These include scientists, physicians, engineers, and technicians whose work entails the preparation, utilization, or disposal of radioactive materials and the measurement of radioactivity in the environment. Among these are persons working in the fields of radiopharmacy, radiobiology, radiochemistry, industrial applications, nuclear medicine, health physics, nuclear power, nuclear waste management, etc.

Gamma radiation is the basis of gammaspectrometry, a central nuclear method for the quantitative and qualitative analysis of radioactive materials in all ambient environments. Gamma rays can be used to examine metallic castings or welds in oil pipelines for weak points. The rays pass through the metal and darken a photographic film at places opposite the weak points. In industry, gamma rays are used for detecting internal defects in metal castings and in welded structures. Gamma rays are also used in nuclear reactors and atomic bombs.

Gamma rays are used in medicine to kill and treat certain types of cancers and tumors. In radionuclide imaging diagnostics, metastable gamma emitters are most favorable for use, while the beta radiation is absorbed by the body, causing an unnecessary radiation overload. Gamma rays passing through tissue of the body produce ionization in tissue, and thus can harm the cells in our body.

Gamma rays are often used in the food industry. The radioisotopes preserve foods. Gamma radiation is a modern, efficient, economical and fast method for the commercial sterilization and microbial decontamination. Gamma radiation is used in many ways in the food industry and agriculture. In the pharmaceutical industry, gamma radiation is used for cold sterilization.

The book contains 15 chapters which are grouped into five areas: Nuclear physics, Environmental sciences, Materials science, Radiation Biology, Agriculture and Forestry.

From the most chapters of this book can be seen great progress in applying gamma radiation in certain scientific disciplines, which will undoubtedly provide new guidelines for research in these and other scientific fields. And that is the aim of this project. The book which treats such important issues, required the contribution of scientists from a wide range of scientific disciplines. I noted with satisfaction the participation of scientists from different parts of the world, who have made special contributions to the quality of the book. I believe that with such a concept it will bring benefit to scientific workers who have contacts with gamma radiation, as well as those who intend to enter the world of applications and problems of gamma radiation. I hope that this extensive research field will be continued with new contents and integrated in future releases of Intech.

I am grateful to the authors, scientific workers from all parts of our beautiful but agitated planet, who have made valuable contributions to the concept of this book, and to Ms Daria Nahtigal (Publishing Process Manager) who brilliantly coordinated activities on this project, and the visionary editorial team of Intech which promotes global scientific solidarity in best way.

**Prof. Feriz D. Adrovic**
Faculty of Science
Univerzity of Tuzla
Bosnia and Herzegovina

# Part 1

## Nuclear Physics

# 1

# Air Kerma Rate Constants for Nuclides Important to Gamma Ray Dosimetry and Practical Application

Marko M. Ninkovic[1] and Feriz Adrovic[2]
[1]Institute of Nuclear Sciences – Vinca, Belgrade,
[2]University of Tuzla, Faculty of Science, Tuzla,
[1]Serbia
[2]Bosnia and Herzegovina

## 1. Introduction

It is often necessary to estimate the exposure rate at a distance from radionuclide emitting gamma or X rays. Such calculations may be required for planning radiation protection measures around radioactive sources, for calibration radiation monitoring instruments, for patient containing radionuclides or for estimating the absorbed dose to patients receiving brachytherapy. The factor relating activity and exposure rate has been various names: the k factor (Johns, 1961), the specific gamma ray constant (ICRU Rep. 10a, 1962), exposure rate constant (Parker et al., 1978) and gamma rate constant (Kereiakes & Rosenstein, 1980). Conversion to SI units required that this factor be replaced by the air kerma rate constant $\Gamma_{\delta,}$ which is now defined as:

$$\Gamma_\delta = \frac{l^2}{A} \left( \frac{dK_{air}}{dt} \right)_\delta \tag{1}$$

where $(dK_{air}/dt)_\delta$ is the air kerma rate due to photons of energy >δ, at a distance $l$ from a point source of activity $A$. The SI unit for $\Gamma_\delta$ is J m$^2$ kg$^{-1}$ which, when the terms gray and becquirel are used, becoms Gy m$^2$ s$^{-1}$ Bq$^{-1}$.

In the process of analysing accessible data on the are kerma rate constants and its precursos for many radionuclides often used in practice (Nachtigal, 1969; Ninkovic & Mladenovic, 1970; NCRP Rep. 49, 1976; Ungar & Trabey, 1982; Aird et al., 1984; Attix, 1986; Ninkovic, 1987; Wasserman & Groenwald, 1988; Ninkovic & Raicevic, 1992,1993; Sabol & Weng, 1995; Ninkovic et al., 2005) it was concluded that published data are in strong disagreement. That is the reason we decided to recalculate this quantities on the basis of the latest data on gamma ray spectra and on the latest data for mass energy-transfer coefficients for air.

## 2. Derivation of the equation for calculation of $\Gamma_\delta$

The kerma $K_{air}$, for interaction of X-rays and gamma rays with air is given by:

$$K_{\text{air}} = \Phi \frac{\mu_k}{\rho} E \qquad (2)$$

where $\Phi$ is the flunce, $E$ the photon energy, and $\mu_\kappa/\rho$ the energy-dependent mass energy-transfer coefficient for air.

The kerma rate, d$K$ /d$t$, is obtained from the kerma by substituting the flux density $\psi$ for the fluence $\Phi$ in Equation 2:

$$\frac{dK_{air}}{dt} = \psi \frac{\mu_k}{\rho} E \qquad (3)$$

where $\psi$ is expressed in $m^{-2}\,s^{-1}$. The quantity $\psi$ is derived from the activity $A$, of a radiation source in accordance with inverse square low:

$$\psi = \frac{A}{4\pi l^2} \qquad (4)$$

By inserting Equation 4 in Equation 3, the following equation is obtained:

$$\frac{dK_{air}}{dt} = \frac{A}{4\pi l^2} \frac{\mu_k}{\rho} E \qquad (5)$$

If photons with energy $E_i$ are emitted per decay event with yield $p_i$, Equation 5 becomes:

$$\frac{dK_{air}}{dt} = \frac{A}{4\pi l^2} \sum_i \left(\frac{\mu_k}{\rho}\right)_i p_i E_i \qquad (6)$$

By inserting Equation 6 in Equation 1, the following equation is obtained for $\Gamma_\delta$:

$$\Gamma_\delta = \frac{1}{4\pi} \sum_i \left(\frac{\mu_k}{\rho}\right)_i p_i E_i \qquad (7)$$

## 3. Calculation of $\Gamma_\delta$

Starting from Equation 7, the air kerma rate constants, $\Gamma_\delta$, were calculated using data on mass energy-transfer coefficients for air (Hubbell, 1969; Hubbell & Seltzer, 2001) and data on photon emission yield in the process of decay of the radionuclides (Firestone, 1996; Stabin & Luz, 2002). The subscript $\delta$ , implies that only photons with energy $>\delta$ , in MeV are included in the calculation.

Concerning the radiation spectra emitted per decay of a radionuclide, there are three types of photons: the gamma ray photons, those characteristic X-ray photons, those from internal conversion of gamma rays and electron capture and those accompanying bremsstrahlung processes of electrons from $\beta^-$ decay and internal conversion of gamma rays and X rays. In this calculation gamma rays and characteristic X-ray photons with energies >20 keV as $\delta$ – value are only ones to have been taken into account. The contribution of bremsstrahlung radiation has not been included.

In the calculation, instead of gamma ray total transition intensities, the gamma ray intensities corrected for internal conversion of gamma rays were used.

The particular air kerma rate constants were calculated for each discrete line of the photon spectrum of the radionuclide, with effective yield per decay >0.01% and energy >20 keV. Since the energy structure of the photon spectra and accessible discrete numerical values of the mass energy-transfer coefficient for air are not the same, the cubic spline interpolation was used to calculate the coefficient , where photon spectrum data are available.

## 4. Results

### 4.1 New recalculated values of $\Gamma_\delta$

Table 1, lists recalculated air kerma rate constants for the 35 radionuclide used most often in gamma ray dosimetry and practical applications. For every radionuclide in the table are given the following data:

in column 1 the symbol of gamma-emitting nuclide,
in column 2 the half-life,
in column 3 the low- energy photon spectra limit,
in column 4 the high-energy photon spectra limit ,
in column 5 the calculated value of the constant in basic SI units, and finaly
in column 6 the calculated value of the constant in practical units (μGy $m^2$ $GBq^{-1}$ $h^{-1}$)

The last unit, for air kerma rate constant, is the practical one especially, for radiation protection and safety calculations in nuclear medicine laboratories, industrial radiography and many others applications of point gamma radiation sources.

The accuracy of calculation of air kerma rate constants is not more than three significant figures. The major portion of the standard error associated with these calculated values of $\Gamma_\delta$ arise from uncertainties in relative intensity measurements of the X ray and gamma ray photon spectra and intensity of omitted bremsstrahlung radiation.

Bremsstrahlung radiation contributes to the total air kerma rate constant by, for exam≤≤le, for $^{60}$Co, not more than 0.4%, and this decreases markedly with decreasing photon energy (BCRUM, Br.J.Rad., 55, 1982). The contribution to $\Gamma_\delta$ from the omitted photons of energies < 20 keV, varies from radionuclide to radionuclide, this is not interesting for the purposes of practical health physics, but is of interest in specific nuclear medicine radionuclide applications.

### 4.2 Examples of our previous measurements of photon spectra and calculation of $\Gamma_\delta$ for selected radionuclide

The next section of the text shows, as examle, the data of our previous measurement of the photon spectrum and the results of calculating the air kerma rate constants for the three selected radionuclides ( $^{182}$Ta, $^{192}$Ir and $^{226}$Ra in equilibrium with its decay products).

#### 4.2.1 Photon spectra and recalculated of $\Gamma_\delta$ for $^{182}$Ta radionuclide

As can be seen from Table 2, the entire of photon ray spectrum of $^{182}$Ta is divided into five characteristic groups of photon lines. The air kerma rate constant was calculated for every

| Radionuclide | Half – life | Energy interval (MeV) | | Air kerma rate constant | |
|---|---|---|---|---|---|
| | | from | to | aGy $m^{-2}$ $Bq^{-1}$ $s^{-1}$ | μGy $m^{-2}$ $GBq^{-1}$ $h^{-1}$ |
| $^{11}C$ | 20.38 min | - | 0.5110 | 38.7 | 139.3 |
| $^{13}N$ | 9.965 min | - | 0.5110 | 38.7 | 139.4 |
| $^{15}O$ | 2.037 min | - | 0.5110 | 38.7 | 139.5 |
| $^{18}F$ | 109.8 min | 0.0005 | 0.5110 | 37.5 | 135.1 |
| $^{24}Na$ | 14.96 h | 1.3690 | 3.8660 | 121.3 | 436.7 |
| $^{42}K$ | 12.36 h | 0.3126 | 2.4240 | 9.10 | 32.8 |
| $^{43}K$ | 22.3 h | 0.2206 | 1.3940 | 35.5 | 127.8 |
| $^{51}Cr$ | 27.70 h | 0.0005 | 0.3201 | 1.17 | 4.22 |
| $^{52}Fe$ | 8.275 h | 0.0006 | 1.0399 | 27.01 | 97.24 |
| $^{59}Fe$ | 44.50 d | 0.0069 | 1.4817 | 40.54 | 145.9 |
| $^{57}Co$ | 271.74 d | 0.0007 | 0.6924 | 3.92 | 14.11 |
| $^{58}Co$ | 70.86 d | 0.0007 | 1.6747 | 35.84 | 129.0 |
| $^{60}Co$ | 5.271 a | 1.1732 | 1.3325 | 85.82 | 309.0 |
| $^{67}Ga$ | 3.261 d | 0.0010 | 0.8877 | 5.40 | 19.45 |
| $^{68}Ga$ | 1.127 h | 0.0010 | 1.8830 | 35.84 | 129.0 |
| $^{75}Se$ | 119.79 d | 0.0013 | 0.5722 | 13.40 | 48.25 |
| $^{99}Mo$ | 65.94 h | 0.0024 | 0.9608 | 5.49 | 19.77 |
| $^{99m}Tc$ | 6.01 h | 0.0024 | 0.1426 | 3.92 | 14.10 |
| $^{111}In$ | 67.31 h | 0.0031 | 0.2454 | 23.09 | 83.13 |
| $^{113m}In$ | 99.49 min | 0.0033 | 0.3917 | 12.22 | 44.00 |
| $^{123}I$ | 13.27 h | 0.0038 | 0.7836 | 10.0 | 36.1 |
| $^{125}I$ | 59.4 d | 0.0038 | 0.0355 | 10.48 | 37.73 |
| $^{131}I$ | 8.021 d | 0.0041 | 0.7229 | 14.50 | 52.20 |
| $^{127}Xe$ | 36.4 d | 0.0039 | 0.6184 | 14.19 | 51.09 |
| $^{133}Xe$ | 5.243 d | 0.0043 | 0.1606 | 3.98 | 14.33 |
| $^{137}Cs/^{137}Ba$ | 30.04 a | 0.0045 | 0.6617 | 22.80 | 82.10 |
| $^{152}Eu$ | 13.537 a | 0.0056 | 1.7691 | 41.36 | 148.9 |
| $^{154}Eu$ | 8.593 a | 0.0061 | 1.5965 | 44.23 | 159.2 |
| $^{170}Tm$ | 128.6 d | 0.0070 | 0.0843 | 0.154 | 0.554 |
| $^{182}Ta$ | 114.43 d | 0.0084 | 1.4531 | 44.45 | 160.0 |
| $^{192}Ir$ | 73.827 d | 0.0089 | 1.0615 | 30.30 | 109.1 |
| $^{197}Hg$ | 2.672 d | 0.0097 | 0.2687 | 3.159 | 11.37 |
| $^{198}Au$ | 2.695 d | 0.0100 | 1.0877 | 15.15 | 54.54 |
| $^{201}Tl$ | 3.038 d | 0.0058 | 0.1674 | 2.84 | 10.22 |
| $^{241}Am$ | 432.2 a | 0.0139 | 0.1030 | 1.102 | 3.97 |

Table 1. Air kerma rate constant for some radionuclide considering photon energy above 20 keV

discrete photon line with yield per decay event >0.01 % and starting with energy of 0.03174 MeV as the delta value. That means that four characteristic X-ray lines are included. The group and total air kerma rate constant are obtained then by addition of partiale or single photon lines constant. Finally, a value of ( 44.8 $\pm$ 0.9 ) aGy $m^2$ $s^{-1}$ $Bq^{-1}$ for an unshielded $^{182}Ta$ source has been obtained. That value is in good agreement with a new recalculated value given in Table 1.

Bearing in mind that standard tantalum sources are usually packed into 0.1 mm of platinum, it was calculated the constant for this type of source also. For that goal, it was calculated the absorption of tantalum photons into 0.1 mm of platinum and obtained that in this way the air kerma rate constant is reduced by 4,46 %. After this correction, a value of (42.8 $\pm$ 0.9) aGy m$^2$ s$^{-1}$ Bq$^{-1}$ was obtained for air kerma rate constant for standard packaged encapsulated tantalum source (Ninkovic & Raicevic, 1992).

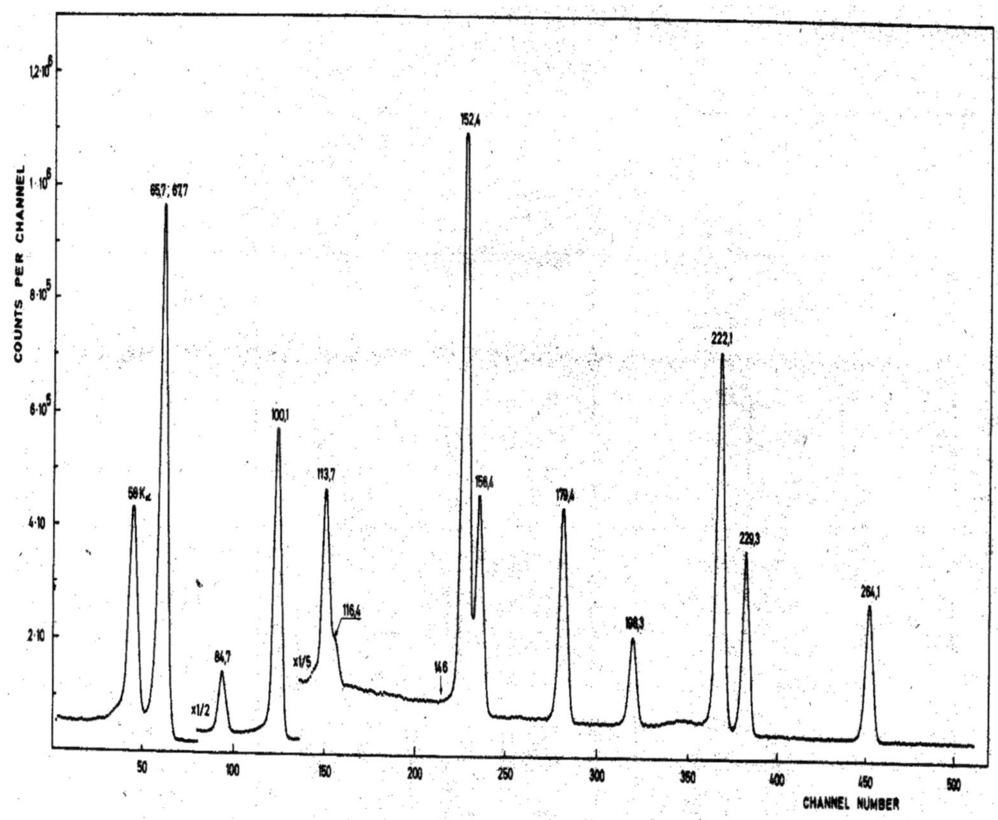


Fig. 1a. Low energy region (50-300 keV) of the photons spectrum emitted in decay of $^{182}$Ta radionuclide

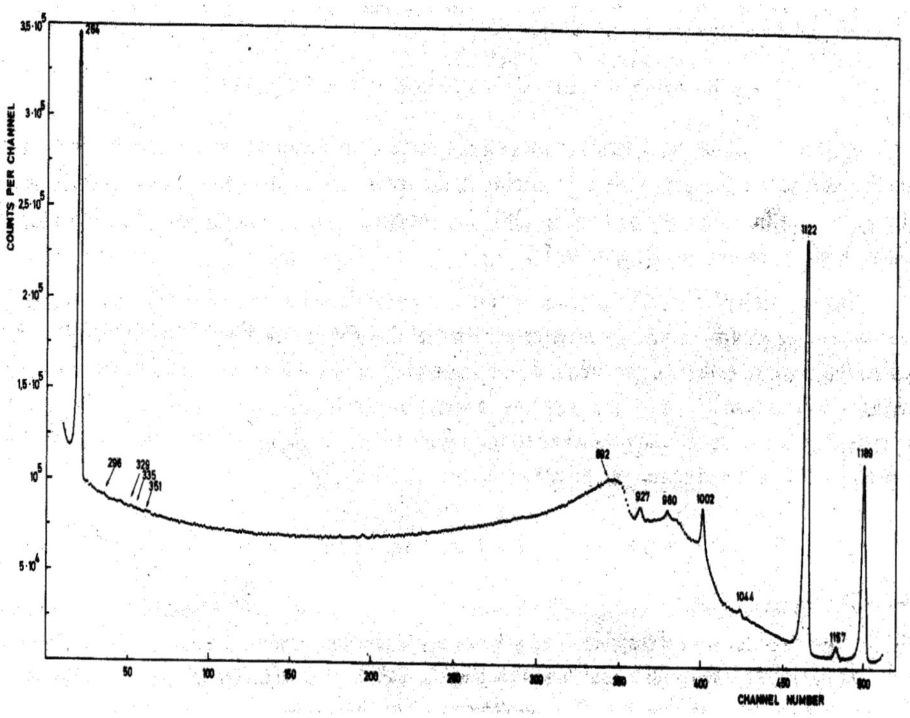


Fig. 1b. Middle energy region (250-1200 keV) of the photons spectrum emitted in decay of $^{182}$Ta radionuclide

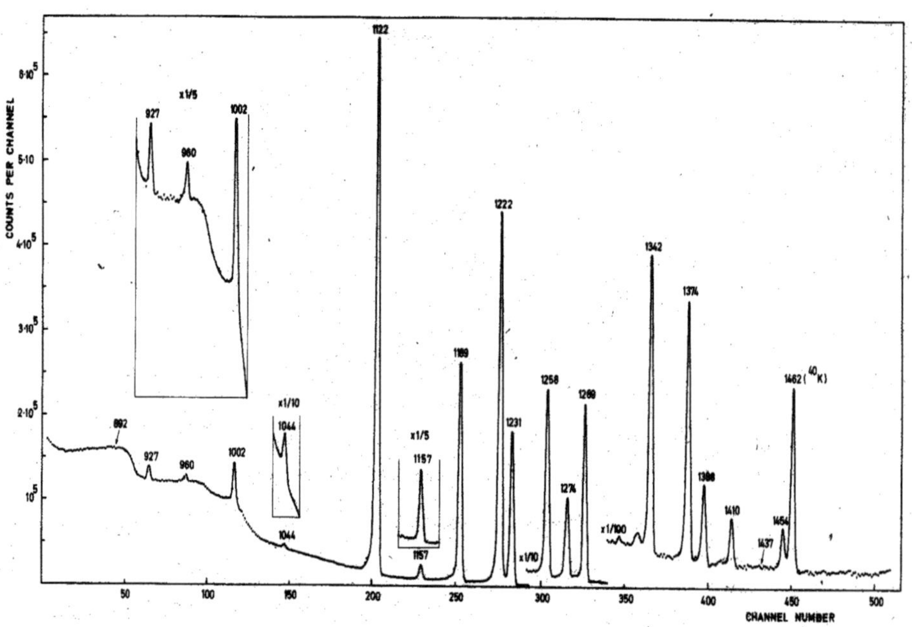


Fig. 1c. High energy region (800-1500 keV) of the photons spectrum emitted in decay of $^{182}$Ta radionuclide

| Group of lines | Energy [MeV] | Yield per decay [%] | Air mass energy transfer coeff. [$10^{-3}$ m$^2$ kg$^{-1}$] | Air Kerma-rate const., $\Gamma_\delta$ [aGy m$^2$ s$^{-1}$ Bq$^{-1}$] | Yield to total $\Gamma_\delta$ [%] |
|---|---|---|---|---|---|
| 1 | 2 | 3 | 4 | 5 | 6 |
| I | 0.03174 | 0.46 ± 0.07 | 12.780 | 0.02 ± 0.003 | 0.05 |
| | 0.04272 | 0.24 ± 0.02 | 5.650 | 0.01 ± 0.001 | 0.02 |
| | 0.05798(*) | 10.1 ± 0.4 | 3.145 | 0.24 ± 0.01 | 0.53 |
| | 0.05932(*) | 17.6 ± 0.6 | 3.045 | 0.41 ± 0.02 | 0.91 |
| | 0.06572 | 3.00 ± 0.07 | 2.720 | 0.07 ± 0.002 | 0.16 |
| | 0.06695(*) | 2.1 ± 0.1 | 2.760 | 0.05 ± 0.002 | 0.11 |
| | 0.0672(*) | 7.5 ± 0.3 | 2.665 | 0.17 ± 0.01 | 0.38 |
| | 0.06775 | 41.0 ± 1.0 | 1.645 | 0.94 ± 0.02 | 2.10 |
| | 1.91 ± 0.1 | | | | 4.26 |
| II | 0.08468 | 2.6 ± 0.3 | 2.352 | 0.07 ± 0.01 | 0.16 |
| | 0.10011 | 14.0 ± 0.4 | 2.320 | 0.41 ± 0.01 | 0.92 |
| | 0.11367 | 1.90 ± 0.04 | 2.345 | 0.06 ± 0.01 | 0.13 |
| | 0.11642 | 0.43 ± 0.01 | 2.350 | 0.02 ± 0.001 | 0.04 |
| | 0.15243 | 6.9 ± 0.1 | 2.500 | 0.34 ± 0.01 | 0.76 |
| | 0.15639 | 2.7 ± 0.2 | 2.520 | 0.14 ± 0.01 | 0.31 |
| | 0.17639 | 3.1 ± 0.2 | 2.595 | 0.18 ± 0.01 | 0.40 |
| | 0.19835 | 1.5 ± 0.1 | 2.665 | 0.10 ± 0.01 | 0.22 |
| | 0.22211 | 7.4 ± 0.2 | 2.735 | 0.57 ± 0.12 | 1.27 |
| | 0.22932 | 3.7 ± 0.1 | 2.750 | 0.30 ± 0.01 | 0.67 |
| | 0.26408 | 3.6 ± 0.1 | 2.830 | 0.34 ± 0.01 | 0.76 |
| | 2.53 ± 0.08 | | | | 5.64 |
| III | 0.92798 | 0.63 ± 0.02 | 2.830 | 0.21 ± 0.01 | 0.47 |
| | 0.95872 | 0.36 ± 0.05 | 2.815 | 0.12 ± 0.02 | 0.27 |
| | 1.00170 | 2.1 ± 0.1 | 2.810 | 0.75 ± 0.04 | 1.67 |
| | 1.04443 | 0.24 ± 0.01 | 2.775 | 0.09 ± 0.004 | 0.20 |
| | 1.17 ± 0.22 | | | | 2.61 |
| IV | 1.11341 | 0.38 ± 0.07 | 2.752 | 0.15 ± 0.03 | 0.33 |
| | 1.12130 | 35.0 ± 0.7 | 2.750 | 13.76 ± 0.27 | 30.71 |
| | 1.1575 | 0.98 ± 0.06 | 2.730 | 0.39 ± 0.02 | 0.87 |
| | 1.18905 | 16.3 ± 0.3 | 2.710 | 6.70 ± 0.12 | 14.96 |
| | 1.22141 | 27.2 ± 0.5 | 2.700 | 11.46 ± 0.21 | 25.58 |
| | 1.23102 | 11.6 ± 0.4 | 2.750 | 4.92 ± 0.17 | 10.98 |
| | 1.25742 | 1.50 ± 0.05 | 2.690 | 0.65 ± 0.03 | 1.45 |
| | 1.27373 | 0.65 ± 0.01 | 2.670 | 0.28 ± 0.01 | 0.63 |
| | 1.28916 | 1.35 ± 0.03 | 2.665 | 0.59 ± 0.02 | 1.32 |
| | 38.90 ± 0.82 | | | | 86.83 |
| V | 1.34273 | 0.27 ± 0.01 | 2.645 | 0.12 ± 0.01 | 0.27 |
| | 1.37384 | 0.22 ± 0.01 | 2.635 | 0.10 ± 0.01 | 0.22 |
| | 1.38740 | 0.09 ± 0.01 | 2.625 | 0.04 ± 0.01 | 0.09 |
| | 1.41010 | 0.05 ± 0.01 | 2.618 | 0.02 ± 0.01 | 0.04 |
| | 1.45305 | 0.04 ± 0.01 | 2.600 | 0.02 ± 0.01 | 0.04 |
| | 0.30 ± 0.003 | | | | 0.66 |
| Total air-kerma rate constant 44.8 ± 0.9 | | | | | 100.0 |

(*)- Characteristic X-ray $K\alpha_2$, $K\alpha_1$, $K\beta$ and $K\beta_1$ respectively

Table 2. Data for calculation and calculated Partial, Groups and Total Air kerma rate constant of $^{182}$Ta radionuclide (Ninkovic & Raicevic, 1992)

### 4.2.2 Photon spectra and calculated of $\Gamma_\delta$ for $^{192}$Ir radionuclide

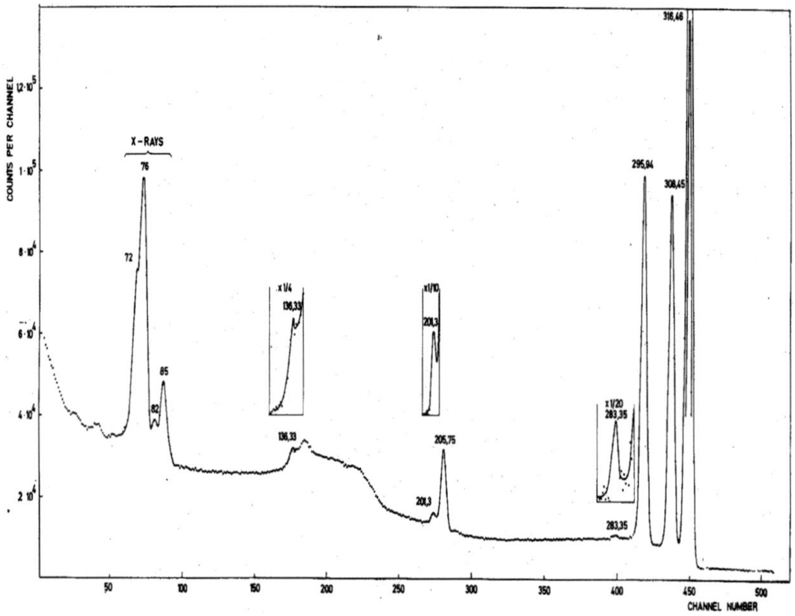


Fig. 2a. Energy spectrum of the photons emitted in decay of $^{192}$Ir radionuclide in energy interval from 50 to 350 keV

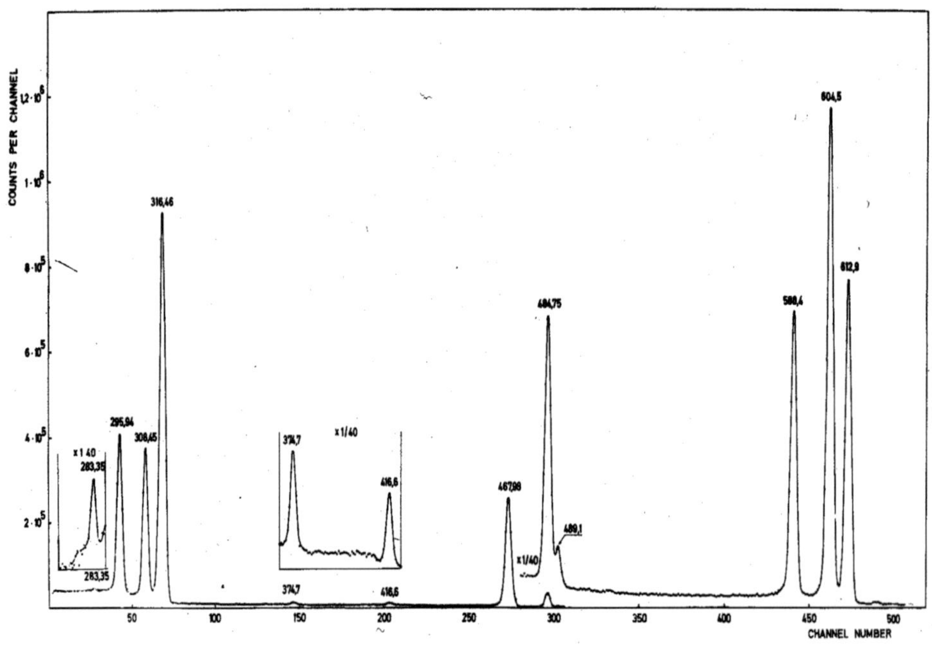


Fig. 2b. Energy spectrum of the photons emitted in decay of $^{192}$Ir radionuclide in energy interval from 250 to 650 keV

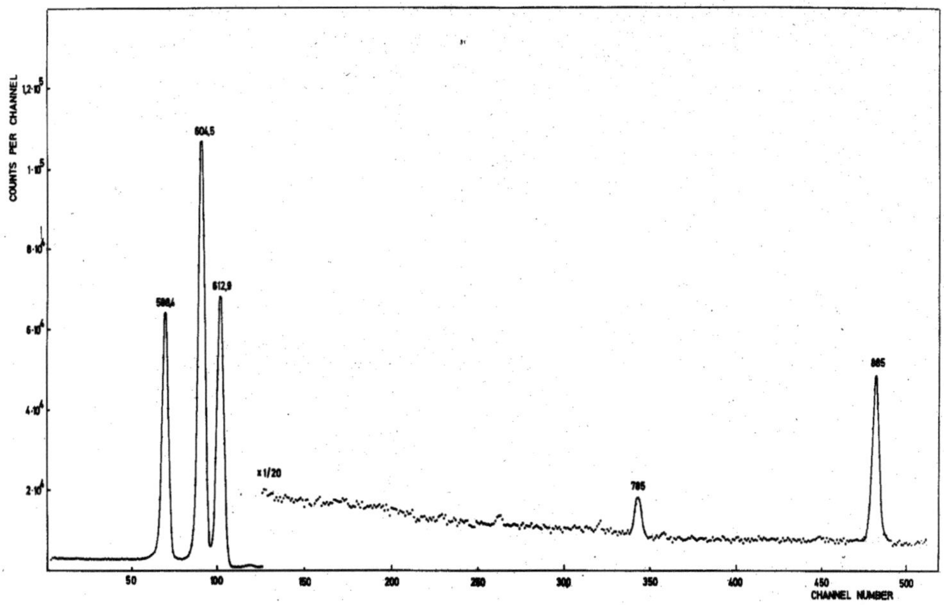


Fig. 2c. Energy spectrum of the photons emitted in decay of $^{192}$Ir radionuclide in energy interval from 550 to 900 keV

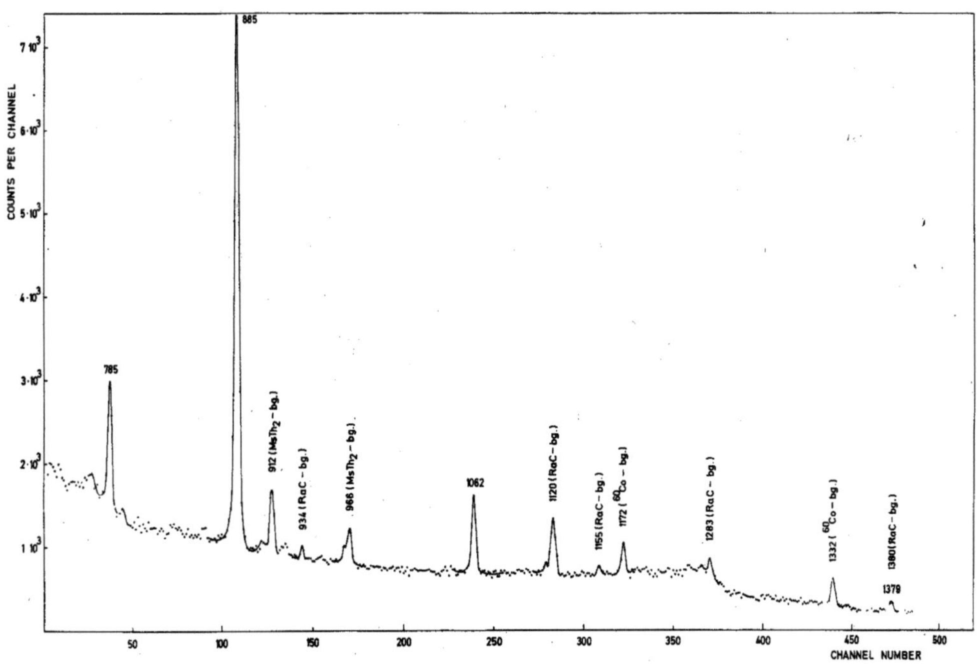


Fig. 2d. Energy spectrum of the photons emitted in decay of $^{192}$Ir radionuclide in energy interval from 750 t0 1400 keV. This part of the spectrum contains many background lines from the RaC, MsTh and $^{60}$Co

| Group of lines | Energy [MeV] | Yield per decay [%] | Air mass energy transfer coeff. [$10^{-3}$ m$^2$ kg$^{-1}$] | Air Kerma-rate const., $\Gamma_\delta$ [aGy m$^2$ s$^{-1}$ Bq$^{-1}$] | Yield to total $\Gamma_\delta$ [%] |
|---|---|---|---|---|---|
| 1 | 2 | 3 | 4 | 5 | 6 |
| I | 0.1363 | $0.17 \pm 0.02$ | 2.47 | $0.01 \pm 0.001$ | <0.1 |
| | 0.2013 | $0.47 \pm 0.04$ | 2.67 | $0.03 \pm 0.003$ | 0.1 |
| | 0.2058 | $3.32 \pm 0.18$ | 2.68 | $0.23 \pm 0.002$ | 0.8 |
| | 0.2832 | $0.27 \pm 0.02$ | 2.85 | $0.03 \pm 0.003$ | 0.1 |
| | $0.30 \pm 0.03$ | | | | <1.1 |
| II | 0.08468 | $28.67 \pm 0.50$ | 2.86 | $3.10 \pm 0.12$ | 10.3 |
| | 0.10011 | $29.65 \pm 0.45$ | 2.88 | $3.35 \pm 0.12$ | 11.2 |
| | 0.11367 | $82.90 \pm 0.4$ | 2.88 | $9.63 \pm 0.28$ | 32.1 |
| | $16.08 \pm 0.44$ | | | | 53.6 |
| III | 0.3745 | $0.70 \pm 0.03$ | 2.94 | $0.10 \pm 0.01$ | 0.3 |
| | 0.4165 | $0.64 \pm 0.04$ | 2.95 | $0.10 \pm 0.01$ | 0.3 |
| | 0.4681 | $47.94 \pm 0.80$ | 2.97 | $8.50 \pm 0.31$ | 28.4 |
| | 0.4846 | $3.18 \pm 0.15$ | 2.98 | $0.58 \pm 0.04$ | 1.9 |
| | 0.4891 | $0.40 \pm 0.04$ | 2.98 | $0.08 \pm 0.01$ | 0.2 |
| | $9.36 \pm 0.34$ | | | | 31.1 |
| IV | 0.5886 | $4.47 \pm 0.35$ | 2.97 | $1.00 \pm 0.10$ | 3.3 |
| | 0.6044 | $8.25 \pm 0.45$ | 2.96 | $1.88 \pm 0.14$ | 6.3 |
| | 0.6125 | $5.26 \pm 0.27$ | 2.96 | $1.22 \pm 0.09$ | 4.1 |
| | $4.10 \pm 0.30$ | | | | 13.7 |
| V | 0.785 | $0.05 \pm 0.01$ | 2.88 | $0.01 \pm 0.003$ | <0.1 |
| | 0.8845 | $0.29 \pm 0.05$ | 2.88 | $0.09 \pm 0.02$ | 0.3 |
| | 1.062 | $0.06 \pm 0.01$ | 2.76 | $0.02 \pm 0.004$ | 0.1 |
| | $0.12 \pm 0.03$ | | | | 0.5 |
| Total air-kerma rate constant: $30.0 \pm 0.9$ | | | | | 100.0 |

Table 3. Data for calculation and calculated Partial, Groups and Total Air kerma rate constant of $^{192}$I radionuclide (Ninkovic & Raicevic, 1993)

As can be seen from Table 3, the entire of photon ray spectrum of $^{192}$Ir is divided into five characteristic groups of photon lines. The air kerma rate constant was calculated for each discrete photon line with yield per decay event >0.05 % and starting with energy of 0.1363 MeV as the lowest energy. That means X-ray were not included. The air kerma rate constant for the groups and for the total were obtained by addition of partial or single photon lines constant. Finally, a value of ( $30.0 \pm 0.9$ ) aGy m$^2$ s$^{-1}$ Bq$^{-1}$ for an unshielded $^{192}$Ir source has been obtained. That value is in good agreement with a new recalculated value given in Table 1.

Keeping in mind that standard iridium sources are usually packed into 0.15 mm of platinum, the constant for that type of source was also calculated. For that goal, it was calculated the absorption of iridium photons into 0.15 mm of platinum and found that in the air kerma rate constant is reduced by 7.33 %. After this correction, a value of ($27.8 \pm 0.9$)

aGy m² s⁻¹ Bq⁻¹ was obtained for the air kerma rate constant for standard packaged iridium source (Ninkovic & Raicevic, 1993).

### 4.2.3 Results of $\Gamma_\delta$ calculation for $^{226}$Ra (in equilibrium with its decay product) radionuclide

| Group of lines | Energy [MeV] | Yield per decay [%] | Air mass energy transfer coeff. [$10^{-3}$ $m^2$ $kg^{-1}$] | Air Kerma-rate const., $\Gamma_\delta$ [aGy $m^2$ $s^{-1}$ $Bq^{-1}$] | Yield to total $\Gamma_\delta$ [%] |
|---|---|---|---|---|---|
| 1 | 2 | 3 | 4 | 5 | 6 |
| I | 0.1857 | 4.83 ± 0.26 | 2.63 | 0.301 ± 0.020 | 0.51 |
| | 0.2419 | 8.56 ± 0.42 | 2.78 | 0.737 ± 0.033 | 1.25 |
| | 0.2588 | 0.49 ± 0.04 | 2.28 | 0.046 ± 0.004 | 0.08 |
| | 0.2748 | 0.38 ± 0.15 | 2.84 | 0.038 ± 0.013 | 0.06 |
| | 0.2952 | 19.74 ± 1.00 | 2.86 | 2.125 ± 0.104 | 3.61 |
| | 0.3520 | 38.27 ± 2.00 | 2.92 | 5.015 ± 0.262 | 8.51 |
| | 0.3868 | - - | 2.94 | - - | - |
| | 0.3888 | 0.76 ± 0.14 | 2.94 | 0.111 ± 0.020 | 0.19 |
| | 0.4550 | 0.28 ± 0.07 | 2.98 | 0.048 ± 0.013 | 0.08 |
| | 0.4621 | 0.16 ± 0.06 | 2.98 | 0.028 ± 0.006 | 0.06 |
| | 0.4805 | - - | 2.98 | - - | - |
| | 0.4872 | 0.38 ± 0.05 | 2.98 | 0.080 ± 0.007 | 0.12 |
| | 0.6094 | 46.46 ± 1.42 | 2.96 | 10.685 ± 0.334 | 18.14 |
| | 19.20 ± 0.96 | | | | 32.60 |
| II | 0.6656 | 1.68 ± 0.05 | 2.945 | 0.420 ± 0.013 | 0.71 |
| | 0.7031 | 0.58 ± 0.05 | 2.935 | 0.135 ± 0.013 | 0.26 |
| | 0.7199 | 0.40 ± 0.05 | 2.925 | 0.107 ± 0.013 | 0.18 |
| | 0.7684 | 4.88 ± 0.05 | 2.892 | 1.383 ± 0.026 | 2.35 |
| | 0.7860 | 1.10 ± 0.05 | 2.89 | 0.319 ± 0.013 | 0.54 |
| | 0.8062 | 1.31 ± 0.04 | 2.89 | 0.389 ± 0.013 | 0.66 |
| | 0.8212 | 0.10 ± 0.04 | 2.88 | 0.030 ± 0.012 | 0.05 |
| | 0.8392 | 0.52 ± 0.05 | 2.875 | 0.160 ± 0.013 | 0.27 |
| | 0.9340 | 3.20 ± 0.15 | 2.83 | 1.078 ± 0.033 | 1.83 |
| | 0.9641 | 0.38 ± 0,05 | 2.82 | 0.132 ± 0.020 | 0.22 |
| | 0.0520 | 0.35 ± 0.04 | 2.78 | 0.130 ± 0.013 | 0.22 |
| | 0.1040 | 0.15 ± 0.04 | 2.76 | 0.058 ± 0.012 | 0.10 |
| | 0.1204 | 16.70 ± 0.42 | 2.75 | 6.560 ± 0.231 | 11.14 |
| | 10.92 ± 0.40 | | | | 18.54 |
| III | 1.1338 | 0.28 ± 0.04 | 2.745 | 0.111 ± 0.013 | 0.19 |
| | 1.1553 | 1.58 ± 0.10 | 2.735 | 0.636 ± 0.040 | 1.08 |
| | 1.2078 | 0.42 ± 0.05 | 2.715 | 0.176 ± 0.020 | 0.30 |
| | 1.2382 | 6.03 ± 0.16 | 2.695 | 2.565 ± 0.066 | 4.35 |
| | 1.2811 | 1.52 ± 0.05 | 2.675 | 0.664 ± 0.020 | 1.13 |

| Group of lines | Energy [MeV] | Yield per decay [%] | Air mass energy transfer coeff. [$10^{-3}$ $m^2$ $kg^{-1}$] | Air Kerma-rate const., $\Gamma_\delta$ [aGy $m^2$ $s^{-1}$ $Bq^{-1}$] | Yield to total $\Gamma_\delta$ [%] |
|---|---|---|---|---|---|
| 1 | 2 | 3 | 4 | 5 | 6 |
| | 1.3038 | 0.084 ± 0.035 | 2.70 | 0.048 ± 0.024 | 0.08 |
| | 1.3777 | 4.30 ± 0.10 | 2.63 | 1.986 ± 0.060 | 3.37 |
| | 1.4016 | 1.31 ± 0.05 | 2.62 | 0.613 ± 0.027 | 1.04 |
| | 1.4081 | 2.47 ± 0.05 | 2.61 | 1.157 ± 0.033 | 1.96 |
| | 1.5093 | 2.26 ± 0.05 | 2.57 | 1.118 ± 0.033 | 1.90 |
| | 1.5387 | - - | 2.56 | - - | - |
| | 1.5433 | 0.29 ± 0.06 | 2.555 | 0.146 ± 0.027 | 0.25 |
| | 1.5833 | 0.76 ± 0.05 | 2.54 | 0.390 ± 0.027 | 0.66 |
| | 1.5948 | 0.33 ± 0.04 | 2.535 | 0.170 ± 0.020 | 0.29 |
| | 1.6055 | 0.33 ± 0.05 | 2.53 | 0.171 ± 0.027 | 0.29 |
| | 1.6614 | 1.10 ± 0.04 | 2.505 | 0.584 ± 0.020 | 0.99 |
| | 1.6841 | 0.22 ± 0.0,04 | 2.495 | 0118 ± 0.020 | 0.20 |
| | 1.7298 | 2.91 ± 0.11 | 2.48 | 1.592 ± 0.060 | 2.70 |
| | 1.7646 | 16.70 ± 0.32 | 2.46 | 9.243 ± 0.231 | 15.69 |
| | 21.49 ± 0.79 | | | | 36.49 |
| IV | 1.8386 | 0.34 ± 0.05 | 2.43 | 0.194 ± 0.026 | 0.33 |
| | 1.8476 | 2.14 ± 0.09 | 2.425 | 1.222 ± 0.043 | 2.07 |
| | 1.8735 | 0.20 ± 0.03 | 2.415 | 0.115 ± 0.019 | 0.20 |
| | 1.8967 | 0,06 ± 0.03 | 2.405 | 0.035 ± 0.014 | 0.06 |
| | 2.011 | 0.05 ± 0.02 | 2.36 | 0.030 ± 0.012 | 0.05 |
| | 2.017 | - - | 2.355 | - - | - |
| | 2.119 | 1.24 ± 0.02 | 2.325 | 0.779 ± 0.020 | 1.32 |
| | 2.2043 | 5.25 ± 0.03 | 2.305 | 3.401 ± 0.119 | 5.77 |
| | 2.294 | 0.34 ± 0.01 | 2.28 | 0.227 ± 0.007 | 0.38 |
| | 2.448 | 1.73 ± 0.03 | 2.235 | 1.207 ± 0.033 | 2.05 |
| | 7.21 ± 0.30 | | | | 12.24 |
| V | 2.700 | 0.034 ± 0.005 | 2.16 | 0.025 ± 0.026 | 0.042 |
| | 2.770 | 0.028 ± 0.004 | 2.14 | 0.021 ± 0.043 | 0.036 |
| | 2.788 | 0.0058 ± 0.0010 | 2.135 | 0.001 ± 0.019 | 0.02 |
| | 2.885 | 0.0095 ± 0.0010 | 2.11 | 0.007 ± 0.014 | 0.012 |
| | 2.922 | 0.0162 ± 0.0030 | 2.10 | 0.001 ± 0.0005 | 0.002 |
| | 2.979 | 0.0162 ± 0.0010 | 2.08 | 0.001 ± 0.0002 | 0.002 |
| | 3.000 | 0.0089 ± 0.0010 | 2.075 | 0.001 ± 0.0001 | 0.002 |
| | 3.054 | 0.022 ± 0.003 | 2.065 | 0.018 ± 0.0.003 | 0.030 |
| | 3.082 | 0.0047 ± 0.00005 | 2.06 | 0.004 ± 0.001 | 0.007 |
| | 3.142 | - - | 2.05 | - - | - |
| | 0.08 ± 0.01 | | | | 0.14 |
| Total air-kerma rate constant: 58.9 ± 2.4 | | | | | 100.0 |

Table 4. Data for calculation and calculated partial, proup`s and total air kerma rate constant of $^{226}$Ra radionuclide in equilibrium with its decay products (Ninkovic, 1987)

As it can be seen from this table , the entire of photon ray spectrum of $^{226}$Ra (in equilibrium with its decay products) are divided into five characteristic groups of photon lines. The air kerma rate constant was calculated for each discrete photon line with yield per decay event >0.05 % and starting with energy of 0.1857 MeV as $\delta$ value. That means X-ray were not included. The air kerma rate constant for the groups and for the total were obtained by addition of partial or single photon lines constant. Finally, a value of ( $56.9 \pm 2.4$ ) aGy m$^2$ s$^{-1}$ Bq$^{-1}$ for an unshielded $^{226}$Ra source has been obtained.

Having seen that standard radium sources are usually packed into 0.5 mm of platinum, the constant for that type of source was also calculated. For that goal it was used analyses of Shalek and Stoval (Shalek & Stovall, 1969), which is in good accordance with the earlier estimate of Aglincev et al. (Aglincev et al., 1960), that 0,5 mm of Pt by absorption of gamma radiation of radium and its decay products, reduce the air kerma rate constant with 9.25 %. After this correction, a value of ( $53.4 \pm 2.2$ ) aGy m$^2$ s$^{-1}$ Bq$^{-1}$ was obtained for the air kerma rate constant for standard packaged radium sources (Ninkovic, 1987). On the basis of this calculated value and experimentally measured value of Aglincev et al. (Aglincev et al., 1960) it was concluded (Ninkovic, 1987) that the real value of air kerma rate constant of $^{226}$ Ra in equilibrium with its decay product is smaller by about 1 to 2 %, than the value recommended by ICRU (ICRU, Handbook 86, 1963).

## 5. Conclussion

Presented process of recalculation the values for air kerma rate constants, for 35 of the most often used radionuclide in practice, was based on the newest appropriate decay data for every radionuclide and latest numerical data for mass energy-transfer coefficient. That is the reason why, according to the authors opinion, obtained values for $\Gamma_{\delta,}$ listed in the table 1, are the most accurate data that can be found in the literature available at present.

It has to be pointed out that to calculate the absorbed dose to soft tissue the air kerma rate has to be multiplied by the ratio of the mass energy-absorption coefficient of soft tissue to that of air, which can be taken as 1,11 between 2 and 0,1 MeV and drops to 1,04 at 0,02 MeV. Also, since the radiation-waiting factor for gamma rays and X rays is 1, by multiplying air kerma rate constants by a factor 1,11, the soft tissue-equivalent dose constant can be obtained.

## 6. References

Aglincev, K.K, Ostromuhova, G.P. and Holnova, E.A (1960). Izm. Techn. 12, 40

Aird, E.G.A, Williams, S. & Glover, C. (1984). *SI Units and radionuclides: Teaching problems,* Radiography, 50, 174-176

Attix, F.H. (1986). *Introduction to radiological physics and radiation dosimetry,* New York: John Wiley and Sons, Ink

British Committee on Radiation Units and Measurements (1982). *Memorandum from British Committee on Radiation Units and Measurements,* Br. J. Radiol., 55, 375-377

Fireston,R.B. (1996). *Tables of Isotopes, eight edn,* New York: John Wiley and Sons

Hubbell, J.H. (1969). *Photon Cross Sections, Attenuation Coefficients and Energy Absorption Coefficient from 10 keV to 100 GeV,* Washington DC: NBS Publication NSRDS-NBS 29

Hubbell, J.H. & Seltzer, S.M. (2001). *Tables of X-Ray Mass Attenuation Coefficient and Mass Energy Absorption Coefficient from 1 keV to 20 MeV for Elements Z = 1 to 92 and 48 Additional Substances of Dosimetric Interest,* NISTIR 5632, Available at: http://physics.nist.gov./PhysRefData/XrayMassCoef/cover.htmlICRU (1963). *Handbook 86,* Natl. Bur. Stand. (US)

International Commission on Radiation Protection Units and Measurements. (1962). *Radiation Quantities and Units,* ICRU Report 10a, Washington DC : ICRU

International Commission on Radiation Protection Units and Measurements (1980). *Radiation Quantities and Units,* ICRU Report 33, , Washington DC : ICRU

Kereiaces, J.G. & Rosenstein, M. (1980), *Handbook of Radiation Doses in Nuclear Medicine and Diagnostic X-ray* Arai, Boca Raton, FL: CRC Press

Nachtigall, D. (1969). *Table of Specific Gamma Ray Constants,* Muenchen: Verlag Karl Thiemig KG

National Council on Radiation Protection and Measurements (1976). *Structural Shielding Design and Evaluation for Medical Use X-rays and Gamma Rays of Energies UP TO 10 MeV.* NCRP Report 49, Washington DC: NCRP

Ninkovic, M.M. & Mladenovic, M. (1970). *Specific gamma ray constant and gamma spectrum of* $^{192}Ir$. Atompraxis, 16, 5-8

Ninkovic, .M.M. (1987). *The air kerma rate constant of* $^{226}Ra$ *in equilibrium with its decay products,* Nucl. Instr. Methods, A255, 334-337

Ninkovic, M.M. & Raicevic J.J. (1992). *Air kerma rate constant for the* $^{182}Ta$ *radionuclide,* Trans. Am. Nucl. Soc., V. 65, Suppl. 1, No.9, 70-72

Ninkovic, M.M. & Raicevic J.J. (1993). *The air kerma rate constant of* $^{192}Ir$, Health Phys. 64(1), 79-81

Ninkovic, M.M, Raicevic, J.J. and Adrovic F. (2005). *Air kerma rate constants for gamma emitters used most often in practice,* Radiation Protection Dosimetry, 115, 1-4, 247-250

Parker, P.R.. Smith, S.H. & Taylor, M.D. (1978), *Basic Science of Nuclear Medicine,* London: Churghill Livingston

Sabol, J. & Weng, P.S. (1995). *Introduction to Radiation Protection Dosimetry,* Singapore: World Scientific Publishing Co. Pte. Ltd

Shalek, R.S. & Stoval M.(1969). in: *Radiation Dosimetry Vol.III: Sources, Fields, Measurements and Applications,* Attix, F.H., Roesch, W.C. and Tochilin, E., p. 748, Academic Press, New York and London

Stabin, M.G. & Da Luz L.C.Q.P. (2002). *Decay data for internal and external dose assessment,* Health Physics, 83(4), 471-475

Ungar, L.M. & Truby, D.K. (1982). *Specific Gamma Ray Dose Constants for Nuclides Important to Dosimetry and radiological assessment,* Report No. ORNL/RSIC- 45/R1, Oak Ridge National Laboratory, Oak Ridge TN, USA

Wasserman, H. & Groenwald, W. (1988). *Air kerma rate constants for radionuclides,* Eur. J. Nucl. Med., 14, 569-571

2

# Material Analysis Using Characteristic Gamma Rays Induced by Neutrons

Alexander P. Barzilov, Ivan S. Novikov and Phillip C. Womble
*Western Kentucky University*
*USA*

## 1. Introduction

Neutron interrogation based methods of non-destructive analysis are well established techniques employed in the field of bulk material analysis. These methods utilize a source of neutrons (a neutron probe) to irradiate objects under scrutiny. Nuclear reactions initiated by neutrons in the volume of the irradiated sample include the following: inelastic neutron scattering, thermal neutron capture, and neutron activation. As a result of nuclear reactions with the material inside the object, the "fingerprint" $\gamma$-rays are emitted with characteristic energies. These characteristic gamma rays are used for the elemental identification. By measuring and counting the number of $\gamma$-rays emitted with a specific energy, one can deduce the amount of the associated chemical element in the sample. The amounts of chemical elements measured allow specifying the chemical composition of the analyzed sample.

Neutron technique is an excellent choice to rapidly determine elemental content of the sample *in situ* in non-intrusive manner. It is a great fit for in situ applications that involve samples that are hard to reach or unsafe to handle, and that require the analysis to be performed rapidly, in real time.

Accelerator based neutron sources such as deuterium - deuterium (d-d) and deuterium - tritium (d-t) fusion neutron generators provide the electronic control of neutron emission including its time structure. The pulse mode of neutron production allows the use of coincidence methods to segregate prompt and delayed gamma ray signatures emitted from neutron induced nuclear reactions. The kinematics of fusion reactions allows "tagging" of outgoing neutrons using the associated particles.

The pulse neutron systems are used in industry for analysis of coal (Dep et al., 1998; Sowerby, 2009), cement (Womble et al., 2005), metal alloys (James & Fuerst, 2000), in geological and soil analysis (Wielopolski et al., 2008), and oil well logging (Nikitin & Bliven, 2010). Security applications of neutron based systems are for chemical and explosive threats detection (Vourvopoulos & Womble, 2001; Aleksandrov et al., 2005; Lanza, 2006), including the search for threats in cargo containers (Barzilov & Womble, 2003) and vehicles (Reber et al., 2005; Koltick et al., 2007), humanitarian demining and confirmation of unexploded ordinance (Womble et al., 2002; Holslin et al., 2006). Such technologies are considered in astrochemistry applications for in situ analysis of planetary samples (Parsons et al., 2011).

The use of the pulse neutron based analysis of nitrogen and oxygen content *in vivo* is discussed in nutrition research (Shypailo & Ellis, 2005) and in cancer diagnostics (Maglich & Nalcioglu, 2010).

In the presented chapter we discuss the components of pulse neutron based material analysis systems, nuclear reactions induced by neutrons, characteristic gamma radiation emitted in these nuclear reactions, gamma ray spectral analysis methods for elemental characterization, and "neutrons in - photons out" methods that utilize the characteristic gamma radiation.

## 2. Pulse neutron sources and system components

A pulse neutron based material analysis system consists of a neutron source, gamma and particle radiation detector(s), a shadow radiation shielding to cover detectors from direct source neutrons, and associated hardware and software for system control, data acquisition and processing. Fig.1 shows the scheme of a typical system. The system operates as follows. Emitted by a source neutrons induce nuclear reactions in the irradiated object and excite nuclei. Excited nuclei emit photons due to various de-excitation processes that are measured by a gamma ray detector. The gamma ray spectrum is analyzed providing information on the chemical composition of the irradiated sample.

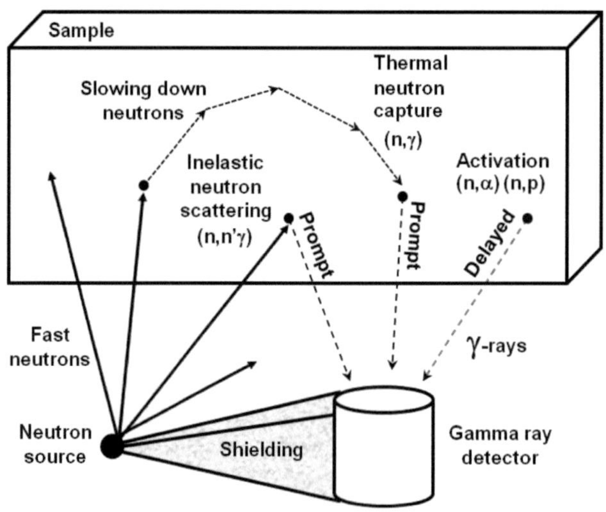


Fig. 1. Pulse neutron based elemental analysis scheme

Various radioisotopes, neutron beams from a nuclear reactor core, or accelerator-based devices are used as neutron sources. Radioisotopes used in neutron sources include $^{252}$Cf, $^{239}$Pu, $^{241}$Am, and others. Californium-252 undergoes spontaneous fission with emission of neutrons with the average energy 2.5 MeV. The $^{252}$Cf neutron emission spectral distribution is described by the semi-empirical Watt formula. Plutonium and americium based sources emit neutrons using alpha decay of $^{239}$Pu or $^{241}$Am and (α,n) reactions in the matrix of light elements such as beryllium or lithium. The neutron energy spectrum of these sources is wide (up to ~11 MeV) with the average neutron energy ~4.5 MeV. Radioisotope sources may require radiation shielding while not in use.

Pulse structure of neutron emission from isotopic source or reactor is usually controlled with a chopper system. Some reactors provide the pulse periodic operation mode. The neutron beam's energy follows $^{235}U$ fission spectrum distribution, or depends on the moderator type used inside the core or in neutron beam optics. Nuclear reactors are bulky, expensive, and require significant radiation shielding. That makes them impossible for use as a neutron source for portable material analysis systems.

Accelerator-based neutron sources are widely used in material analysis. These sources utilize charged particle beams to create fast neutrons in nuclear reactions induced in various targets. Some examples of such neutron producing reactions are the following: $T(d,n)^4He$, $D(d,n)^3He$, $^9Be(d,n)^{10}Be$, $^7Li(d,n)^8Be$, $^7Li(p,n)^7Be$, $^7Be(p,n)^7B$. The pulse neutron emission scheme allowing high repetition rates is provided by controlling acceleration parameters electronically. Such sources can be turned off thus simplifying radiation shielding requirements. The neutron sources based on fusion reactions are compact systems due to a large reaction resonance at low deuteron energy (approximately 3.4 barn for 100 keV for d-t fusion). The d-t fusion neutron generators are widely used as sources of neutrons for portable probe and industrial applications. These isotropic neutron sources are rugged, low maintenance, and relatively inexpensive systems.

### 2.1 Accelerator based fusion neutron generators

Neutron generators utilize d-t and d-d fusion reactions that produce mono energetic neutrons. The d-t reaction shows greater energy release. At the incident particle's small energies, $^4He$ and neutron share 17.59 MeV with conservation of linear momentum, and mono energetic 14.1 MeV neutrons are emitted out of reaction

$$^2H + {}^3H \rightarrow {}^4He + n \; (Q=17.59 \text{ MeV}, E_n=14.1 \text{ MeV}, E_{He}=3.49 \text{ MeV}) \tag{1}$$

In 50% of events, d-d reaction produces mono energetic 2.45 MeV neutrons and $^3He$ that share 3.27 MeV:

$$^2H + {}^2H \rightarrow {}^3He + n \; (Q=3.27 \text{ MeV}, E_n=2.45 \text{ MeV}, E_{He}=0.82 \text{ MeV}) \tag{2}$$

With 50% probability, $^3T$ and proton may be also produced in d-d reaction (Q=4.03 MeV, $E_p$=3.02 MeV, $E_T$=1.01 MeV).

Neutrons produced in the d-t reaction are emitted isotropically. Neutron emission in d-d reaction is slightly peaked forward along the direction of ion beam. The yield of d-d reaction at low energies of deuterons (reaction cross section is $3.3\times10^{-2}$ barn at 100 keV) is approximately two orders of magnitude lower than in d-t fusion. Therefore, higher deuteron current is required to achieve d-d neutron yields comparable to a d-t source. Because of that d-t neutron generators are more common in applications requiring small size neutron sources of higher energy neutrons. The d-d systems may be preferable in applications where only the lower energy neutrons are required and where 14.1-MeV neutrons may cause unnecessary interference in the analysis due to many reactions channels open for high energy neutrons.

Another reaction which can be used for neutron production is the t-t fusion

$$^3H + {}^3H \rightarrow {}^4He + n + n \; (Q=11.3 \text{ MeV}) \tag{3}$$

The reaction cross section is $3.4\times10^{-2}$ barn at 100 keV, which is similar to the cross section of the d-d reaction. The distribution of energy between reaction products varies producing the wide neutron spectrum with maximum energy up to ~11 MeV. Wide neutron spectrum may be useful for material analysis applications that require both low and high energy neutrons.

The compact "sealed tube" neutron generator design includes an ion source, a positive ion accelerator, and a target. Commonly used source to generate positive ions is the cold cathode Penning source. This source has a cylindrical anode under ~1-2 kV potential applied to it, and grounded cathodes on the ends of the anode. The magnet surrounds the anode cylinder setting up the coaxial magnetic field inside it. The tritium or deuterium gas is introduced into the volume of the anode cylinder. The electric field between anode and cathodes causes ionization of gas molecules creating the cold plasma. Trapped inside the anode electrons are moving in the volume and ionizing gas molecules which helps to maintain the plasma quality. Ions are transferred into the acceleration region through the exit port of the cathode. This region supplies the electric field (up to ~100-120 kV) to accelerate the positive ions. Neutron generator target is a metal hydride loaded with deuterium or tritium or the mixture of both. The ions interact with a target, producing neutrons in fusion reactions. Typical neutron output levels of sealed tube neutron generators are ~$10^8$-$10^9$ n/s (d-t) and ~$10^6$ n/s (d-d). Higher output usually shortens the sealed tube's life time. Sealed tube neutron generators are produced by Thermo Fisher Scientific, Schlumberger, Baker Hughes, EADS SODERN in France, and VNIIA in Russia.

Other designs of neutron generators utilize ion sources such as hot cathode source, radiofrequency ion source, or inertial electrostatic confinement (IEC) based source. For example, the pulse d-d system produced by Adelphi Technology Inc. (Williams et al., 2008) using the microwave driven plasma source technique provides the 2.45-MeV neutron output up to $8\times10^9$ n/s. The similar technique using tritium provides the yield of 14.1-MeV neutrons ~$10^{11}$ n/s. The d-t neutron generator developed by NSD-Fusion GmbH for irradiation of extended samples uses the IEC technique producing ~$10^{10}$ n/s. These neutron generator designs have the longer life time compared to sealed tube sources.

## 2.2 Gamma ray detectors

Physical parameters and limitations of the gamma ray detectors used in the system govern parameters of the entire system. The choice of gamma ray detectors is important for neutron based system to be effective. Improper solution can generate both false positive and false negative results.

The gamma ray detectors must be suitable for operation in mixed radiation fields where neutrons and gamma rays present. The detector material must have a high Z value to effectively detect characteristic photons with energies up to 10.8 MeV. The detection medium must also provide the energy resolution that allows resolving peaks of interest. Ideally, the detector should provide minimum interference with the signal emitted from a sample when the detector material is irradiated with neutrons. Thus, if possible the detector material should avoid isotopes that are anticipated in the analyzed samples. The neutron induced gamma ray peaks for elements of the detector material should not interfere with the sample's spectral signatures. In addition, neutrons may produce radioactive activation products with the time delayed decay inside the detector volume. These decays (for example, the beta decay) may produce photons or charged particles that interfere with the characteristic gamma ray spectra adding the noise and overloading the data acquisition electronics. It is a complicated task to satisfy all these requirements, especially with added cost limitations. Usually, the trade-off between various detector parameters including its cost is considered for a particular application (Barzilov & Womble, 2006).

Standard gamma ray detector solutions for spectroscopy are high purity germanium detectors (HPGe) with liquid nitrogen dewar or mechanical cooling subsystems (Sangsingkeow et al., 2003), and scintillation detectors such as NaI(Tl), $Bi_4Ge_3O_{12}$, $LaBr_3$(Ce) (van Loef et al., 2001), etc. Noble gas scintillation or ionization detectors and gamma ray telescopes can also be used for neutron induced photon measurements in the MeV energy range (April et al., 2006).

The NaI(Tl) scintillator material has the light yield 38 photons/keV, 1/e decay time 250 ns, and density 3.67 g/cc. Atomic numbers are 53 and 11 for iodine and sodium, respectively. Under neutron irradiation, the NaI(Tl) scintillator is activated by neutrons showing the delayed beta decay spectral continuum with the endpoint energy ~2 MeV.

The BGO scintillator has the light yield 9 photons/keV, and 1/e decay time 300 ns. Due to the high atomic number of bismuth, 83, and the crystal's high density of 7.13 g/cc, the BGO scintillator is very effective for detection of high energy photons. Its energy resolution is lower than NaI(Tl) resolution: ~10% FWHM versus ~7% FWHM for 662-keV $\gamma$-rays. The BGO demonstrates excellent behaviour under neutron irradiation without delayed decay issues. Significant downside of the BGO detector is its sensitivity to the environmental temperature (Womble et al., 2002).

The $LaBr_3$(Ce) scintillator has the ~3%-resolution of the 662-keV peak, and density 5.08 g/cc. The lanthanum atomic number is 57. This scintillator has high light yield 63 photons/keV, fast 1/e decay time 16 ns, and better timing properties than NaI(Tl). The $LaBr_3$(Ce) material contains small quantities of the radioactive lanthanum-138 isotope ($t_{1/2}=1.02\times10^{11}$ years) producing the 1.47-MeV gamma ray peak that is always visible in the spectrum; it can be used for calibration purposes. The $LaBr_3$(Ce) is affected by neutrons showing the delayed beta decay spectral continuum with endpoint energy ~3 MeV when irradiated with a d-t neutron source. The measured $\beta$-decay curve exhibits cumulative nature: two isotopes decay at the same time. The $^{80}Br$ decays with a half-life 17.68 minutes. The $^{82}Br$ isotope decays with the half-life approximately 35 hours. Lanthanum halide demonstrates stable gamma ray spectrum parameters in the mixed field under d-t neutron irradiation, when properly shielded. The good energy resolution under the room temperature, the high brightness, and the high scintillation decay speed pose this material as a promising candidate for active neutron interrogation applications, if the crystal's neutron activation issues are properly addressed.

The HPGe detector has superior energy resolution comparing to scintillation detectors. The atomic number of germanium is 32. The HPGe crystal density is 5.35 g/cc. HPGe crystal is sensitive to the high energy neutrons, which cause detector damage (Tsoulfanidis & Landsberger, 2010). High energy neutrons produce charges in the germanium crystal which are adding noise to the collected gamma ray spectrum (Ljungvall & Nyberg, 2005). Neutron collisions with the crystal cause atom displacements into interstitial positions creating a vacancy pair. These crystal defects behave as trapping centers for holes and electrons, and may create new donor and acceptor states, thus gradually changing the charge collection efficiency, the resolution, and the pulse timing characteristics of the detector. The n-type HPGe detectors are preferable in applications that involve neutron irradiation. They have been shown to be more resistant to damage by fast neutrons (Pehl et al., 1979). The neutron damage problem requires special attention and treatment (Fourches et al., 1991). The speed of the HPGe charge collection is another parameter to be considered in high count rate conditions and applications that require good timing resolution (Cooper & Koltick, 2001).

The comparison of selected gamma ray detectors used in neutron-based material analysis applications is shown in Table 1.

The shielding is required to protect the gamma ray detector from direct hit by the neutrons. Shielding size defines the geometry of the system since a neutron source and a gamma ray detector are separated by the shielding column. The combination of materials with large scattering cross sections for fast neutrons and large low energy neutron capture cross sections, and high Z materials with high stopping power for gamma rays is used. The goal is to keep fast neutrons away from the detector volume either by redirecting their path or moderating them with the subsequent capture. The d-d or d-t targets are in general of the "point source" type, thus the shielding may have a conical shape to minimize the weight. For $10^8$-n/s d-t source, the simplest "shadow" shielding is a layered conical structure of ~50 cm length; the 30-40 cm borated polyethylene layer near the source, and the 10-20 cm lead layer near the gamma ray detector (Womble et al., 2003). The more complex shielding designs are possible using layers of other materials, but the size / weight / cost considerations add design limitations. In addition, the detector may be also shielded from lower energy neutrons scattered from surrounding materials. The two-layer shielding can reduce spectral noise due to low energy neutron interactions with the detector crystal. The outer layer of borated resin is effective as a thermal neutron shielding; the inner lead layer attenuates photons emitted from thermal neutron capture reactions in the outer layer. The lead also attenuates low energy photons that are not of interest in material analysis thus helping to reduce dead time of the gamma ray spectroscopy system.

| Detector | Energy resolution, %FWHM @ 662 keV | Efficiency | Cooling | Neutron activation issues |
|---|---|---|---|---|
| NaI(Tl) | ~7% Fair | Medium | No | Activated, beta-decay |
| $Bi_4Ge_3O_{12}$ | ~10% Fair | High | No, temp. shifts | No |
| $LaBr_3$(Ce) | 2.8% Good | Medium | No | Activated, beta-decay |
| HPGe | 0.4% Excellent | Medium | $LN_2$ Temp. | No |

Table 1. Gamma ray detectors used in neutron-based material analysis applications

Data acquisition electronics used with the gamma ray detectors in such systems should be appropriate for the detector's signal processing and count rates attainable in neutron interrogation. Standard analog and digital spectroscopy solutions are typically used.

## 3. Nuclear reactions induced by neutrons

Neutrons emitted in d-d ($E_n$=2.45 MeV) and d-t ($E_n$=14.1 MeV) fusion reactions are highly penetrating particles. The typical range is several feet into materials commonly utilized in industry and commerce. Nuclear reactions energetically possible under 14.1-MeV fusion neutron's action in the volume of the irradiated object are the following: (n,n'γ), (n,γ), (n,α), (n,p), (n,d), (n,t), (n,2p), (n,n'p), (n,n'α), (n,$^3$He), and (n,2n). If the sample contains heavy nuclei, (n,3n) and nuclear fission reactions may be induced with the low probability. Production of charged particles is prevailing for light nuclei; neutron production is favourable for heavier nuclei. The reactions (n,d) and (n,t) have noticeable cross-section for light mass isotopes, but products produced in such reactions are stable. The (n,d) and (n,t) reaction cross sections for medium and heavier mass nuclei are low.

Widely used in material analysis neutron induced nuclear reactions are inelastic neutron scattering (n,n'γ), thermal neutron capture (n,γ), and neutron activation (n,αγ) and (n,pγ). The only source of fast neutrons is a fusion neutron source. Thermal neutrons are created by slowing down the fast source neutrons in collisions with low Z materials within the sample itself or within the environment around the sample, or by using neutron moderating materials.

| Isotope | $\sigma_{total}$ | $\sigma_{inl}$ | $\sigma_{n\text{-}n'\ 1st}$ | $\sigma_{n\text{-}n'\ 2nd}$ | $\sigma_{n\text{-}n'\ 3rd}$ | $\sigma_{n,\alpha}$ | $\sigma_{n,p}$ |
|---|---|---|---|---|---|---|---|
| $^1$H | 692.0 | 0.0 | 0.0 | 0.0 | 0.0 | 0.0 | 0.0 |
| $^{12}$C | 1303.2 | 426.9 | 184.7 | 0.9 | 9.9 | 72.7 | 0.2 |
| $^{14}$N | 1628.6 | 399.3 | 14.9 | 26.7 | 15.3 | 60.1 | 54.0 |
| $^{16}$O | 1611.1 | 508.5 | 27.0 | 82.5 | 43.0 | 109.0 | 43.7 |
| $^{19}$F | 1740.7 | 164.2 | 0.3 | 36.8 | 0.3 | 21.3 | 14.7 |
| $^{31}$P | 1831.7 | 53.9 | 0.2 | 0.2 | 0.1 | 126.9 | 91.9 |
| $^{32}$S | 1829.7 | 378.9 | 99.3 | 10.3 | 18.0 | 159.6 | 247.4 |
| $^{35}$Cl | 2100.0 | 820.0 | 5.8 | 5.2 | 12.0 | 137.3 | 98.0 |
| $^{75}$As | 3456.2 | 685.1 | 0.8 | 0.5 | 7.3 | 10.1 | 19.0 |

Table 2. 14.1-MeV neutron induced nuclear reaction cross sections (in millibarns): $\sigma_{tot}$ – the total neutron cross-section; $\sigma_{inl}$ – the inelastic neutron cross-section; $\sigma_{n\text{-}n'\ 1st\ level}$ – the (n,n') cross-section which excites the nucleons to the first nuclear level; $\sigma_{n\text{-}n'\ 2nd\ level}$ – the (n,n') cross-section which excites the nucleons to the second nuclear level; $\sigma_{n\text{-}n'\ 3rd\ level}$ – the (n,n') cross-section which excites the nucleons to the third nuclear level; $\sigma_{n,\alpha}$– the (n,α) cross-section; and $\sigma_{n,p}$ – the (n,p) cross-section

| Isotope | $\sigma_{total}$ | $\sigma_{inl}$ | $\sigma_{n\text{-}n'\ 1st}$ | $\sigma_{n\text{-}n'\ 2nd}$ | $\sigma_{n\text{-}n'\ 3rd}$ | $\sigma_{n,\alpha}$ | $\sigma_{n,p}$ |
|---|---|---|---|---|---|---|---|
| $^1$H | 2683.6 | 0.0 | 0.0 | 0.0 | 0.0 | 0.0 | 0.0 |
| $^{12}$C | 1595.3 | 0.0 | 0.0 | 0.0 | 0.0 | 0.0 | 0.0 |
| $^{14}$N | 1512.6 | 0.0 | 0.0 | 0.0 | 0.0 | 70.2 | 22.4 |
| $^{16}$O | 561.4 | 0.0 | 0.0 | 0.0 | 0.0 | 0.0 | 0.0 |
| $^{19}$F | 2763.5 | 995.3 | 246.3 | 346.8 | 99.4 | 0.01 | 0.0 |
| $^{31}$P | 3036.1 | 448.3 | 448.3 | 0.0 | 0.0 | 0.0 | 30.8 |
| $^{32}$S | 3422.6 | 6.9 | 0.0 | 0.0 | 0.0 | 129.9 | 58.2 |
| $^{35}$Cl | 3050.4 | 428.3 | 124.4 | 243.9 | 0.0 | 4.1 | 32.0 |
| $^{75}$As | 3238.3 | 1728.5 | 37.0 | 60.0 | 78.4 | 0.0 | 0.02 |

Table 3. 2.45-MeV neutron induced nuclear reaction cross sections (in millibarns)

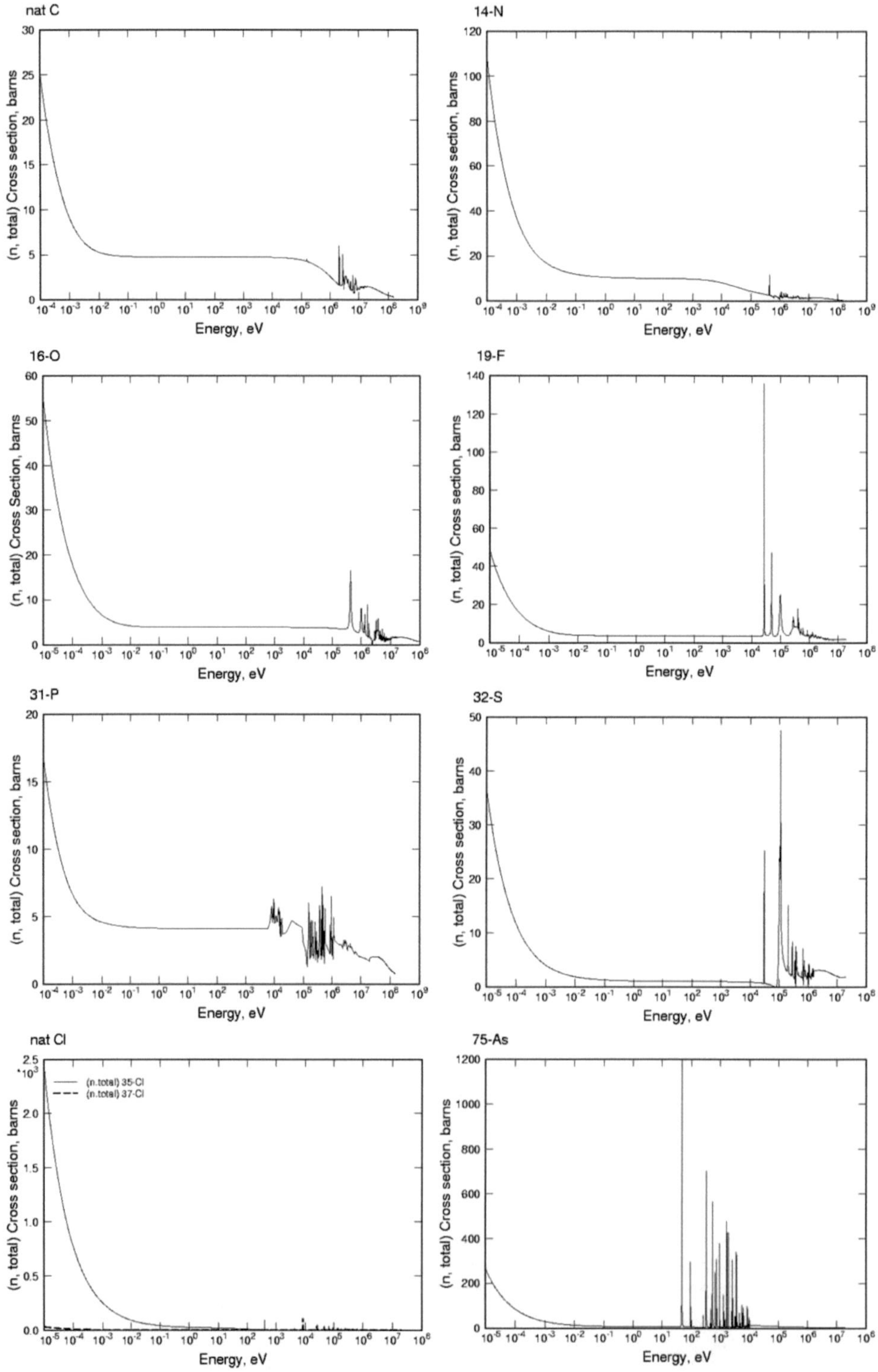


Fig. 2. Neutron cross sections (n, total) for C, N, O, F, P, S, Cl, and As

| Isotope | $^{1}H$ | $^{12}C$ | $^{14}N$ | $^{16}O$ | $^{19}F$ | $^{31}P$ | $^{32}S$ | $^{35}Cl$ | $^{75}As$ |
|---|---|---|---|---|---|---|---|---|---|
| $\sigma_{th}$ | 332.7 | 3.5 | 79.8 | 0.2 | 9.7 | 172.7 | 548.1 | 33070.2 | 4528.3 |

Table 4. Thermal neutron capture reaction cross sections at $E_n$=0.025 eV (in millibarns)

The representative set of elements (H, C, N, O, F, P, S, Cl, and As) is selected as an example of isotopes found in explosive, chemical threats, coal and other materials. The parameters of the d-t neutron induced nuclear reactions are shown in Table 2. Table 3 shows the d-d neutron induced nuclear reactions at $E_n$=2.45 MeV for the same set of elements. The (n,$\gamma$) thermal neutron capture reaction's cross-section at $E_n$=0.025 eV are shown in Table 4. The total neutron cross-sections $\sigma_{tot}$ for C, N, O, F, P, S, Cl, and As are shown in Fig.2. The neutron data for other isotopes are available from website of the Nuclear Information Service of the Los Alamos National Laboratory at http://t2.lanl.gov/data/data.html.

The (n,2n) reaction is also utilized to produce the excited energy states causing the delayed beta decay with associated photon emission that may be non-fingerprint in nature, but it may assist to identify the amount of a parent isotope in the sample. The fast neutron activation of this type can be used for example to measure amount of nitrogen in a sample via reaction $^{14}N(n,2n)^{13}N$. The produced $^{13}N$ isotope has $t_{1/2}$=10.1 minutes emitting positrons. They annihilate immediately with electrons in the sample matrix emitting 511 keV gamma rays. Although it is not characteristic photon energy, it indicates the presence of a positron emitter. If measured correctly in time and associated with $t_{1/2}$ of $^{13}N$ isotope, this signature can be used in material analysis. The issue of such approach is the possibility of the beta annihilation photon's emission by other $^{13}N$-producing parent nuclei. For example, elements that may cause neutron based production of $^{13}N$ are boron and oxygen. The 1.47-MeV alpha particles may be initiated by thermal neutrons via $^{10}B(n,\alpha)^{7}Li$ reaction, producing 511-keV photons through the $^{10}B(\alpha,n)^{13}N$ reaction. Knockout protons of high energy produced by fast neutrons may initiate the 5.5-MeV-threshold $^{16}O(p,\alpha)^{13}N$ reaction. The $^{63}Cu(n,2n)^{62}Cu$ reaction can produce the positron emitter $^{62}Cu \rightarrow {}^{62}Ni + e^{+}$ ($t_{1/2}$=9.8 minutes). So, the 511-keV annihilation photons emitted from copper and nitrogen nuclei have a close half-life values. Thus the use of other gamma ray signatures utilizing other reactions in conjunction with the positron annihilation would be beneficial in the material analysis.

## 4. Characteristic gamma radiation

As a result of nuclear reactions involving the isotopes contained in the object under scrutiny, exited nuclei emit gamma rays with specific energies in the de-excitation process. They act as the "fingerprints" of these isotopes. Most $\gamma$-rays are emitted promptly after the reaction. The "prompt" photon emission from excited nucleus occurs within approximately $10^{-9}$ seconds after initial excitation. However, in some cases, a nucleus with a half-life of a few seconds to a couple of minutes is formed. This radioactive nucleus decays to a daughter nucleus emitting various particles ($\alpha$, $\beta^{+}$, $\beta^{-}$, etc.) and delayed photons. The prompt gamma ray emission occurs either in the single transition as it happens in the case of hydrogen 2.223-MeV gamma rays, or through several transitions emitting many prompt $\gamma$-rays of lower energy. The examples of energy level schemes for $^{12}C$ and $^{16}O$ nuclei are shown in Fig.3.

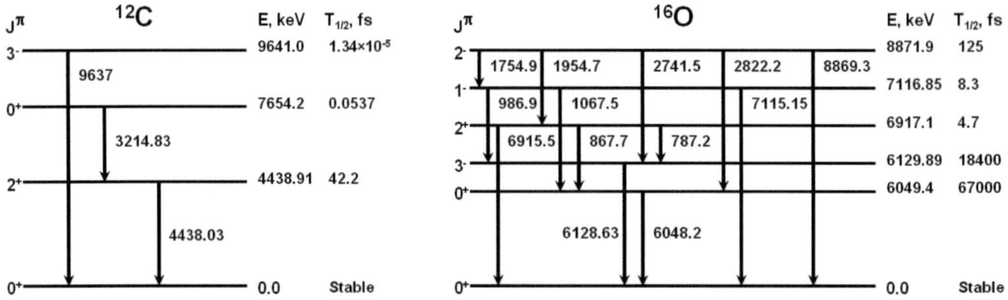


Fig. 3. Energy level schemes for $^{12}C$ and $^{16}O$

| Isotope | Reaction | $E_\gamma$, keV |
|---|---|---|
| $^{1}H$ | (n,γ) | 2223 |
| $^{12}C$ | (n,n′γ) | 4438 |
| $^{14}N$ | (n,n′γ) | 730, 1634, 2313 |
| | (n,γ) | 1885, 5269, 5298, 10829, 10318 |
| $^{16}O$ | (n,n′γ) | 5618, 6129 |
| $^{19}F$ | (n,n′γ) | 197, 1236, 1348, 1357 |
| | (n,γ) | 582, 2453, 3589 |
| $^{31}P$ | (n,n′γ) | 1266, 2028, 2233 |
| | (n,γ) | 636, 2154, 3900, 6785 |
| $^{32}S$ | (n,n′γ) | 1273, 2230 |
| | (n,γ) | 841, 2380, 3221, 5420 |
| $^{35}Cl$ | (n,γ) | 788, 1165, 1951, 1959, 6111, 7414 |
| $^{75}As$ | (n,n′γ) | 199, 265, 280, 573 |
| | (n,γ) | 165, 472, 1534, 6810 |

Table 5. Characteristic gamma rays

The intensities of the obtained specific gamma rays provide information about the number of atoms in the sample. Hence, the information on its chemical composition can be extracted from the measured gamma ray spectrum. The list of isotopes, nuclear reactions, and energies of most prominent characteristic gamma rays are shown in Table 5. Emitted due to neutron induced reactions photons are highly penetrating. For example, energy of gamma rays emitted from nuclear reactions on nuclei of carbon, oxygen, and nitrogen isotopes is between 4 and 11 MeV. Table 5 is not inclusive. The prompt gamma rays for other elements can be found in the following libraries. The prompt gamma rays from thermal neutron captures (n,γ) are catalogued in the library for natural elements (Lone et al., 1981). It lists the prompt gamma ray energies in the range from 23 keV to 10829 keV for all isotopes, in terms of gamma rays emitted per 100 neutron radiative captures. These data are also available online from the National Nuclear Data Center (Brookhaven National Laboratory) at http://www.nndc.bnl.gov/capgam/. Zhou Chunmei compiled the thermal neutron capture data for nuclides with A>190 (Chunmei, 2001) and new evaluation data for thermal neutron capture for elements A=1-25: level properties, prompt gamma rays, and decay scheme properties (Chunmei, 2000). The experimental data on the (n,n′γ) photons are compiled by

Demidov and colleagues in the IAEA document INDC-CCP-120 (Demidov et al., 1978). The gamma ray spectra from inelastic scattering were measured for all elements except unstable isotopes and noble gases.

The gamma spectrum obtained from ammonium nitrate sample irradiated with a d-t source is shown in Fig.4. Spectrum was measured using HPGe detector. We would like to note the spectral feature of Doppler broadening that is specific for photons induced by 14.1-MeV neutrons on light nuclei. The two inset expanded spectra in Fig.4 show the gamma ray from $^{12}C(n,n'\gamma)^{12}C$, 4.438 MeV and $^{16}O(n,n'\gamma)^{16}O$, 6.13 MeV. It is readily apparent that the gamma ray peak from $^{12}C$ is much wider than the gamma ray peak from $^{16}O$. Other causes of this widening such as electronic noise, crystal damage due to neutron irradiation can be dismissed since $^{16}O$ does not have any evidence of the broadening. The broadening of the gamma ray peaks for light nuclei was studied in (Womble et al., 2009). The energy levels of the nucleus have different spins and parities, and the state's life times. For example, in $^{16}O$, the 2nd excited state with the energy of 6.13 MeV has a half-life of 18.4 ps (see Fig.3). The 3rd excited state has a half-life nearly 2000 times shorter (8.3 fs). Energies of these two states are close to each other, but the 2nd excited state to ground state transition is $3^- \rightarrow 0^+$ and the 3rd excited state to ground state transition is $2^+ \rightarrow 0^+$. Thus the difference in half-life is due to the transition probability of producing E3 radiation versus E2 radiation. The half-life time of 4.43-MeV level in $^{12}C$ is 42 fs. Carbon and oxygen nuclei recoiling in inelastic neutron scattering reactions under 14.1-MeV neutrons have similar stopping times moving in the matrix of the sample; for example, approximately 1800 fs for the $NH_4NO_3$ sample. Therefore $^{12}C$ nucleus may emit photon while in motion exhibiting the Doppler broadening effect for the 4.43-MeV peak, but $^{16}O$ nucleus is stopped before the emission of the 6.13-MeV gamma ray and therefore does not experience the peak broadening in the measured spectrum.

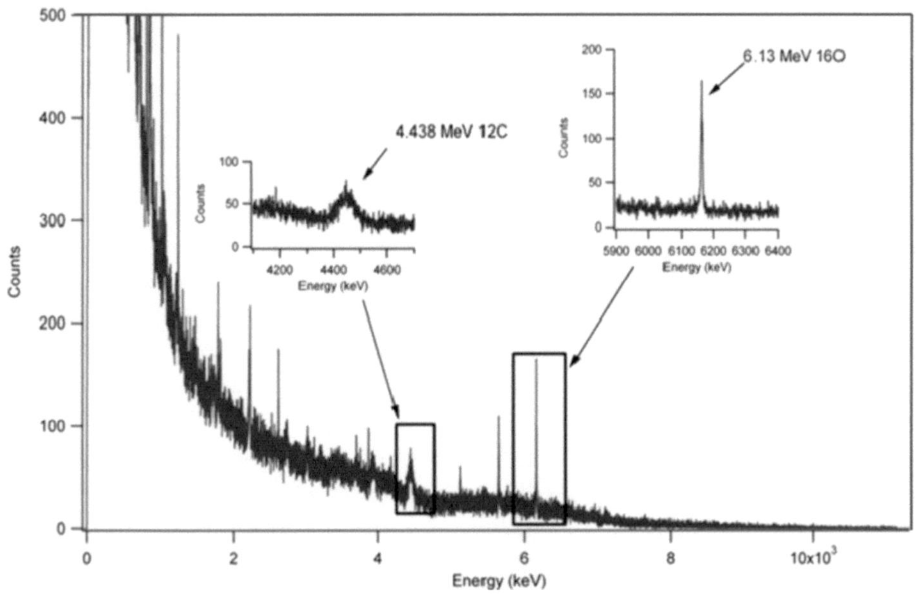


Fig. 4. d-t neutron induced gamma ray spectrum for ammonium nitrate

## 5. Methods and applications

This section covers a number of systems that perform bulk material analysis using neutron induced gamma spectroscopy. Although this is by no means a complete list, it represents systems and techniques that have been utilized in the past twenty years. Some of these systems are still in the market and some never made it to market. Some of the systems have different acronyms or different trade names but rely on the same physical principles:

- PGNAA - Prompt Gamma Neutron Activation Analysis
- PFNA - Pulsed Fast Neutron Analysis
- PFTNA - Pulsed Fast/Thermal Neutron Analysis
- PFNTS - Pulsed Fast Neutron Transmission Spectroscopy
- API – Associated Particle Imaging

### 5.1 Prompt gamma neutron activation analysis

Admittedly, Prompt Gamma Neutron Activation Analysis (PGNAA) is usually thought of as a continuous (or DC) source technique. In this technique, fast neutrons from a radioisotope, a neutron generator or a reactor impinge upon a sample. The sample then emits gamma rays through (1) prompt gamma ray emission from gamma decay, (2) gamma ray emission through short-lived beta decay, or (3) prompt gamma ray emission due to exoergic nuclear reactions. By the latter, we mean gamma ray emission to satisfy conservation of energy (or mass) in a nuclear reaction. The most well-known gamma rays from this type of reaction are the 2.22 MeV gamma ray from the $^{1}H(n,\gamma)$ reaction and the 10.8 MeV gamma ray from the $^{14}N(n,\gamma)$ reaction.

Pulsed neutron sources can be used in PGNAA systems. However, using pulsing neutron generators without taking advantage of the pulsing mechanism is not efficient.

### 5.2 Pulse fast neutron analysis

Tsahi Gozani, Peter Sawa, and Peter Ryge conceived of Pulsed Fast Neutron Analysis (PFNA) in 1987 while they were working at SAIC (Gozani, 1995). In PFNA, a large accelerator creates a deuteron beam which is directed at a deuterium gas target. A chopper, which consists of strong electric field that periodically sweeps the deuteron beam away from the target, creates neutron pulses of a few nanoseconds duration. In described system, user has control of the chopper, so moment when neutrons are created and duration of the neutron pulse are known precisely. The PFNA uses time-of-flight (TOF) methods to obtain a favourable signal-to-noise ratio (SNR) to detect various chemicals. The accelerator used in PFNA is typically an 8 MV Van de Graff accelerator. The D(d,n) reaction has a low Q-value (approximately 100 keV). Thus any energy above the Q-value is mostly transferred to the kinetic energy of the neutron, producing a neutron with a maximum kinetic energy of 8 MeV. The velocity of 8 MeV neutrons is about 6 cm/ns. The user, then, knows exactly when the neutron pulse is made and now can estimate its position at any time after its creation.

Furthermore, due to the high kinetic energy of the deuteron beam, the neutron's momentum is parallel to the momentum of the deuteron. The developers took advantage of this fact and used a sophisticated system to "raster" the neutron beam across the object of interest. The neutron beam also has a small angular divergence and estimates (Strellis, 2009) are that the beam is 9 cm × 12 cm in the center of the object of interest.

PFNA systems can be used to screen very large cargo shipments such as tractor-trailer shipping containers and airport shipping containers. A large, 2-dimensional array of NaI gamma ray detectors covers the cross-section area of the object under scrutiny.

The energies of the gamma rays emitted from the object are plotted against the TOF of the neutron. This creates a two-dimensional array of data that looks similar to a spectrogram in that the intensity of the gamma ray at a particular TOF is represented by a color using the RGB color scheme. In this array of data, color bands parallel to the TOF axis indicate constant gamma ray background such as from normally occurring radioactive materials (NORM). Color bands parallel to the energy axis represent the gamma ray spectra of volume elements ("voxels") within the object of interest. The volume element size is based on the time resolution of the system so the voxels are approximately 5-cm thick. For example, in the center of the container the voxel is 9 cm × 12 cm × 5 cm. The small voxel size increases the SNR of the system. Another benefit is that the lifetime of certain activation products can be measured and this gives more data upon which to identify the material.

At early development stage, price and size were the drawbacks of using PFNA. However, since the 9/11 attacks the main challenge is the system cost. The cost includes installation and maintenance of this complex system. In 2009, there was a single system working at the George Bush Intercontinental Airport (Strellis, 2009). As of this writing, we are aware of no other installations.

### 5.3 Pulsed fast / thermal neutron analysis

The Pulsed Fast/Thermal Neutron Analysis (PFTNA) is a technique used in conjunction with small, portable electronic neutron generators. It was originally developed by George Vourvopoulos, Phillip Womble, and Frederick Schultz and presented in (Womble et al., 1995). Unlike PFNA, which has pulse duration of approximately 2 ns, PFTNA employs pulses with a minimum duration of 5 μs. Longer pulse duration significantly reduces cost of PFTNA systems. The PFNA system can be used in a "macro-pulse" mode, in which the neutron beam is turned off for a period of 100 μs. This "macro-pulse" mode mimics the PFTNA system's mode.

The advantage of the PFTNA systems is an ability to separate the gamma ray spectrum of inelastic scattering reactions (n,n'γ) from thermal neutron capture (n,γ) and activation reactions (e.g. (n,p)) gamma-ray spectra. The data acquisition system collects data during the neutron pulse at one memory address and then switches to another memory address to acquire data between pulses. The data collected during the pulse is primarily from (n,n'γ) reactions and the data collected between pulses is primarily from (n,γ) reactions. Often systems are designed to be shut off for a few minutes to collect short-lived activation products such as $^{16}O(n,p)$ ($t_{1/2} \approx 16$ s). It is a common misconception that the frequency and duration of the neutron pulses is chosen to maximize the data from the (n,n'γ) reactions. In fact these parameters are chosen to maximize the (n,γ) reactions or more precisely the thermal neutron flux. The neutron pulse frequency determines whether the thermal neutron flux is kept near constant or if it is allowed to diffuse. Applications such as differential die-away analysis (DDA), a method of measuring fissile content, allow the thermal neutron flux to completely diffuse and use pulse frequencies less than 1 kHz. PFTNA systems use pulse frequencies greater than 5 kHz to ensure that the thermal neutron flux is nearly constant for the entire period of measurement (Vourvopoulos & Womble, 2001). Our personal experience in this area has shown that frequencies higher than 10 KHz may be desirable as well.

For a d-t or d-d neutron generator, the typical pulsing method is to clamp the so-called source voltage using "clamping circuits". The source voltage causes ionization of the deuterium gas before the ions are accelerated. A consequence of the higher pulse frequency is the shorter pulse duration. This is due to the fact that these clamping circuits operate at a constant duty cycle. The source voltage duration must be a few microseconds (approximately 4 μs) for the deuterium gas to reach a pressure where ionization occurs (the "fill time"). With this condition along with the constant duty cycle, the maximum neutron pulse frequency is about 20 kHz since higher frequencies (>25 kHz) will not have a sufficiently long fill time. Thus PFTNA pulses are typically 10 μs in duration with a pulsing frequency of 10 kHz. The PFTNA scheme with the neutron pulse's time structure is shown in Fig.5.

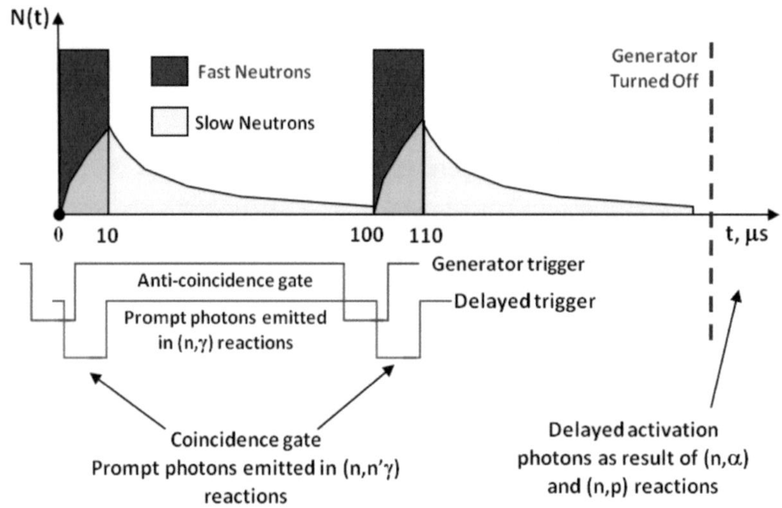


Fig. 5. Pulse fast thermal neutron analysis scheme

As discussed earlier, PFTNA method uses two different memory addresses depending whether the neutron generator is on or off. The use of two memory addresses is sometimes described as "ping-ponging" since the data "bounces" between two addresses. A gate signal is sent from the neutron generator to the data acquisition to indicate whether the generator is on or off. The gate signal is usually delayed from the rise of the source voltage by the fill-time. Furthermore the gate signal will extend past the fall-time of the source voltage by a few microseconds. This lag is due to the processing time of the data acquisition system. One of the reasons that PFTNA systems use fast data acquisition electronics is to minimize this lag.

Another reason to use fast data acquisition electronics in PFTNA systems is high counting rate in the detector when the neutron generator is producing neutrons. Data acquisition rates in the system during this period can exceed 100 kcps. High rates such as these can overwhelm HPGe detectors and analog amplifies. Modern digital electronics can cope with these rates but as rates approach 1 Mcps, scintillation detectors such as sodium iodide or bismuth germanate can be overwhelmed. Count rate limitations force PFTNA system designers to place shielding material between the detector and the neutron generator that adds to system weight.

The d-d and d-t fusion reactions take place at low momentum which means that the neutrons are emitted isotropically. These systems typically consume about 100 W during operation. The average neutron outputs for PFTNA systems are $10^8$ n/s for d-t based systems and $10^6$ n/s for d-d based systems. For d-t systems, this leads to radiological concerns to personnel which can be mitigated by distance (approximately 8 meters stand-off for unshielded operation) or shielding (approximately 30-50 cm shielding).

The benefits of PFTNA systems are their smaller size and relatively low cost. However, these features lead to a lower SNR compared to PFNA systems. Some research teams have suggested combining PFTNA method with the associated particle imaging technique to improve the SNR.

### 5.4 Associated particle imaging

In associated particle imaging (API), the recoiling residual nucleus, e.g. the alpha particle for d-t reaction, is used to perform time-of-flight and direction selectivity. SNR could be greatly improved for (n,n'γ) gamma ray spectra by measuring gamma ray signals that are emitted only from the selected volume. However, application of this technique would have no effect on the SNR of the (n,γ) or the time delayed activation gamma ray spectra.

The scheme of API technique is shown in Fig.6. The d-t fusion reaction produces alpha particle and 14.1-MeV fast neutron that are emitted in opposite directions due to linear momentum conservation. The segmented alpha detector installed inside the sealed neutron generator tube is used for detection of the α-particle event's position and time to "tag" the direction of the 14.1-MeV neutron (Koltick et al., 2009). The geometry of segments of the alpha detector and the neutron's times-of-flight define the geometry of "voxels" for the 3D analysis. ZnO(Ga) detector was used as an alpha detector. It was found that detector's efficiency is about 90% for 3.49-MeV alpha particles. The phosphor coating emits ~15 photoelectrons / alpha; its scintillation emission peaks at 390 nm with ~3.3-ns decay time, allowing up to $\sim 2\times 10^{10}$ n/s output for 2% tagged solid angle without significant pile-up (Cooper et al., 2003).

The alpha particle detection event and gamma ray detection event are both stamped with the timing signals. The DAQ system is set up to produce the logic signal when both events (the alpha particle and the photon detection) are recorded within a short time interval - the "coincidence window". This logic signal is used to select those gamma ray signals in the energy spectrum that arrive from the tagged voxel. The 14.1-MeV neutron travels in air with the velocity ~5 cm/ns. The 3.49-MeV α-particle has the velocity ~1.3 cm/ns. Thus the coincidence window should be in the order of nanoseconds. The quality of the timing signals for both detectors should be very high, without jitter. The width of the coincidence window and the neutron flux are interconnected: the random coincidence rate increases with the higher neutron flux thus limiting neutron yield of the generator.

The API technique was used in such systems as SENNA (Vakhtin et al., 2006), EURITRACK (Perret et al., 2006), and UNCOSS (Eleon et al., 2010).

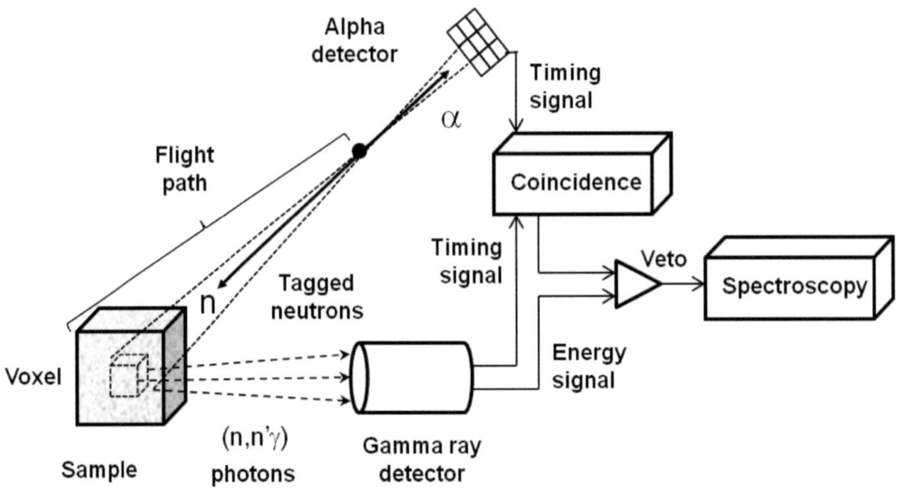


Fig. 6. Associated particle imaging technique

### 5.5 Pulsed fast neutron transmission spectroscopy

Pulsed fast neutron transmission spectroscopy (PFTNS) is the only technique in this section, which examines the resulting neutron spectrum, instead of the gamma ray spectrum. In this technique, a broad energy beam of neutrons is directed at an array of neutron detectors. The object under interrogation is passed through the beam and the resulting attenuated neutron spectrum is measured using the neutron detectors. This method is the same method that researchers use to perform neutron cross-section measurements.

The pulsing in PFTNS allows the system to perform neutron TOF measurements. These TOF measurements are used to determine the energy of the neutrons with flight paths of 4 to 10 m. The resulting neutron spectrum is used to estimate the attenuation of neutrons as function of energy. Light elements such as H, C, N, and O have high cross-sections for neutron attenuation at these energies. Thus the relative amounts of H, C, N, and O can be determined, and the "imaging" of elements is possible. The voxel sizes would be similar to those of PFNA due to the limits of the speed of the neutrons.

Due to the high neutron fluences and precise timing required for PFTNS, this system needs an accelerator similar to the one utilized by PFNA. The TOF path means that the systems take a large amount of space. These are two of the reasons that these systems were not widely adopted by the security community.

PFTNS was proposed as a primary or secondary screening system for airline security. Designs were proposed which would handle a large number of bags per minute. This would be achieved by having the bags ride a carousel around accelerator. The neutron detectors would be placed in a "wall configuration" and the neutron beam would raster through a number of bags. The National Academy report (NNMAB-482-6, 1999), written in 1999, was extremely critical of the utilization of PFTNS for airport security. However, Overley suggests that detection rates of 93% and false alarm rates of 4% are possible with this technique (Overley et al., 2006).

## 6. Gamma radiation spectral analysis

Neutron based material analysis methods generally require a skilled analyst to interpret the gamma ray spectral data collected, and to classify the interrogated object using the elemental parameters extracted from the spectral data. Automatic spectral analysis algorithms and the object's classification algorithms are required for real world applications where access to nuclear spectroscopy expertise is limited, or the autonomous and/or the robotic operation is necessary.

### 6.1 Analysis of neutron induced gamma ray spectra

The first step in the data analysis process is to extract the sample's elemental information from the measured gamma ray spectra. The spectrum analysis algorithms that are used for that purpose should simultaneously provide quick, accurate, and objective analysis of gamma ray spectra by evaluating the intensities of the characteristic photon peaks. For spectra measured with high resolution detectors such as HPGe, the approach can be based on the peak finding algorithm using the regions of interest (ROIs). Usually, the "blank" spectrum (measured with no sample present) is subtracted from the "sample" spectrum (measured with the sample) before the spectral analysis. It takes into account the signatures of the same elements that are present in surrounding materials, and in the sample. The "nuclear" ROI parameters such as the net peak area in counts /second units are proportional to the number of isotopes in the sample that emitted the fingerprint gamma rays. The "nuclear" parameters may be converted into other appropriate units, if needed, using the elemental calibration library (for example, "chemical" parameters accepted in the coal or the cement analysis industry, etc.). These libraries are created for the system using calibration measurements using known samples.

The simple ROI-based method may be appropriate for non-complicated spectra with the peaks that are well resolved. For spectra with many closely positioned peaks, or low resolution spectra with overlapping peaks, the peak-shape fitting algorithms are required. The mathematical method of measured spectrum fitting as the linear combination of single element's detector responses, that are measured experimentally, was developed by George Vourvopoulos and Phillip Womble (Vourvopoulos & Womble, 2001). To use this method, one must first measure the response of the low resolution detector to $\gamma$-rays from pure elements. For example, a block of pure graphite is used to determine the detector's elemental response to the carbon $\gamma$-rays. To determine the detector's elemental response to hydrogen, a response is measured from a water sample, and so on. The counts in i-th channel of the spectrum of a sample S can be represented by the equation:

$$S_i = K \cdot BL_i + \sum_{j=1}^{m} A_j \cdot RF_{i,j} \tag{4}$$

Here: $BL_i$ is the blank spectrum at the i-th channel and K is its coefficient; $RF_{i,j}$ is the detector's elemental response of the j-th element at the i-th channel and $A_j$ is its coefficient, and $m$ is the total number of elements used for this spectrum decomposition procedure. The coefficients K and $A_j$ are found by the least squares algorithm minimizing the $\chi^2$ to find the "best fit". As the result of this decomposition procedure, the intensities of peaks of j elements used in this fitting are found in counts / second.

Another spectral decomposition technique developed by Robin Gardner and colleagues (Shyu et al., 1993) utilizes the detector's elemental responses that are calculated using Monte Carlo methods. The experimental method of detector's elemental response generation provides detailed realistic spectral features (i.e. electronics noise, peak broadening, neutron activation effects, etc.), but it is time intensive, and the set of pure element samples may be limited. The computational method allows generation of responses for the larger set of elements, but it may be problematic to represent detailed spectral features because not all processes in a Monte Carlo code may be taken into account.

Bruce Kessler applied the original mathematical method based on multi-wavelets to analyze the neutron induced photon spectra (Kessler, 2010). In this approach, the set of special scaling vector components was developed for spectrum fitting. Wavelet decompositions ignore signal components up to the approximation space of the basis, so the wavelet analysis is used to look for patterns over the top of spectral "noise". The measured sample spectrum wavelets are decomposed using a variable linear combination of the wavelets from the decompositions of detector's elemental responses providing the intensities of characteristic gamma ray peaks. The algorithm was shown to be effective for both high resolution and low resolution spectra.

### 6.2 Classification algorithms

The object's classification algorithms are responsible for material identification using the characteristic gamma ray peak data that are produced by the spectral analysis algorithms. The classification uses the fact that the amount of particular isotopes varies for different materials (i.e. based on their chemical formula, taking the reaction cross sections into account).

The suitable approach is to represent the measurement result as a "point" in the space of several parameters (elemental intensities). Different materials containing the same isotopes but in different ratios are represented by points that are segregated in such "elemental" space. The dimension of this space is determined by the number of isotopes.

In general, the "nuclear" data obtained with neutron based systems differ from elemental composition evaluations based on chemical formula due to several reasons such as statistical nature of nuclear reactions, short measurement times, presence of radiation shielding, and other environmental conditions. Thus, the chemical compound measured in various conditions is represented not by the single "point" in the elemental space, but rather by a cloud-like set of points, where each point corresponds to one measurement.

Clouds have circular structure, and create specific patterns for classes of different materials. Fig. 7 shows a two-dimensional slice of the elemental space. It represents two isotopes: oxygen and nitrogen. Four materials that contain different amounts of oxygen and nitrogen were used as samples for measurements in various environmental conditions. The (nitrogen, oxygen) points from many measurements shape the 2D clouds on the figure.

The classification decision is made using the boundaries calculated to separate these classes. For the well separated clouds, the boundaries can be found easily, and simple decision making logic "trees" can be constructed. But in many cases, the patterns for threats and innocuous materials are overlapped in the multidimensional space of parameters making the differentiation task challenging for classical decision-tree algorithms.

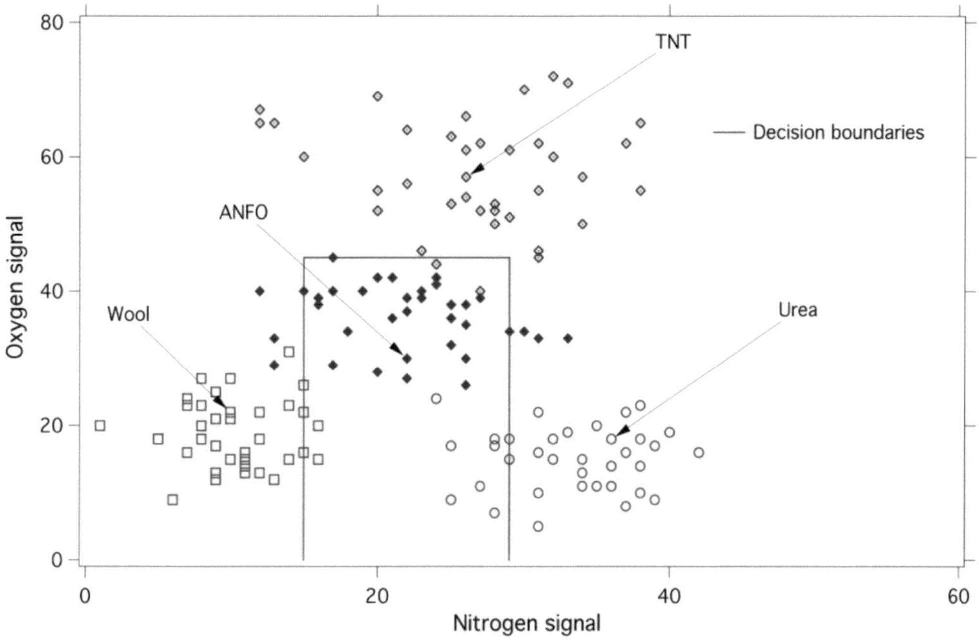


Fig. 7. Data points for four different substances

The decision-tree algorithm for the identification of the particular threat material ("ANFO") amongst four classes of materials is shown in Fig.8. It can be described as the following: if detected oxygen signal lies between lower and higher thresholds, then nitrogen signal is checked. If nitrogen signal lies between lower and higher thresholds, then substance can be identified as "ANFO". The thresholds in the decision tree ($N_{low}$, $N_{high}$, $O_{high}$) can be varied in order to achieve better results. Each threshold value can be represented as a horizontal (oxygen threshold) or vertical (nitrogen threshold) line. In order to pick best low and high thresholds, parameters of the decision making tree were varied and Receiver Operating Characteristics (ROC) curves were plotted. In signal detection theory, a ROC curve is a graphical plot of true positive rate (or sensitivity) versus false positive rate for a binary classifier system as its discrimination threshold is varied. The ROC-analysis provides tools to select possibly optimal decision boundaries. The ROC curve methodologies are discussed elsewhere (Fawcett, 2006).

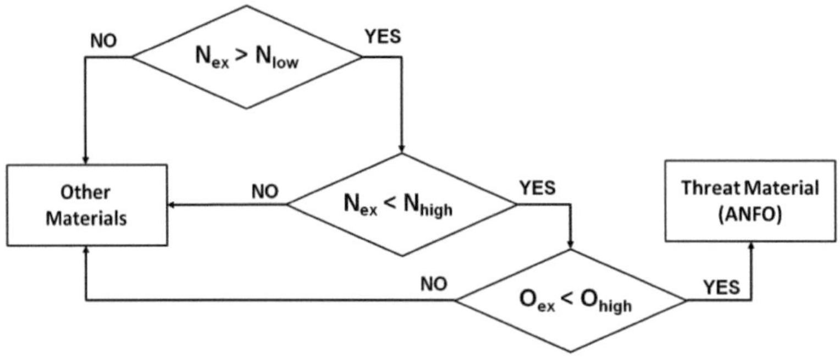


Fig. 8. Example of the decision tree used to analyze data from Fig. 7

Each set of parameters of the decision-making algorithm corresponds to the point on the ROC curve (or ROC surface). Therefore, we define the optimal parameters for the decision-making algorithm as a set of parameters, which allow the minimum "decision" vector magnitude from (0,1) point (left-upper corner of the ROC graph) to the corresponding point on the ROC curve (see Fig.9). The best low and high thresholds were selected by variation of parameters of the decision making tree aimed to determine the ROC curve with the minimal decision vector length. The optimal decision boundaries for identifying the ANFO material are shown as black lines in Fig.7.

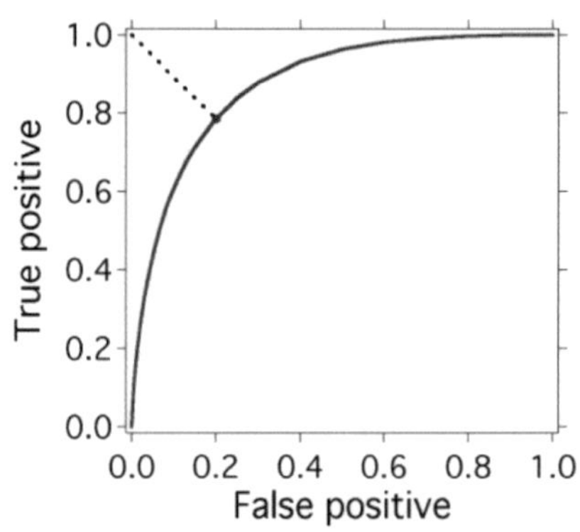


Fig. 9. ROC curve

It is clear that this algorithm does not satisfactory identify substances when classes are overlapping. For example, True Positive rate of the classification between ANFO and urea is only 75%, which is unacceptable for the field deployable system. To improve performance of the classifier, the linear boundary was used. In the case of general linear boundary, the decision making algorithm can be described as the following: if point with experimentally measured nitrogen and oxygen counts ($N_{ex}$, $O_{ex}$) lies below the line that is defined as $O = k \cdot N + \ell$, then this point belongs to the class A, if it lies above that line, then it belongs to the class B. The parameters for the linear boundary (the slope and the offset) were varied, and the ROC curves were generated. The optimal pair of parameters corresponds to the minimal decision vector magnitude. This approach tested with the same data set as shown in Fig. 7 produced better results: true positive rates for all classifiers are better than 95%. Optimal linear boundaries are shown in Fig.10 as black lines.

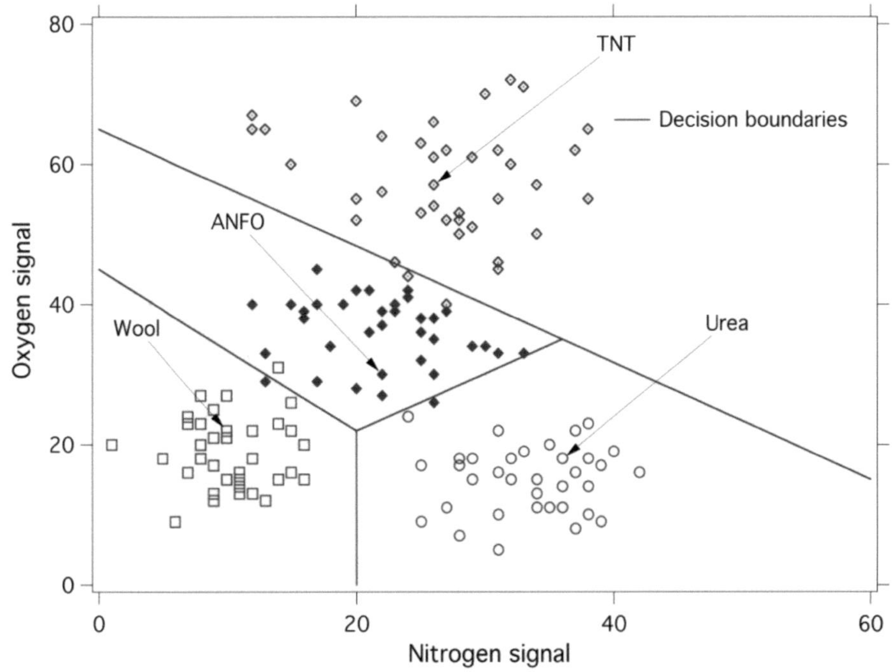


Fig. 10. Linear boundaries between four classes

The use of linear boundaries is significantly improving the material identification capabilities of the neutron based system. The use of polynomial functions is a natural generalization of this approach. Other pattern recognition methods can also be used to construct the decision boundaries of complex shapes and can be applied to analyze the detector signals - for example, methods based on R-functions (Bougaev & Urmanov, 2005).

## 7. Conclusion

This chapter provided an overview of several material analysis methods using different nuclear reactions induced by pulse neutrons: PGNAA, PFNA, PFTNA, PFTNS, and API. These methods utilize the characteristic gamma radiation and other radiation signatures, prompt and delayed in time, to measure the elemental content of unknown bulk samples. The pulse neutron based elemental analysis is the non-intrusive, non-destructive technique that has yielded the development of in situ material characterization systems in many areas: process control in industry, medicine, security, geological and environmental studies, and others. These applications require automatic, rapid spectra analysis and sample classification algorithms to be effective for the real world use. The methods of spectral decomposition using the combination of single element's detector responses proved to be effective. The pattern recognition methods shown true positive rates ~95% in the material classification.

## 8. References

Aleksandrov, V.D.; Bogolubov, E.P.; Bochkarev, O.V.; Korytko, L.A.; Nazarov, V.I.; Polkanov, Yu.G.; Ryzhkov, V.I. & Khasaev, T.O. (2005). Application of Neutron

Generators for High Explosives, Toxic Agents and Fissile Material Detection, *Applied Radiation and Isotopes*, Vol. 63, pp. 537-543, ISSN 0969-8043

April, E.; Bolotnikov, A.E.; Bolozdynya, A.I & Doke, T. (2006). *Noble Gas Detectors*, Wiley-VCH, 1st Edition, ISBN 978-3527405978

Barzilov, A. & Womble, P.C. (2006). Comparison of Gamma-Ray Detectors for Neutron-Based Explosives Detection Systems, *Transactions of American Nuclear Society*, Vol. 94, p. 543, ISSN 0003-018X

Barzilov, A. & Womble, P.C. (2003). NELIS - a Neutron Inspection System for Detection of Illicit Drugs, *AIP Conference Proceedings*, Vol. 680, pp. 939-942, ISSN 1551-7616

Bougaev, A. & Urmanov, A. (2005). R-functions Based Classification for Abnormal Software Process Detection, In: *Computational Intelligence and Security*, J. Carbonell and J. Siekmann, (Eds.), 991-996, Lecture Notes in Computer Science, Vol. 3801/2005, Springer, ISBN 3-540-30818-0, Berlin / Heidelberg, Germany

Chunmei, Z. (2001). Thermal-Neutron Capture Data Update and Revision for Some Nuclides with A>190, INDC-CPR-0055, Nuclear Data Services, International Atomic Energy Agency, Vienna, Austria

Chunmei, Z. (2000). Thermal Neutron Capture Data for A=1-25, INDC-CPR-0051, Nuclear Data Services, International Atomic Energy Agency, Vienna, Austria

Cooper, J.C.; Koltick, D.S.; Mihalczo, J.T. & Neal, J.S. (2003). Evaluation of ZnO(Ga) Coatings as Alpha Particle Transducers within a Neutron Generator, *Nuclear Instruments and Methods in Physics Research A*, Vol. 505, Issues 1-2, pp. 498-501, ISSN 0168-9002

Cooper, J.C. & Koltick, D.S. (2001). Optimization of Time and Energy Resolution at High Count Rates with a Large Volume Coaxial High Purity Germanium Detector, *Nuclear Science Symposium Conference Record*, IEEE, Vol.4, pp. 2420-2423, ISBN 0-7803-7324-3

Demidov, A.M.; Govor, L.I. & Cherepantsev, Yu.K. (1978). *Atlas of Gamma-Ray Spectra from Inelastic Scattering of Reactor Fast Neutrons*, INDC-CCP-120, Nuclear Data Services, International Atomic Energy Agency, Vienna, Austria

Dep, L.; Belbot, M.; Vourvopoulos, G. & Sudar, S. (1998). Pulsed Neutron-Based On-Line Coal Analysis, *Journal of Radioanalytical and Nuclear Chemistry*, Vol. 234, No.1-2, pp. 107-112, ISSN 0236-5731

Eleon, C.; Perot, B.; Carasco, C.; Sudac, D.; Obhodas, J. & Valcovic, V. (2010). Experimental and MCNP Simulated Gamma-Ray Spectra for the UNCOSS Neutron-Based Explosive Detector, *Nuclear Instruments and Methods in Physics Research A*, Vol. 629, pp. 220-229, ISSN 0168-9002

Fawcett, T. (2006). An introduction to ROC analysis, *Pattern Recognition Letters*, Vol. 27, pp. 861–874, ISSN 0167-8655

Fourches, N.; Walter, G. & Bourgoin, J.C. (1991). Neutron-Induced Defects in High-Purity Germanium, *Journal of Applied Physics*, Vol. 69, pp. 2033-2043, ISSN 0021-8979

Gozani, T. (1995). Understanding the Physics Limitations of PFNA - the Nanosecond Pulsed Fast Neutron Analysis, *Nuclear Instruments and Methods in Physics Research B*, Vol. 99, pp. 743-747, ISSN 0168-583X

Holslin, D.T.; Shyu, C.M.; Sullivan, R.A. & Vourvopoulos, G. (2006). PELAN for Non-Intrusive Inspection of Ordnance, Containers, and Vehicles, *Proceedings of SPIE*, Vol. 6213, p. 621307, ISSN 0277-786X

James, W.D. & Fuerst, C.D. (2000). Overcoming Matrix Effects in the 14-MeV Fast Neutron Activation Analysis of Metals, *Journal of Radioanalytical and Nuclear Chemistry*, Vol. 244, No.2, pp. 429-434, ISSN 0236-5731

Kessler, B. (2010). An Algorithm for Wavelet-Based Elemental Spectrum Analysis, 13th International Conference on Approximation Theory, San Antonio, TX, March 7-10, 2010. Available at: http://works.bepress.com/bruce_kessler/65

Koltick, D.S.; Kane, S.Z.; Lvovsky, M.; Mace, E.K.; McConchie, S.M. & Mihalczo, J.T. (2009). Characterization of an Associated Particle Neutron Generator With ZnO:Ga Alpha-Detector and Active Focusing, *IEEE Transactions on Nuclear Science*, Vol. 56, pp. 1301-1305, ISSN 0018-9499

Koltick, D.S; Kim, Y.; McConchie, S.; Novikov, I.; Belbot, M. & Gardner, G. (2007). A Neutron Based Vehicle-Borne Improvised Explosive Device Detection System, *Nuclear Instruments and Methods in Physics Research B*, Vol. 261, pp. 277-280, ISSN 0168-583X

Lanza, R. (2007). Nuclear Techniques for Explosive Detection: Current Prospects and Requirements for Future Development, In: *Combined Devices for Humanitarian Demining and Explosive Detection*, International Atomic Energy Agency, ISBN 978-92-0-157007-9, Vienna, Austria

Ljungvall, J. & Nyberg, J. (2005). A Study of Fast Neutron Interactions in High-Purity Germanium Detectors, *Nuclear Instruments and Methods in Physics Research A*, Vol. 546, pp. 553–573, ISSN 0168-9002

Lone, M.A.; Leavitt, R.A. & Harrison, D.A. (1981). Prompt Gamma Rays from Thermal-Neutron Capture, *Atomic Data and Nuclear Data Tables*, Vol. 26, Issue 6, pp. 511-559, ISSN 0092-640X

Maglich, B.C. & Nalcioglu, O. (2010). ONCOSENSOR for Noninvasive High-Specificity Breast Cancer Diagnosis by Carbogen-Enhanced Neutron Femto-Oximetry, *ASME Conference Proceedings: Congress on NanoEngineering for Medicine and Biology*, pp. 57-58, ISBN 978-0-7918-4392-5

Nikitin, A. & Bliven, S. (2010). Needs of Well Logging Industry in New Nuclear Detectors, *Nuclear Science Symposium Proceedings*, pp. 1214-1219, Knoxville, TN, Oct. 30 - Nov. 6, 2010, ISSN 1082-3654

NNMAB-482-6 (1999). *The Practicality of Pulsed Fast Neutron Transmission Spectroscopy for Aviation Security*, National Academies Press, ISBN 0-309-07367-7

Overley, J.C.; Chmelik, M.S.; Rasmussen, R.J.; Schofield, R.M.S.; Sieger, G.E. & Lefevre, H.W. (2006). Explosives Detection via Fast Neutron Transmission Spectroscopy, *Nuclear Instruments and Methods in Physics Research Section B*, Vol. 251, pp. 470-478, ISSN 0168-583X

Parsons, A.; Bodnarik, J.; Evans, L.; Floyd, S.; Lim, L.; McClanahan, T.; Namkung, M.; Nowicki, S.; Schweitzer, J.; Starr, R. & Trombka, J. (2011). Active Neutron and Gamma-Ray Instrumentation for In-Situ Planetary Science Applications, *Nuclear Instruments and Methods in Physics Research A*, Vol. 652, pp. 674-679, ISSN 0168-9002

Pehl, R.H.; Madden, N.W.; Elliott, J.H.; Raudorf, T.W.; Trammel, R.C & Darken, L.S. (1979). Radiation Damage Resistance of Reverse Electrode Ge Coaxial Detectors, *IEEE Transactions on Nuclear Science*, Vol. 26, pp. 321-323, ISSN 0018-9499

Perret, G.; Perot, B.; Artaud, J-L. & Mariani, A. (2006). EURITRACK Tagged Neutron Inspection System, *Journal of Physics: Conference Series*, Vol. 41, pp. 375-383, ISSN 1742-6596

Reber, E.L.; Blackwood, L.G.; Edwards, A.J.; Jewell, J.K.; Rohde, K.W.; Seabury, E.H. & Klinger, J.B. (2005). Idaho Explosives Detection System, *Nuclear Instruments and Methods in Physics Research B*, Vol. 241, Issue 1-4, pp. 738-742, ISSN 0168-583X

Sangsingkeow, P.; Berry, K.D.; Dumas, E.J.; Raudorf, T.W. & Underwood, T.A. (2003). Advances in Germanium Detector Technology, *Nuclear Instruments and Methods in Physics Research A*, Vol. 505, pp. 183-186, ISSN 0168-9002

Shypailo, R.J. & Ellis, K.J. (2005). Design Considerations for a Neutron Generator-Based Total-Body Irradiator, *Journal of Radioanalytical and Nuclear Chemistry*, Vol. 263, No. 3, pp. 759-765, ISSN 0236-5731

Shyu, C. M.; Gardner, R.P. & Verghese, K. (1993). Development of the Monte Carlo Library Least-Squares Method of Analysis for Neutron Capture Prompt Gamma-Ray Analyzers, *Nuclear Geophysics*, Vol. 7, No. 2, pp. 241-268, ISSN 0969-8086

Sowerby, B.D. (2009). Nuclear techniques for the On-Line Bulk Analysis of Carbon in Cola-Fired Power Stations, *Applied Radiation and Isotopes*, Vol. 67, pp. 1638-1643, ISSN 0969-8043

Strellis, D. A.; Gozani, T. & Stevenson, J. (2009). Air Cargo Inspection Using Pulsed Fast Neutron Analysis, In: *IAEA Proceedings Series: Topical Meeting on Nuclear Research - Applications and Utilization of Accelerators*, Paper SM/EN-05, International Atomic Energy Agency, ISBN 978-92-0-150410-4, Vienna, Austria

Tsoulfanidis, N. & Landsberger, S. (2010). *Measurement and Detection of Radiation*, 3rd Edition, CRC Press, ISBN 1420091859

Vakhtin, D.N.; Gorshkov, I.Yu.; Evsenin, A.V.; Kuznetsov, A.V. & Osetrov, O.I. (2006). Senna - Portable Sensor for Explosives Detection Based on Nanosecond Neutron Analysis, In: *Detection and Disposal of Improvised Explosives*, H. Schubert and A. Kuznetsov, (Eds.), 87-96, NATO Security through Science Series B: Physics and Biophysics, Vol. 6, Springer, ISBN 1-4020-4886-6, Dordrecht, The Netherlands

van Loef, E.V.D. , Dorenbos, P; van Eijk, C.W.E; Gudel, H.U. & Kraemer, K.W. (2001). High-Energy-Resolution Scintillator: $Ce^{3+}$ activated $LaBr_3$, *Applied Physics Letters*, Vol. 79, No. 10, pp. 1573-1575, ISSN 0003-6951

Vourvopoulos, G. & Womble, P. (2001). Pulsed Fast Thermal Neutron Analysis: A Technique for Explosives Detection, *Talanta*, Vol. 54, pp. 459-468, ISSN 0039-9140

Wielopolski, L.; Hendrey, G.; Johnsen, K.; Mitra, S.; Prior, S.; Rogers, H. & Tolbert, H. (2008). Nondestructive System for Analyzing Carbon in the Soil, *Soil Science Society of America Journal*, Vol. 72, pp. 1269-1277, ISSN 0361-5995

Williams, D. L.; Vainionpaa, J. H.; Jones, G.; Piestrup, M. A.; Gary, C. K.; Harris, J. L.; Fuller, M. J.; Cremer, J. T.; Ludewigt, B. A.; Kwan, J. W.; Reijonen, J.; Leung, K.-N. & Gough, R. A. (2008). High Intensity, Pulsed, D-D Neutron Generator, *AIP Conference Proceedings*, Vol. 1099, pp. 936-939, ISSN 1551-7616

Womble, P.C.; Barzilov, A.; Novikov, I.; Howard, J. & Musser, J. (2009). Evaluation of the Doppler-Broadening of Gamma-Ray Spectra from Neutron Inelastic Scattering on Light Nuclei, *AIP Conference Proceedings*, Vol. 1099, pp. 624-627, ISSN 1551-7616

Womble, P.C.; Paschal, J. & Moore, R. (2005). Cement Analysis Using d + D neutrons, *Nuclear Instruments and Methods in Physics Research B*, Vol. 241, No.1-4, pp. 765-769, ISSN 0168-583X

Womble, P.C.; Vourvopoulos, G.; Paschal, J.; Novikov, I. & Chen, G. (2003). Optimizing the Signal-to-Noise Ratio for the PELAN System, *Nuclear Instruments and Methods in Physics Research A*, Vol. 505, Issue 1-2, pp. 470-473, ISSN 0168-9002

Womble, P.C.; Vourvopoulos, G.; Paschal, J.; Novikov, I. & Barzilov, A. (2002). Results of Field Trials for the PELAN System, *Proceedings of SPIE*, Vol. 4786, pp. 52-57, ISSN 0277-786X

Womble, P.C.; Schultz, F.J. & Vourvopoulos, G. (1995). Non-Destructive Characterization Using Pulsed Fast-Thermal Neutrons, *Nuclear Instruments and Methods in Physics Research B*, Vol. 99, pp. 757-760, ISSN 0168-583X

# Part 2

# Environmental Sciences

# 3

# Gamma Radiation

Richard Stalter and Dianella Howarth
*St. John's University*
*USA*

## 1. Introduction

The content of this chapter includes a brief history of gamma radiation, units of radiation measurement, ecological importance, tables including the half life of gamma emitting nuclides, comparative sensitivity of living organisms to gamma radiation, biological magnification of radioactive and nuclear materials, and brief descriptions of case studies of Woodwell (1962), Stalter and Kincaid 2009), and nuclear power plant disasters (Three Mile Island, USA, 1980, Chernobyl 1986, Japan 2011).

Gamma radiation is somewhat similar to x-rays in that both pass through living materials easily. Also referred to as "photons" they travel at the speed of light. Gamma rays have sufficient energy to ionize matter and therefore can damage living cells. The damage produced in the cell or tissue is proportional to the number of ionizing paths produced in the absorbing material. Isotopes of elements that are emitters are radionuclides important in fission products from nuclear testing, nuclear power plant disasters or waste.

The injurious affect of gamma rays depends on (1) their number (2) their energy and (3) their distance from the source of radiation. Radiation intensity decreases exponentially with increasing distance. Radiation damage on vascular plant species was demonstrated by Woodwell (1962) who subjected a mature pine oak forest at Brookhaven National Laboratory to gamma radiation from a cesium 137 source (Figure 1).

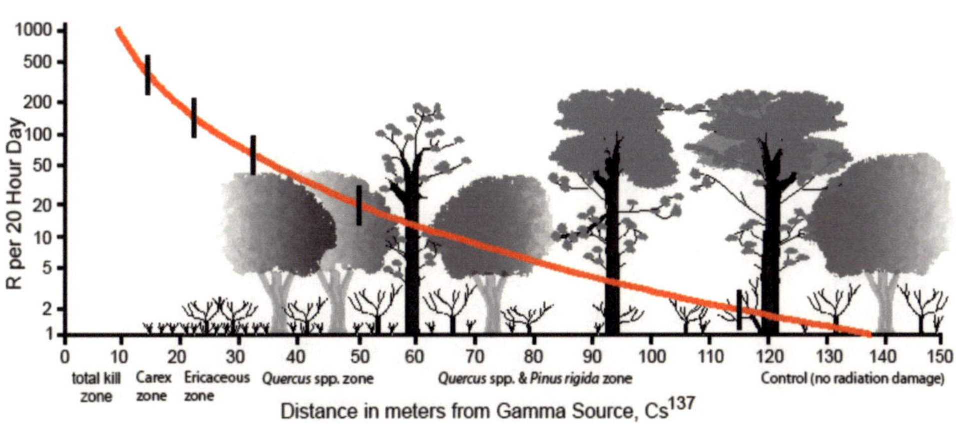


Fig. 1. Radiation dose and damage to a pine-oak forest, Brookhaven National Laboratory, 1961. Zones delineated by vertical lines (Woodwell 1962, Stalter and Kincaid 2009).

Gamma rays are external emitters that penetrate biological materials easily and produce their insidious effects without being taken internally. Alpha and beta particles are internal emitters; their damage to organisms is greatest when taken internally. Odum (1971) summarizes this concept best, "the alpha beta gamma series is one of increasing penetration but decreasing concentration of ionization and local damage." Alpha and beta radiation, unlike gamma radiation, are corpuscular in nature. While alpha particles travel but a few centimeters, and can be stopped by a layer of dead skin, they are dangerous because they produce a large amount of local ionization which can cause mutations disrupting cell processes. Beta particles are high speed electrons. While much smaller than alpha particles, they are able to travel up to a couple of centimeters in living tissue, giving up their energy over a large path. Beta particles, like alpha particles can damage tissue, and like alpha particles, can cause mutations that affect the functioning of cells.

## 2. The history of gamma radiation as applied to biological systems

Most are familiar with the discovery of x-radiation by Roentgen in 1895 and the isolation of radium by the Curies in 1898 (Goodspeed and Uber 1939). Researchers soon learned that both x-rays and radioactive substances such as radium produced similar effects on biological materials. Koernicke (1905) noted that cell division was delayed on x-ray and radium treated cells. Both Koernicke (1905) and Gager (1907) described "striking chromosomal disruptions" after cells were dosed with x-rays or exposed to radium, a gamma emitter. Gamma irradiated cells were also broken or fragmented by radiation treatment (Gager 1907, 1908). For additional historical work on radiation and plant cytogenetics the reader is directed to a review article by Goodspeed and Uber (1939). Smith (1958) compiled a paper on the use of radiation in the production of useful mutations based on papers presented in three symposia in the United States from August 1956 to January 1957. A more recent review article on ionizing radiation damage to plants was prepared by Klein and Klein (1971).

There are numerous studies applying gamma radiation to biological systems. Several investigations involving botanicals follow. Nuttall et al (1961) found that yellow sweet Spanish onions exposed to 4000 or 8000 rad prevented sprouting in 97% of their experimental group suggesting that irradiation might be a viable method of prolonging storage life for onions. This study, while intriguing, has not been generally accepted by a public concerned with the problems of radiation. A second article by Heeney and Rutherford (1964) examined the effects of gamma radiation on the storage life of fresh strawberries. A dose of 330,000 rad prevented fungal development of the redcoat strawberry variety stored at 40 degrees F for 26 days. The fugal free period was sharply reduced at lower radiation doses and/or at higher temperatures. Pritchard et al (1962) studied the effect of gamma radiation on the utilization of wheat straw by rumen microorganisms. They concluded that, "high levels of gamma radiation were needed to release nutrients trapped in wheat straw needed by microbes. However, the levels of gamma irradiation necessary for nutrient release were well above what was practical for commercial purposes."

Baumhover et al (1955) investigated the use of gamma irradiation on male sterilization on the control of screw-worm flies in the southern United States while Bushland (1960) Cutcomp (1967) and Lawson (1967) discussed this practice as a general way of controlling certain insect pests. Gambino and Lindberg (1964) examined the response of the pocket mouse to ionizing radiation. McCormick and Golley (1966) presented data on irradiation of natural vegetation in the southeastern United States while Monk (1966) published a similar study on the effects of short-term gamma radiation on an old field. Witherspoon (1965, 1969) examined radiation damage to a forest surrounding an unshielded fast reactor in 1965, and followed this study with a report in 1969 on radiosensitivity of forest tree species to acute fast neutron radiation. Odum and Pigeon (1970) researched the effect of irradiation and ecology of a tropical rain forest in Puerto Rico.

## 3. Units of measurement

Three units, the gigabecquerel (GBq), gray (GY), and roentgen (R) are used to measure radiation. The GBq measures the number of gamma rays emitted from a source of radiation and is a unit of radioactivity that is defined as 1.37 x $10^{-12}$ atomic decays each second. The weight of the material comprising a GBq varies. One gram of radium is 37 GBq while $10^{-7th}$ of a gram of newly formed radio-sodium is also 37 GBq since both release 3.7 x $10^{-10}$ disintegrations/second (Odum 1971). In dealing with biological systems, smaller units are generally used such as the millicurie microcurie and picocurie which are $10^{-3}$, $10^{-6}$ and $10^{-12}$ respectively.

A second measurement of radiation is the GY. The absorbed dose of 1 GY means the absorption of 1 joule of radiation energy per kg of tissue. The third, the roentgen is nearly the same as the GY, and is used as a unit of measurement for exposure to gamma and x rays. Both are units of the total dose of radiation received by an organism. The dose rate is the amount of radiation received per unit time.

## 4. Ecological importance of radionuclides

There are different kinds of atoms of each element; these are referred to as isotopes. Some isotopes are radioactive, some not. Radioactive isotopes are unstable. These decay into other isotopes releasing radiation. Each radioactive isotope, radionuclide, have a specific rate of disintegration, its half life.

Radionuclides fall into well defined groups (Tables 1 and 2). Naturally occurring nuclides are included in Table 1 while those from fallout produced by fission or uranium and other elements are found in Table 2. Fission isotopes are produced from nuclear explosions which have for the most part been eliminated and from "controlled" reactions that produce nuclear power. While most of the aforementioned nuclides are not essential for the growth of organisms, they may be incorporated in biogeochemical cycles and become concentrated in food chains, especially strontium and cesium. Thus Woodwell (1962) used cesium as a gamma radiation emitter in his well published study of an irradiated pine oak forest at Brookhaven National Laboratory, Long Island, New York. More will be said about this research later in this paper.

## 5. Sensitivity of organisms to radioactivity

There is a wide range of sensitivity of organisms to radioactivity. Mammals are most sensitive while bacteria are most resistant especially as spores. Moreover there is a wide range of tolerance to radiation during the life cycle of an organism. Radiation sickness in humans can be caused by as little as 0.35 Gy while a dose of 6-8 Gy is lethal to nearly 100% of individuals (Donnelly et al 2010). A dose of 2 Gy may kill some insect embryos while a dose of 100 Gy is necessary to kill all adult individuals (Odum 1971). Dividing cells are generally more susceptible to radiation than resting cells. The toxicity of radionuclides depends on the absorption, distribution in the body, half-life, elimination half-time, type of radiation emitted, and their energy.

| Element | Half-Life | Radiations Emitted | |
|---|---|---|---|
| Uranium-235($^{235}$U) | 7 x $10^8$ yrs. | Alpha$^3$ | Gamma$^0$ |
| Radium-226 ($^{226}$Ra) | 1620 yrs. | Alpha$^3$ | Gamma$^0$ |
| Potassium-40 ($^{40}$K) | 1.3 x $10^9$ yrs. | Beta$^2$ | Gamma$^2$ |
| Carbon-14 (See Table 3.) | | | |

Table 1. Naturally occurring gamma emitting isotopes which contribute to background radiation (Odum 1971).

| Element | Half-Life | Radiations Emitted | |
|---|---|---|---|
| The cesium group | 33 yrs. | Beta$^2$ | Gamma |
| Cesium-137 ($^{137}$Cs) and | 2.6 min | Beta | Gamma$^1$ |
| daughter barium-137 ($^{137}$Ba) | 2.3 yrs. | Beta$^1$ | Gamma$^2$ |
| Cesium-134 ($^{134}$Cs) | | | |
| The cerium group | 285 days | Beta$^1$ | Gamma$^0$ |
| Cerium-144 ($^{144}$Ce) and | 17 min. | Beta$^2$ | Gamma$^2$ |
| daughter praseodymium-144 ($^{137}$Pr) | 33 days | Beta$^1$ | Gamma$^1$ |
| Cerium-141 ($^{141}$Ce) | | | |
| | 1 yr. | Beta$^2$ | |
| The ruthenium group | 30 sec. | Beta$^3$ | Gamma$^2$ |
| Ruthenium-106 ($^{106}$Ru) and | 40 days | Beta$^1$ | Gamma$^1$ |
| daughter rhodium-106 ($^{106}$Rh) | 65 days | Beta$^1$ | Gamma$^1$ |
| Ruthenium-103 ($^{103}$Ru) | 35 days | Beta$^0$ | Gamma$^1$ |
| Zirconium-95 ($^{95}$Zr) and daughter | 12.8 days | Beta$^1$ | Gamma$^1$ |
| niobium-95($^{95}$Nb) | 40 hrs | Beta$^2$ | Gamma$^2$ |
| Barium-140 ($^{140}$Ba) and daughter | 11.3 days | Beta$^1$ | Gamma$^1$ |
| lanthanium-140($^{140}$La) | 2.6 yrs. | Beta$^1$ | Gamma |
| Neodymium-147 ($^{147}$Nd) and | 61 days | Beta$^2$ | Gamma$^1$ |
| daughter promethium-147($^{147}$Pm) | 2.4 x $10^4$ yrs. | Alpha$^3$ | Gamma$^1$ |
| Yttrium-91 ($^{91}$Y) | 8 days | Beta$^1$ | Gamma$^1$ |
| Plutonium-239 ($^{239}$Pu) | 7 x $10^8$ yrs. | Alpha$^3$ | Gamma$^0$ |
| Iodine-131 ($^{131}$I) | | | |
| Uranium-235 ($^{235}$U) | | | |

Table 2. Elements important in fission products entering the environment through fallout or waste disposal.

Sparrow (1962), Sparrow and Evans (1961), Sparrow and Woodwell (1962), and Sparrow et al (1963) have demonstrated that sensitivity of ionizing radiation is directly proportional to the size of the cell nucleus or chromosome volume. The larger the chromosome volume the more sensitive the material is to radiation. There are also differences in radiation tolerance between wild and laboratory rodent populations. Gambino and Lindberg (1964) and Golley et al (1965) have reported that the lethal dose for 50% of some wild rodent populations is roughly twice that of laboratory white mice or white rats, likely due to the reduced variation in the latter.

Radioactivity has been successfully used to sterilize certain male insect pests. Sterile males are introduced to natural populations in large numbers which mate with females. A female mates only once, and once mated with a sterile male produces no young. Introducing radiated sterile male screw-worm flies in areas where they occur successfully reduced the number of screw-worm flies, a major pest in the southern United States. For those seeking more general information on this topic see Baumhover et al (1955) Bushland (1960), Cutcomp (1967), Knipling ( 1960,1964, 1965, 1967) and Lawson (1967).

## 6. Radiation effects on ecosystems

Since the early 1960's there have been numerous studies on the effect of gamma radiation on ecosystems. These studies were fueled by the arms race between the Soviet Union and the United States (Stalter and Kincaid 2009). After lengthy negotiations between the two powers the SALT (Strategic Arms Limitation Treaty) was signed in 1971 and extended in 1977. With the signing of the treaty, less funding for irradiation studies was available (Stalter and Kincaid 2009). Thus most studies cited in this paper are those conducted prior to the SALT agreement of 1971. The gamma source that has been used has been either cesium 137 or cobalt 60. These include the studies of Woodwell (1962, 1965a) at Brookhaven National Laboratory, Long Island, New York, a tropical rain forest , Puerto Rico (Odum and Pigeon 1970) and the desert of Nevada (French 1965). Additional studies have been conducted in the fields and forests of Georgia (Odum and Kuenzler 1963) (Platt 1965), and Oak Ridge, Tennessee (Witherspoon 1965, 1969). Much additional work involving a portable gamma source on plant communities has been conducted at the Savanna River Ecology Laboratory, Aiken, South Carolina (McCormick and Platt 1962, McCormick and Golly 1966, Monk 1966, McCormick 1969).

Stalter and Kincaid (2009) investigated community development following gamma radiation at a pine-oak forest, Brookhaven National Laboratory, Long Island, New York. The objective of this study was to compare vascular plant community change at five vegetation zones the site of Woodwell's (1962) gamma irradiated forest (Figure 1). The zones were: the dead zone where all vegetation was killed; a gramminoid *Carex pensylvanica* zone; an ericaceous zone; an oak dominated zone; and a control, the original oak pine forest. Radiation greater than 63,000 roentgens killed all vegetation. *Carex* dominated the zone receiving 27,000 to 63,000 roentgens, ericaceous shrubs, *Vaccinium* spp. and *Gaylussacia baccata* were dominant at the zone receiving 11,000 to 27,000 roentgens while oaks survived at the zone receiving 3600 to 11,000 roentgens. Upon completion of the Woodwell study in the 1970's, pitch pine (*Pinus rigida*) has invaded the total kill zone as bare mineral soil favors pine regeneration (Stalter and Kincaid 2009). *Carex* remained the dominant taxon in the original *Carex* zone demonstrating again that different plant species vary in their tolerance of radiation.

Herbaceous plant communities may be more resistant to radiation than mature forests because many early successional species have small nuclei (Sparrow and Evans 1961) and also because herbaceous taxa like *Carex pensylvanica* have more below ground plant material which is shielded from gamma radiation. Sparrow (1962), Sparrow and Evans (1961), and Sparrow et al (1963) present detailed information on the relationship between nuclear volumes, chromosome numbers and relative radiosensitivity.

## 7. Biological magnification of radioactive material

Radioactive material may become concentrated or "biologically magnified" during food chain transfer. Numerous biology and ecology text books include information on how living organisms take up nutrients pesticides and radioactive material and concentrate them. Because this concept is well known, we direct the reader to several early studies involving the concentration of radioactive material (See the work of Foster and Rostenbach, 1954; Hanson and Kornberg 1956; Davis and Foster 1958). Ophel (1963) reported a concentration of strontium 90 in perch flesh as 5x that of lake water while that in perch bone was 3000x! Additional information on radioecological concentration can be found in Auberg and Crossley (1958), Auberg and Hungate (1967) and Polikarpov (1966).

## 8. Radioactive fallout

Radioactive particles that fall to the earth after above ground nuclear tests and nuclear power plant accidents are called radioactive fallout. Radioactive particles mix with the dust in the atmosphere and eventually fall to earth often thousands of miles from the initial explosion.

There are two types of nuclear weapons, the fission bomb and fusion bomb or thermonuclear weapon. In thermonuclear devices, deuterium fuses to form a heavier element with the release of energy and neutrons. A fission bomb is needed to trigger the fusion reaction. The thermonuclear weapon produces more neutrons which induce radioactivity in the environment than a fission device per unit of energy released. Roughly ten percent of the energy of a nuclear weapon is in residual radiation which may become dispersed in the atmosphere (Glasstone 1957). The amount of fallout produced depends on the type of weapon, size of the weapon and also on the amount of naturally occurring material that is mixed with the radioactive material released in the explosion. Fallout patterns and intensity depend upon the direction of the wind, speed and direction of the jet stream, presence and amount of precipitation.

Atomic explosions carry radioactive material high in the atmosphere where the radioactive material becomes fused with silica dust and other material present in the vicinity of the explosion. These particles are largely insoluble. The fallout particles may adhere to vegetation where they enter food chains at the primary consumer level. Fallout from Chernobyl in 1986 was deposited in Lappland (Sweden) where caribou consumed contaminated vegetation. Shifting winds also carried Chernobyl radiation particles to northern Italy where rabbit growers fed their rabbits vegetation contaminated with radioactive fallout from Chernobyl. Ultimately the rabbits were destroyed because of the high concentration of radioactive material in their flesh.

There are differences in the kind of radionuclides that enter terrestrial and marine food chains. Soluble fission products, strontium 90 and cesium 137, are generally found in the highest amounts in land plants and animals. In marine systems fallout that forms strong complexes with organic matter such as cobalt 60, iron 59, zinc 65, and manganese 54 are most likely to be concentrated in marine organisms. In addition, those found in colloidal form such as cesium 134 and zirconium 95 are also found in high concentration in marine organisms. Cesium 134 is mostly from the fission products of a power reactor whereas cesium 137 can be formed during atomic power plant accidents or as a product of nuclear bomb explosions.

There are additional considerations/problems associated with concentrating radioactive material entering food chains as the concentration of radioactivity is also a function of nutrient richness, and the exchange and storage capacity of soils. Nutrient poor soils and thin soils such as those found on granite outcrops act as a nutrient trap providing more radionuclides to the vegetation. For example, sheep grazing on hill pastures in England accumulated 20x as much strontium 90 in their bones than sheep pastured in deep valleys where calcium content of the soil was higher and the grasses taller (Bryant et al 1957). For additional radiological work on tracers in food chains and trophic levels see Odum and Golley (1963), Odum and Kuenzler (1963), de la Cruz (1963), Ball and Hooper (1963), Foster (1958), and Foster and Davis (1956).

## 9. Nuclear power plant accidents

Brief descriptions of three power plant accidents in the United States the Soviet Union and Japan follow. The first nuclear power plant accident occurred at 4 am on March 28, 1979, near Harrisburg, Pennsylvania, USA, the state's capital. A malfunction in the cooling system resulted in a portion of the core to melt in the Number 2 reactor. The approximately 2 million people who lived near the plant had an average dose of 0.14 Gy (Rogovin 1980). Although some radioactive gas was released from the plant on the 29th and 30th of March there was, "not enough to cause any radiation dose above background levels in the neighborhood of the accident "(http://www.world-nuclear.org/info/info/info/inf36.html). Fortunately, there were no reported injuries or health issues emanating from the Three Mile Island accident.

A more serious nuclear accident occurred at the Chernobyl power plant located 80 miles north of the city of Chernobyl in the Ukraine, one of the original Soviet Republics. A "routine" shut down and test that began on the 25th of April, 1986, led to this disaster. At one in the morning, 26 April, the reactor's power source dropped and when the backup safety system failed, the reactor, Reactor Four, exploded. Shortly after the initial explosion at Chernobyl, the Swedish government reported high levels of radiation at their Forsmark nuclear power plant at Stockholm. When additional European nuclear power plants also experienced higher than normal levels of radiation, they contacted the USSR for an explanation. Although initially denying the nuclear disaster, on the 28th of April the USSR acknowledged that one of their reactors had been compromised.

Group A. Naturally occurring isotopes which contribute to background radiation.

| NUCLIDE | HALF-LIFE | RADIATIONS EMITTED | |
|---|---|---|---|
| Uranium-235 ($^{235}$U) | 7 x $10^8$ yrs. | Alpha$^3$ | Gamma$^0$ |
| Radium-226 ($^{226}$Ra) | 1620 yrs. | Alpha$^3$ | Gamma$^0$ |
| Potassium-40 ($^{40}$K) | 1.3 x $10^9$ yrs. | Beta$^2$ | Gamma$^2$ |
| Carbon-14 ($^{14}$C) | 5568 yrs. | Beta$^0$ | |

$^0$ Very low energy, less than 0.2 Mev; $^1$ relatively low energy, 0.2-1 Mev; $^2$ high energy, 1-3 Mev; $^3$ very high energy, over 3 Mev.

Group B. Gamma emitting nuclides of elements which are essential constituents of organisms. Modified from Odum (1971).

| NUCLIDE | HALF-LIFE | | RADIATIONS EMITTED | |
|---|---|---|---|---|
| Cobalt-60 ($^{60}$Co) | 5.27 | yrs. | Beta$^1$ | Gamma$^2$ |
| Copper-64 ($^{64}$Cu) | 12.8 | hrs. | Beta$^1$ | Gamma$^2$ |
| Iodine-131 ($^{131}$I) | 8 | days | Beta$^1$ | Gamma |
| Iron-59 ($^{59}$Fe) | 45 | days | Beta$^1$ | Gamma$^2$ |
| Manganese-54 ($^{54}$Mn) | 300 | days | Beta$^2$ | Gamma$^2$ |
| Potassium-42 ($^{42}$K) | 12.4 | hrs. | Beta$^3$ | Gamma$^2$ |
| Sosium-22 ($^{22}$Na) | 2.6 | yrs. | Beta$^1$ | Gamma$^2$ |
| Sodium-24 ($^{24}$Na) | 15.1 | hrs. | Beta$^2$ | Gamma$^2$ |
| Zinc-65 ($^{65}$Zn) | 250 | days | Beta$^1$ | Gamma$^2$ |

Also barium-140 ($^{140}$Ba), bromine-82 ($^{82}$Br), molybdenum-99 ($^{99}$Mo) and other trace elements.

Group C. Nuclides important in fission products entering the environment through fallout or waste disposal.

| NUCLIDE | HALF-LIFE | | RADIATIONS EMITTED | |
|---|---|---|---|---|
| The strontium group | | | | |
| Strontium-90 ($^{90}$Sr) and | 28 | yrs. | Beta$^1$ | |
| daughter yttrium-90 ($^{90}$Y) | 2.5 | days | Beta$^2$ | |
| Strontium-89 ($^{89}$Sr) | 53 | days | Beta$^2$ | |
| The cesium group | | | | |
| Cesium-137 ($^{137}$Cs) and | 33 | yrs. | Beta$^2$ | Gamma |
| daughter barium-137 ($^{137}$Ba) | 2.6 | min. | Beta | Gamma$^1$ |
| Cesium-134 ($^{134}$Cs) | 2.3 | yrs. | Beta$^1$ | Gamma$^2$ |
| The cerium group | | | | |
| Cerium-144 ($^{144}$Ce) and | 285 | days | Beta$^1$ | Gamma$^0$ |
| daughter praseodymium-144 ($^{144}$Pr) | 17 | min. | Beta$^2$ | Gamma$^2$ |
| Cerium-141 ($^{141}$Ce) | 33 | days | Beta$^1$ | Gamma$^1$ |
| The ruthenium group | | | | |
| Ruthenium-106 ($^{106}$Ru) and | 1 | yr. | Beta$^0$ | |
| daughter rhodium-106 ($^{106}$Rh) | 30 | sec. | Beta$^3$ | Gamma$^2$ |
| Ruthenium-103 ($^{103}$Ru) | 40 | days | Beta$^1$ | Gamma$^1$ |
| Zirconium-95 ($^{95}$Zr) and daughter | 65 | days | Beta$^1$ | Gamma$^1$ |
| niobium-95 ($^{95}$Nb) | 35 | days | Beta$^0$ | Gamma$^1$ |
| Barium-140 ($^{140}$Ba) and daughter | 12.8 | days | Beta$^1$ | Gamma$^1$ |

| | | | | |
|---|---|---|---|---|
| lanthanum-140 ($^{140}La$) | 40 | hrs. | Beta[2] | Gamma[2] |
| Neodymium-147 ($^{147}Nd$) and | 11.3 | days | Beta[1] | Gamma[1] |
| daughter promethium-147 ($^{147}Pm$) | 2.6 | yrs. | Beta[1] | Gamma |
| Yttrium-91 ($^{91}Y$) | 61 | days | Beta[2] | Gamma |
| Plutonium-239 ($^{239}Pu$) | 2.4 x $10^4$ | yrs. | Alpha[3] | Gamma[1] |
| Iodine-131 (see Group B) | | | | |
| Uranium (see Group A) | | | | |

Table 3. Radionuclides of Ecological Importance

Scientists estimate that the radiation from the Chernobyl accident was 100x that of the two atom bombs dropped on Hiroshima and Nagasaki. It is estimated that the total atomospheric release was 5200 PBq (petabecquerel, $10^{15}$ Bq). The immediate death toll was 31 individuals though many more may die from the long term effects of radiation. The Soviets battled blazes at the Chernobyl power plant for two weeks. Those battling the fires were heroes in this author's eyes because they knew they were exposing themselves to dangerous levels of radiation. Ultimately the Soviet authorities encased the Chernobyl reactor in concrete. A second more stable sarcophagus is currently being constructed over the original; its scheduled completion date is 2013.

There may have been additional unreported nuclear power plant accidents in the Soviet Union. Radioactive monitoring stations in Europe have picked up higher levels of radiation at various times which may have been the result of other Soviet nuclear power plant accidents.

The third and most recent nuclear power plant crisis occurred at the Fukushima Daiichi power plant in Japan. The cause of this disaster was a severe earthquake and tsunami on the 11th of March, 2011. The earth quake, which registered approximately 9 on the Richter Scale, was the event that set this tragedy in motion. The earthquake and resulting tsunami damaged the power plant compromising the cooling systems to the reactors causing the fuel rods to overheat. This disaster was rated greater than that at Three Mile Island. As of June 2011, the Fukushima disaster has released approximately one tenth the total amount of radiation as was released at Chernobyl. Unfortunately, the damaged Japanese reactor continues to spew forth radiation so the ultimate amount of radiation released from the plant cannot be determined with certainty.

## 10. Acknowledgements

The authors extend their thanks to the following individuals who assisted in the preparation of this chapter: Huizhong Xu, Associate professor of Physics, St. John's University, who reviewed the physics section, to Natacha Lamarre, undergraduate biology major, St. John's University, who typed the paper, prepared the tables and checked the references.

## 11. References

Auberg B., Hungate, F. P. (eds.). 1967. *Radioecological Concentration Processes*. Pergamon Press, Oxford. 1040 pp.

Auerbach, S. I., and Crossley, D. A. 1958. Strontium-90 and cesium-137 uptake under natural conditions. Proc. Int. Conf. Peaceful Uses Atomic Energy, Geneva Paper No. 401.

Ball, R. C., and Hooper, F. F. 1963. Translocation of phosphorus in a trout stream ecosystem. In: *Radioecology* (V. Schultz and A. W. Klement, eds.). Reinhold Publishing Company, New York. pp. 217-228.

Baumhover, A. H., Graham A. J., Hopkins, D. E., Dudley, F. H., New, W. D., and Bushland, R. C. 1955. Screw-worm control through release of sterilized flies. J. Econ. Entomol., 48:462-466.

Bryant, F. J., Chamberlain, A. C., Morgan, A., and Spicer, C. S. 1957. Radiostrontium in soil, grass, milk and bone in U.K.; 1956 results. J. Nuc. En., 6:22-40.

Bushland, R. C. 1960. Male sterilization for the control of insects. In.: *Advances in Pest Control Research* (R. L. Metcalf, ed.), Vol. III. John Wiley & Sons, Inc., New York.

de la Cruz, A. A., and Wiegert, R. G. 1967. 32-Phosphorus tracer studies of a horse weed aphid-ant food chain. Amer. Midl. Nat., 77:501-509.

Cutcomp, L. K. 1967. Progress in insect control by irradiation induced sterility. Pans, 13:61-70.

Davis, J. J., and Foster, R. F. 1958. Bioaccumulation of radioisotopes through aquatic food chains. Ecology, 39:530-535.

Donnelly, E. H., Nemhauser, J. B., Smith, J. M., Kazzi, Z. N. Farfan, E. B., Chang, A. S., and Naeem, S. F. 2010. Acute radiation syndrome: assessment and management. Southern Medical Journal. 103(6):541-546.

Foster, R. F. 1958. Radioactive tracing of the movement of an essential element through an aquatic community with specific reference to radiophosphorus. Publ. della Stazione Zool. di Napoli.

Foster, R. F., Davis, J. J. 1956. The accumulation of radioactive substances in aquatic forms. Proc. Int. Conf. Peaceful Uses Atomic Energy, Geneva, 13:364-367.

Foster, R. F., and Rostenbach, R. E. 1954. Distribution of radioisotopes in the Columbian River. J. Amer. Water Works Assoc., 46:663-640.

French, N. R. 1965. Radiation and animal population: problems, progress and projections. Health Physics, 11:1557-1568.

Gager, C. S. 1907. Some effects of radioactivity on plants. Science 25:264.

Gager, C. S. 1908. Effects of the rays of radium on plants. Mem. N. Y. Bot. Gard. 4:1-278.

Gambino, J. J., and Lindberg, R. G. 1964. Response of the pocket mouse to ionizing radiation. Read. Res., 22:586-597.

Glasstone, S. 1957. *The Effects of Nuclear weapons*. U. S. Atomic Energy Commission, Washington, D. C.

Golley, F. B., Gentry, J. B., Menhinick, E., and Carmon, J. L. 1965. Response of wild rodents to acute gamma radiation. Rad. Res., 24:350-356.

Goodspeed, T. H., and Uber, F. M. 1939. Radiation and plant cytogenetics. Botanical Review 5(1):1-48.

Hanson, W. C. and Kornberg, H. A. 1956. Radioactivity in terrestrial animals near an atomic energy site. Proc. Int. Conf. Peaceful Uses Atomic Energy, Geneva 13:385-388.

Heeney, H. B., and Rutherford, W. M. 1964. Some effects of gamma radiation on the storage life of fresh strawberries. Canadian Journal of Plant Science 44:188-194.

Hungate, R. E. 1966. The Rumen and Its Microbes. Academic Press, New York. 533 pp.

Klein, R. M., and Klein, D. T. 1971. Post-irradiation modulation of ionizing radiation damage to plants. Botanical Review 34(4):397-436.

Knipling, E. F. 1960. The eradication of the screwworm fly. Scient. Amer., 203(4):54-61.

Knipling, E. F. 1963. The sterility principle. Agr. Sci. Rev., 1(1):2.

Knipling, E. F. 1965. The sterility method of pest population control. In: *Research in Pesticides* (G. O. Chichester, ed.). Academic Press, New York. Pp. 233-249.

Knipling, E. F. 1967. Sterile technique, principles involved, current application, limitationsand future applications. In: *Genetics of Insect Vectors of Disease* (Wright and Pal, eds.). Elsevier Publishing Co., Amsterdam. Pp. 587-616.

Koernicke, M. 1905. Uber die Wirkung von Rontgen- und Radiumstrahlen auf pflanzliche Gewebe und Zellen. Ber. Deut. Bot. Ges. 23:404-415.

Lawson, F. R. 1967. Theory of control of insect population by sexually sterile males. Ann. Entomol. Soc. Amer., 60:713-722.

McCormick, F. J. 1969. Effects of ionizing radiation on a pine forest. In: *2nd Nat. Sym. Radioecology* (D. Nelson and F. Evans, eds.). Clearinghouse Fed. Sci. Tech. Info., U. S. Dept. Commerce, Springfield, VA. Pp. 78-87.

McCormick, F. J., and Platt, R. B. 1962. Effects of ionizing radiation on a natural plant community. Rad. Bot., 2:161-204.

McCormick, F. J., and Golley, F. B. 1966. Irradiation of natural vegetation-an experimental facility, procedures and dosimetry. Health Physics, 12:1467-1474.

Monk, C. D. 1966. Effects of short-term gamma irradiation on an old field. Rad. Bot., 6:329-335.

Nuttall, V. W.; Lyall, L. H.; and McQueen, K. F. 1961. Some effects of gamma radiation on stored onions. Canadian Journal of Plant Science, 41:805-813.

Odum, E. P. 1971. Fundamentals of Ecology. W. B. Saunders Company, Philadelphia. Pp. 451-467.

Odum, E. P. and Golley, F. B. 1963. Radioactive tracers as an aid to the measurement of energy flow at the population level in nature. In: *Radio ecology* (V. Shultz and A.W. Klement, eds.). Reinhold Publishing Company, New York. Pp. 403-410.

Odum, E. P., and Kuenzler, E. J. 1963. Experimental isolation of food chains in an old field ecosystem with use of phosphorus-32. In: *Radio ecology* (V. Shultz and A.W. Klement, eds.). Reinhold Publishing Company, New York. pp. 113-120.

Odum, H. T., and Pigeon, R.F. (eds.). 1970. A tropical rainforest. A study of irradiation and ecology at El Verde, Puerto Rico. Nat. Tech. Info, Service, Springfield, VA. 1678 pp.

Odum, H. T., and Pigeon, R.F. 1965. Ionizing radiation and homeostasis of ecosystems. In: *Ecological Effects of Nuclear War* (Woodwell, ed.). Brookhaven National Laboratories, Publ. No. 917. Pp. 39-60.

Ophel, I. L. 1963. The fate of radiostrontium in a freshwater community. In: *Radioecology* (V. Scultz and W. Klement, eds.), Reinhold Publishing Company, New York. Pp. 213-216.

Platt, R. B. 1965. Ionizing Radiation and homeostasis of ecosystems. In: *Ecological Effects of Nuclear War* (Woodwell, ed.) Brookhaven National Library, Publ. No. 917:39-60.

Polikarpou, G. C. 1966. Radioecology of aquatic organisms. (Translated from Russian by S. Technica and edited by Schultz and Klement).

Pritchard, G. I., W. J. Pigden and D. J. Minson. 1962. Effect of gamma radiation on the utilization of wheat straw by rumen microorganisms. Canadian Journal of Animal Science 42:215-217.

Rogovin, M. 1980. Three Mile Island: A report to the Commissioners and to the Public, Volume I. Nuclear Regulatory Commission, Special Inquiry Group.

Smith, H. H. 1958. Radiation in the production of useful mutations. Botanical Review 24(1): 1-24.

Sparrow, A. H. 1962. The role of the cell nucleus in determining radiosensitivity. Brookhaven Lecture Series No. 17. Brookhaven Nat. Lab. Publ. No. 766.

Sparrow, A. H., and Evans, H. J. 1961. Nuclear factors affecting radiosensitivity. 1. The influence of nuclear size and structure, chromosome complement and DNA content. In: *Fundamental Aspects of Radiosensitivity*. Brookhaven Nat. Lab. Pp. 76-100.

Sparrow, A. H.; Shairer, L. A.; and Sparrow, R. C. 1963. Relationship between nuclear volumes, chromosome numbers, and relative radiosensitivity. Science, 141:163-166.

Sparrow, A. H., and Woodwell, G. M. 1962. Prediction of the sensitivity of plants to chronic gamma irradiation. Rad. Bot., 2:9-26.

Stalter, R., and Kincaid, D. T. 2009. Community development following gamma radiation at a pine-oak forest, Brookhaven National Laboratory, Long Island, New York. American Journal of Botany 96(12):2206-2213.

Witherspoon, J. P. 1965. Radiation Damage to forest surrounding an unshielded fast reactor. Health Physics, 11:1637-1642.

Witherspoon, J. P. 1969. Radiosensitivity of forest tree species to acute fast neuron radiation. In: *Proc. 2nd Natl. Sym. Radioecology* (D. Nelson and F. Evans, eds.). Clearinghouse Fed. Sci. Tech. Info., Springfield VA. Pp. 120-126.

Woodwell, G. M. 1962. Effects of ionizing radiation on terrestrial ecosystems. Science, 138:572-577.

Woodwell, G. M. (ed.). 1965. *Ecological Effects of Nuclear War.* Brookhaven National Laboratory Publ. no. 917, 72 pp.

Woodwell, G. M. 1965a. Effects of ionizing radiation on ecological systems. In: *Ecological Effects of Nuclear War* (Woodwell, ed.). Brookhaven National Laboratory Publ. no. 917. Pp. 20-38.

# 4

# Environmental Gamma-Ray Observation in Deep Sea

Hidenori Kumagai, Ryoichi Iwase, Masataka Kinoshita, Hideaki Machiyama, Mutsuo Hattori and Masaharu Okano
*JAMSTEC (Japan Marine Science and Technology Center, Japan Agency for Marine-Earth Science and Technology)*
*Japan*

## 1. Introduction

Deep-Sea *in-situ* radioactivity measurements were initiated in 1964 for investigation of sunken USN atomic power submarine Thresher by using Geiger counters (Wakelin, 1964). Since then, *in-situ* radioactivity monitoring utilizing a towed-fish and/or remote sea bottom stations has been carried out in various countries. But they are mainly used for surveys of artificial radio activities to evaluate potential risks around atomic power plants, radioactive waste disposal sites and sunken nuclear objects etc. (Jones et al., 1988). Thus, underwater gamma ray measurement has been rather limitedly performed compared to the sub aerial ones, which has been widely utilized for explorations of uranium or other valuable minerals for mining or much wider environmental analysis (Bristow, 1983), due to very effective shielding by seawater. In underwater environment, Yoshida and Tsukahara (1987) reported environmental gamma ray characteristics around an active cold seepage associating with an active fault in Sagami Bay, southern coast of mainland Japan. Besides that, deep-sea gamma ray surveys for scientific use are intensively carried out by ODP-IODP logging for determination of rock types.

This chapter intends to demonstrate some usage of environmental gamma-ray measurement, through the studies in the deep sea environment, e.g. deep sea hydrothermal vent, cold seeps, active faults. Further, this chapter also tries to document relationship with tectonic settings and geological events. For this purpose, the specification and assemblage of the gamma ray sensor systems of JAMSTEC are introduced firstly. Next, an examples of the field measurement will be shown as a temporal variation that reflecting the uniqueness of deep sea. Although the sensitivities of the current model of sensor are significantly reduced by the thick Al or Ti pressure hull to endure high pressure in deep sea, the collected gamma ray spectra and distributions of intensities indicated some linkages with tectonic settings to date.

## 2. Instrumental

The apparatus collected to the world-wide gamma ray spectra in the deep sea environment, through hundreds dives of manned submersibles Shinkai6500/2000 and Remotely Operated

Vehicles (ROVs). In addition, an on-line real-time environmental gamma ray observatory has operated at deep seafloor of 1174m water depth for more than ten-years at Hatsushima Observatory, Sagami-Bay, southern coast of mainland Japan.

All the deep-sea systems developed are based on an on-land system originally developed at RIKEN (the Institute of Physical and Chemical Research; Okano et al., 1980; Kumagai and Okano, 1982). Since 1984, Deep-Sea Research department of JAMSTEC had developed deep-sea gamma ray sensor available down to 6500m of water depth (Hattori et al., 1997; 2000; 2002). To achieve required sensitivity for the system, NaI(Tl) scintillation counter has applied to the measurement that stored into Al-pressure vessel (Plate 1). Figure 1 is a schematic block diagram of the system. During the development, various assemblages of high voltage supply, data transfer, and sizes or shapes of scintillator were tried; three or two inches spherical, or three inches cylindrical scintillators. As the current model, three inches

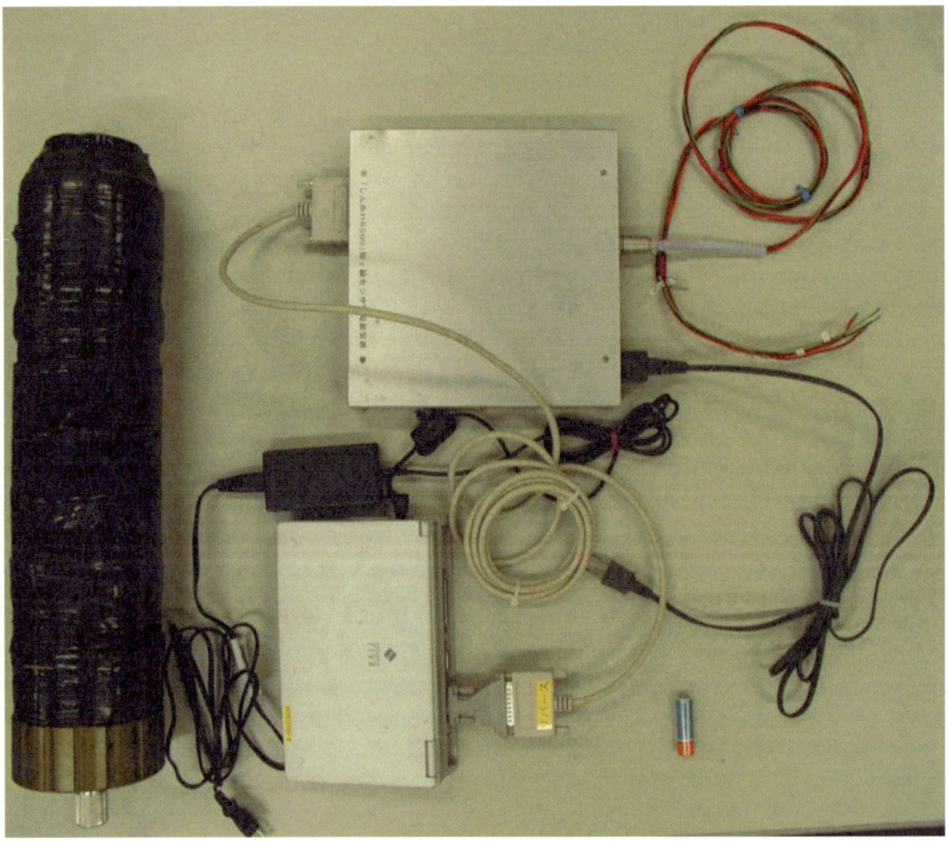

Plate 1. Deep sea gamma-ray sensor for a manned submersible *Shinkai6500*. NaI(Tl) scintillator are in the Al-pressure container (dark green colour, wrapped by black plastic tape, left). Under an operation, the sensor unit is connected thorough an underwater connector on the pressure container (silver coloured cover) to a feed-thorough on the pressure hull of submersible by underwater cable. The data signals are transferred to PC via power supply box (silver coloured) where +12V DC power is converted from AC power supplied from submersible to feed to the sensor unit.

spherical NaI(Tl) scintillator has been adopted and stored into Al-pressure container with high voltage supply and with multi-channel pulse height analyzer (PHA). The power supply for the underwater part is +12 V or +24 V according to the specifications of platforms (submersibles, towed-fish, etc.). When the γ-ray was detected by the system, its energy level was determined by PHA and transferred via serial communication of 9600 baud of RS-232C. Under this constraint, the channels of PHA are limited to 256. Firstly, the system was equipped on a towed-fish (Deep-tow system) and then applied on submersibles and other platforms, including a real time observatory (Hattori et al., 2000; Iwase et al., 2001). The approximate sensitivity of the apparatus is 0.25 nGy/h/cps, which is significantly reduced by the shielding by their pressure vessels.

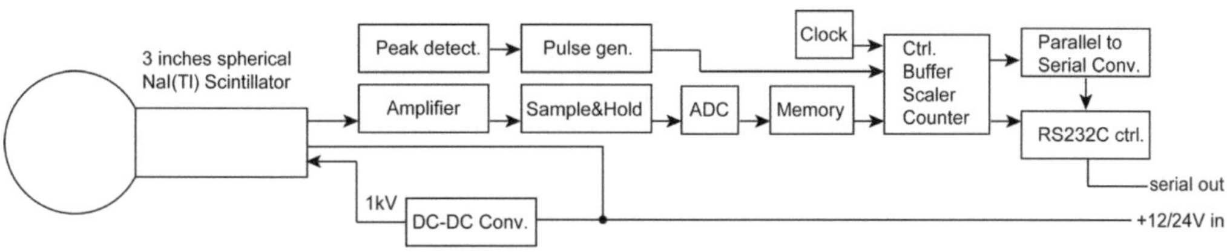


Fig. 1. Schematic block diagram of the Deep-sea gamma-ray sensor.

Once NaI(Tl) scintillation system rolled out, Ge-semiconductor detector had also tried to use. Instead of its high resolution to resolve the energy distribution of γ-ray radiations in environment, the sensitivity of Ge-semiconductor system is one-third or one order lower than that of NaI(Tl) scintillation system. Thus, the Ge-semiconductor system has limited to a trial production in JAMSTEC although another institutions, e.g. Japan Atomic Energy Research Institute (JAERI), tried further development for environmental monitoring (Ito et al., 2005).

### 2.1 Protocols of analysis

In analysis, the peel-off (stripping) method is applied to resolve the contributions from different nuclides to the total gamma-ray intensity. The method identifies the total energy peaks from approx. 3MeV to lower. In this energy region, the scintillation spectra are composed of photopeaks and their Compton continuums. Here, each of Compton continuums is assumed to have flat distribution. Thus, to resolve the photon energy distribution, the Compton continuums were sequentially peeled off from higher energy level to lower (Okano et al., 1982). These protocols are partly available on commercial applications, e.g. Seiko EG&G Co. Energy calibrations are performed for all time-series data typically for individual dives by using frequently identified three photopeaks: 609 keV of $^{214}$Bi, 1460 keV of $^{40}$K and 2614 keV of $^{208}$Tl.

## 3. Intensities of gamma-radiation in seawater and on seafloor

Due to the effective shield by seawater, total count rate in water is rather lower than that of sub aerial measurement (Yoshida and Tsukaraha, 1987). According to the geological environment, total count rate of sub aerial environment is 20-100 nGy/h in Japan, which is equivalent to approx. 80-400 counts per second (cps) in our system: 1 nGy/h is equivalent to 4cps. Figure 2 is an example of the time-series record of total count rate in submersible dive:

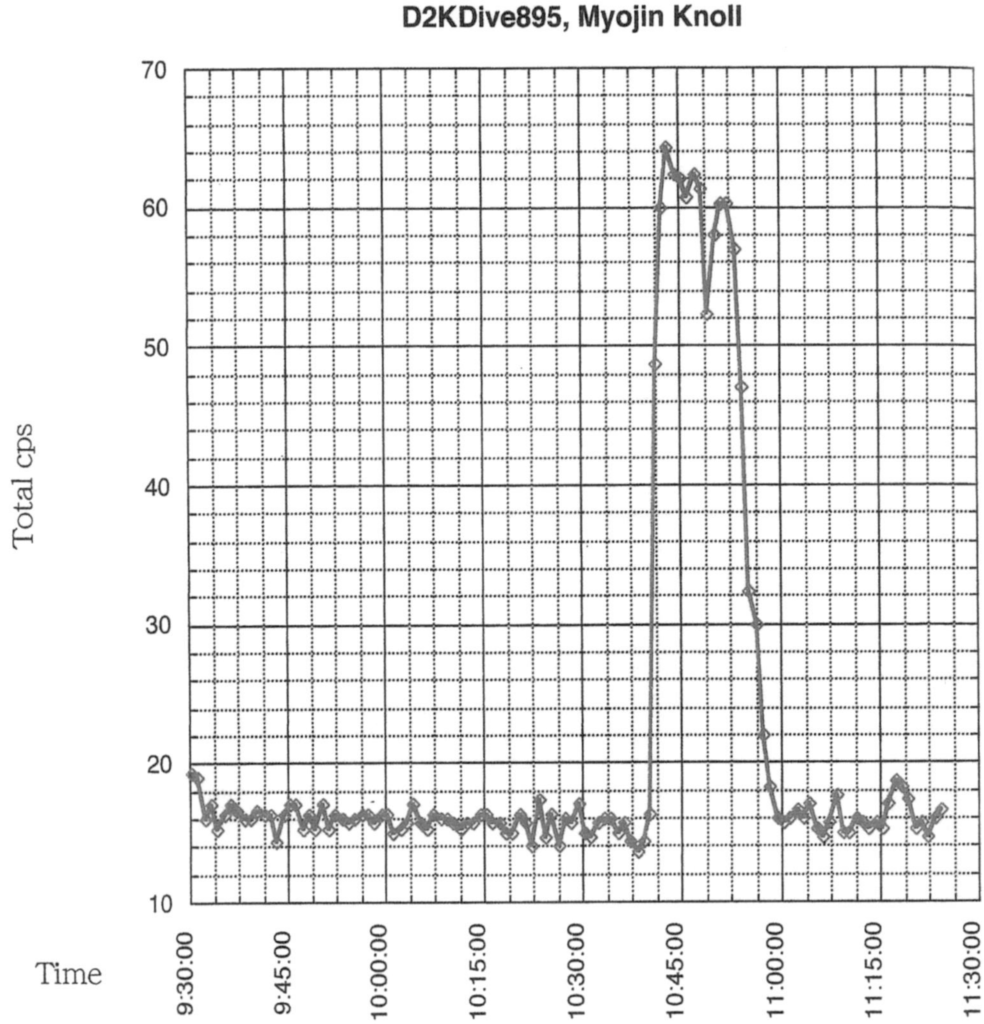


Fig. 2. A typical time sequential record obtained by submersible dives; 895th dive of *Shinnkai2000* at Myojin Knoll, Izu-Bonin Arc. Original Fig. was in Hattori & Okano (1999).

collected at the 895th dive of Shinkai2000. The cosmic-ray induced high energy gamma ray, > 3MeV as its energy level, still penetrate into shallow water, which raises total count rate approx. 20 cps. Descending to deeper, the total count rate decreases down to 10 cps or lower in typical; in this example, the minimum count rate is approx.14 cps. In the area with little suspended material in water column, such a minimum decreases down to 5 cps, e.g. in the vicinity of Mid Atlantic Ridges (Hattori et al., 2001). Approaching to the seafloor, total count rate raises up again to a few tens cps in most cases, which indicate significant contribution from seabed. Figure 3 shows the distributions of the maximum total count rates recorded when submersibles or ROVs were on bottom, touching on the seabed. The mode of the recorded value is in the class of 10-20 cps in linear scale with 10 cps in intervals, however, the frequency of the count rate gradually decreases to much higher count rate; there is no significant gap in the distribution. Even in this context, the area around the hydrothermal vents or cold seeps associated with active fault showed significantly anomalous count-rate, up to $10^4$ cps; the highest total count rate was recorded at Izena Hole caldron, >8900 cps. Other hydrothermally active areas also showed rather high count-rates, e.g. Iheya ridges, Hatoma-knoll, Ishigaki-knolls of Nansei-shoto and Kagoshima-bay (Southwest Japan), Myojin-knoll of Shichito-Iwojima Ridge. To judge the apparent anomaly from the above described general distribution, here after 100 cps is adopted as a tentative threshold to identify geological activities. Here also pseudo-logarithmic scales were applied in Figure 3b to show the entire variation of total count-rate. In this figure, distribution peak of the γ-ray intensity is on the class 50-100 cps, which also support this tentative threshold.

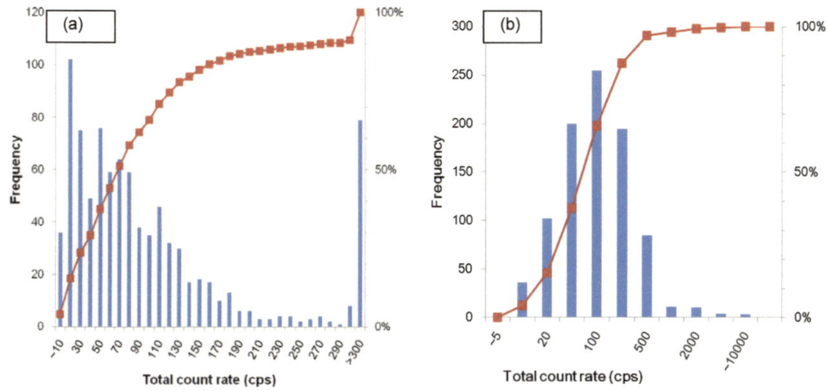


Fig. 3. Maximum total count rate distribution recorded at individual dives of submersibles: (a) Linear scale classification of every 10 cps in intervals, (b) Data classification under logalithmic-like scale. Blue bars: frequency in the class; Red sequential line: cumulative relative frequencies from lower count rates.

### 3.1 Relationships between tectonic settings and gamma-ray signature

Hereafter, some examples of the intensities of environmental γ-ray obtained in various geological settings are discussed. Figure 4 summarizes the measured distributions of total count rate maxima of γ-radiation around Japan. It is clear that the anomalous values were recorded all the area around Japan even in the area rather old geologic edifices; e.g. Komahashi-daini Knoll of Kyushu-Palau Ridge, or Annei Smt. of Nishi-Shichito Ridge.

The localities where very high count-rates were observed, > 1000 cps, were limited to the active hydrothermal sites developing on the arc-backarc volcanoes: volcanism developing above trench-arc system relating to seafloor subduction. In contrast to the hydrothermal sites, fore-arc cold seepages showed moderately high cont-rates, up to 500 cps; mostly not exceeding to 200 cps. It is notable that four of five localities where the very high count-rate recorded (>1000 cps) were in Okinawa Trough. It is a back arc basin developing between Nansei-Shoto and Asian continent in East China Sea, where thick terrigeneous sediments have been accumulated. Generally, major radioactive elements, K, Th and U[1], are rich in

[1] All these elements are geochemically classified to incompatible elements that concentrated into continental crust due to their incompatibility to the rock forming minerals. Thus, oceanic crust or magmas in ocean are relatively poor in such elements.

such terrigeneous sediments, which potentially cause the high count-rate of γ-radiation. The hydrothermal circulation within the sedimentary layer enhanced by the magmatic heat sources may scavenge and concentrate such radioactive species from thick sediments in those areas. Therefore, it is natural that the hydrothermal activities in oceanic environment far from continents or large land masses, e.g. at Mid-Ocean Ridges, did not show any very high count-rates. The vicinity of Hawaiian Island is also the area of low γ-ray intensities. In those areas, maximum count rates did not exceed 200 cps.

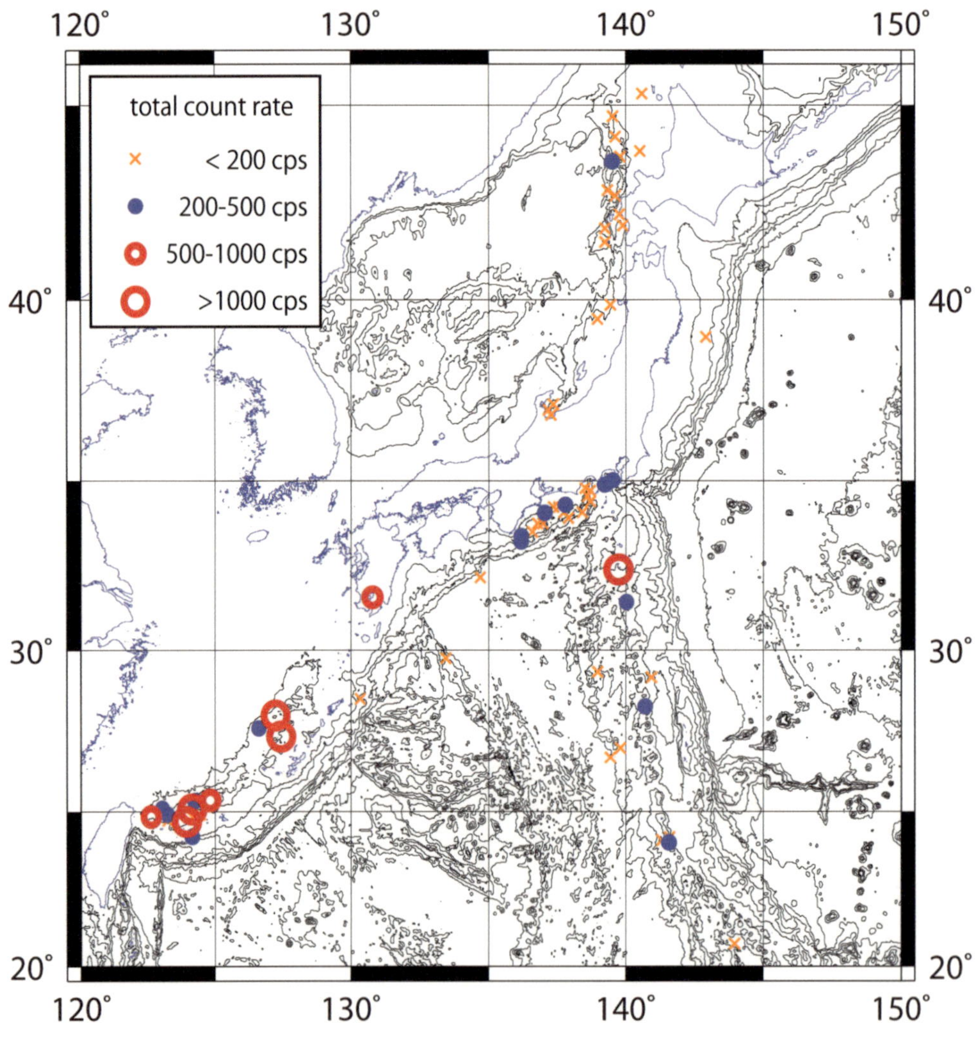


Fig. 4. The regional distribution of anomalous total count rate maxima, 100 cps, in the vicinity of Japan. Bathymetric contours are drawn as 1000 m in intervals. Original figure was taken from Hattori & Okano (2002) and redrawn.

Even by using NaI(Tl) scintillator with 7% resolution, the sources of γ-ray could be resolved by its energies. In this purpose, three characteristic γ-ray of 609 keV of $^{214}$Bi, 1460 keV of $^{40}$K and 2614 keV of $^{208}$Tl were used in this study. Under some reasonable assumptions, either radiation dose rate or concentrations of γ-ray source nuclei in the environment could be calculated. Hereafter the radiation dose rates are applied because it is not obvious the spatial distribution of the source nuclei at the various locations; geometric complexities of the seafloor having tall hydrothermal chimneys abruptly standing up > 10m in some cases and contribution from vigorous flow out from hydrothermal chimneys or seepages may cause unexpected increase of gamma radiation.

Figure 5 shows the statistics of γ-ray dose rate from three possible sources, K, U-series and Th-series, respectively. K is major element in seafloor sediment; the recorded maximum of dose rate was 7.0 μR/h regardless of its mode at 0.28 μR/h. Entire variation was within three orders of magnitude, which was narrower than those of U- and Th-series radiations (Figure 5(a)).

Contrary to K, dose-rate of U-series varies more than five orders of magnitude caused by its maximum of ~200 μR/h regardless of its relatively low mode (0.54 μR/h; Figure 5(b)). Dose rate distribution of Th-series shows the intermediate nature between K and U-series, ranging within four orders of magnitude (Figure 5(c)). It is notable that its mode of distribution is 1.38μR/h, which is much higher than that of U-series. It may relate to the rather broad distribution tailing to the higher side of dose rate. This nature will be discussed later.

Under comparison with the recorded enormous total count-rate and the large variation of U-series dose rate, high concentration of U-series in environment should cause very high intensities of γ-ray, > 1000 cps, equivalent to >250 nGy/h. To confirm this view, the relationship of total count rates with U-series dose rate is plotted in Figure 6. It is clear that the observed high count rates were tightly associated with the high dose rate of U-series, > 100 cps as total count rate. Considering the progeny radio-nuclei of U-series and their half lives, Ra fed from hydrothermal vent fluids causes such radio activities[2]. Even in the lower count-rate environment, rough relationship between total count rate and U-series dose rate still found, which suggests a ubiquity of U-series controlled γ-radiation environment; that is represented by the areas of hydrothermal activities in the higher dose rate end of correlation line. In addition, the trends was slightly above from the extrapolation defined in >100cps of total count rate. It is suggested significant contributions caused from Th-series nuclei, which built up the total count-rate.

To investigate such Th-series contribution, measured dose rates of U-series and Th-series were plotted (Figure 7). As predicted the above described relationship between total count rate and U-series dose rate, the data were dominantly plotted around the correlation line of slope 1. This also supports the view of a ubiquity of U-series controlled γ-radiation environment. Above the trend, a few tens of data obtained from the fore-arc seepage areas were plotted; e.g. in Suruga Bay or in Sagami Bay (both are in southern coast of mainland Japan), or off Kamaishi-city near the Japan Trench. In such areas, the U-series dose rate is not so high regardless of the moderately high total count rate, up to 300 cps. Instead, dose rate of Th-series reaches approx. 5 μR/h regardless of the dose rate of U-series of <1 μR/h. As these areas are in the vicinities of active faults, thus, the contribution from Th-series nuclei is significant in the tectonics controlled environment. The data points from other tectonic active area either on southern coast of mainland Japan (e.g. Zenisu-ridge) or in

[2] Ra is in the group of an alkaline earth elements, which shows high solubility to water. Ra also frequently replaces Ba in minerals as the same group element.

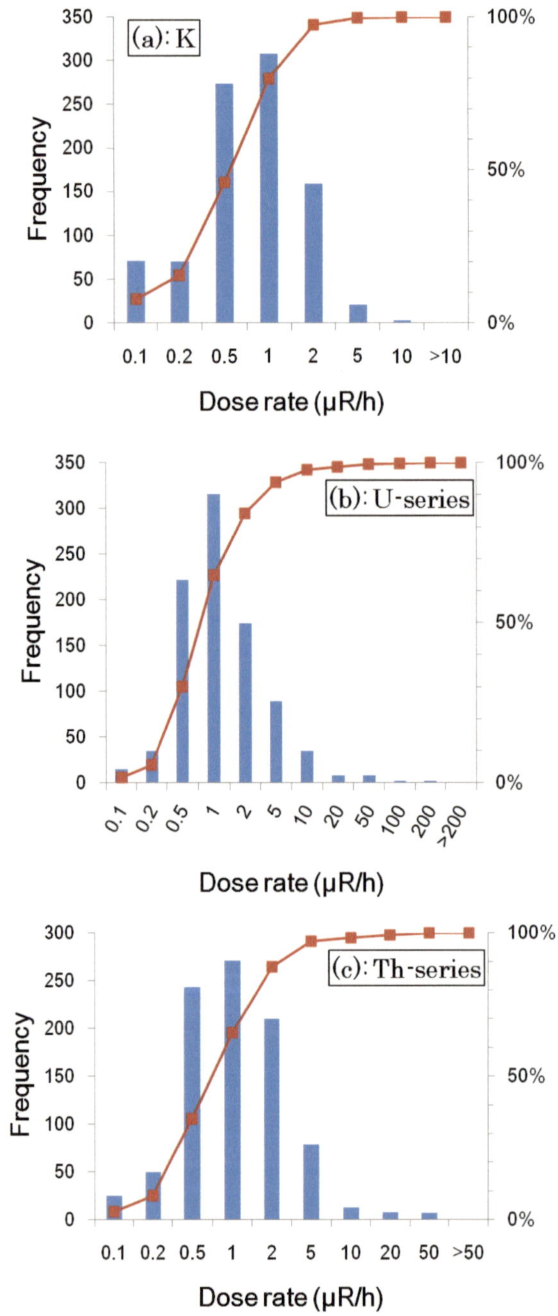


Fig. 5. The maximum dose rate distribution of individual sources; (a): K, (b): U-series, and (c): Th-series.

Japan sea-side (e.g. off Rebun Isl.) are also plotted above the line regardless of rather lower total count rates. These signatures may relate to squeezing of pore fluids by the compaction of sediments, which supplies Ra or other soluble species in seafloor environment.

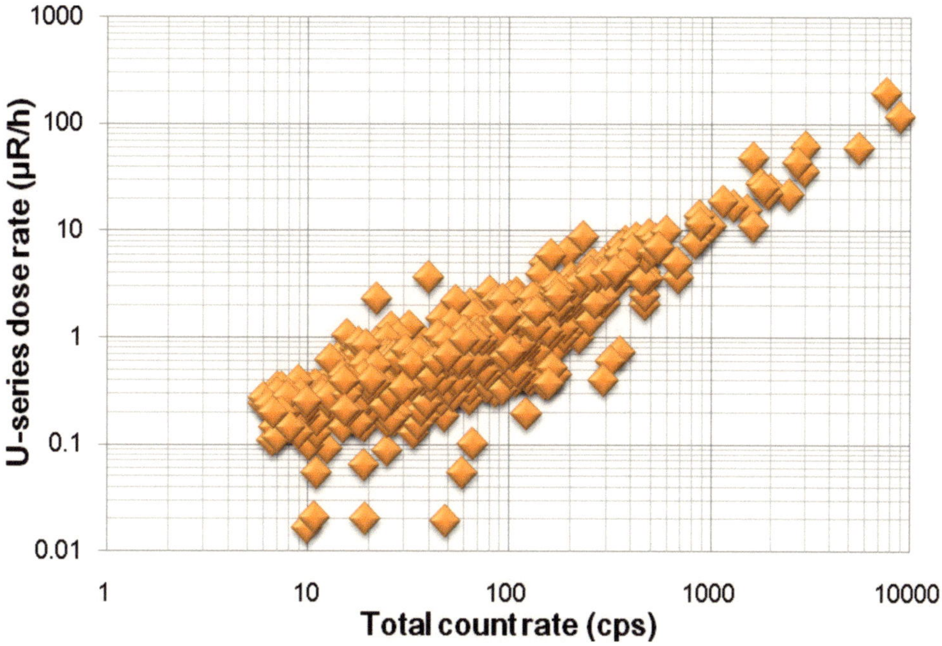


Fig. 6. The relationship between maximum total count rate and maxim dose rate from U-series nuclei. Most of data were tightly scattered around slope-1 correlation line.

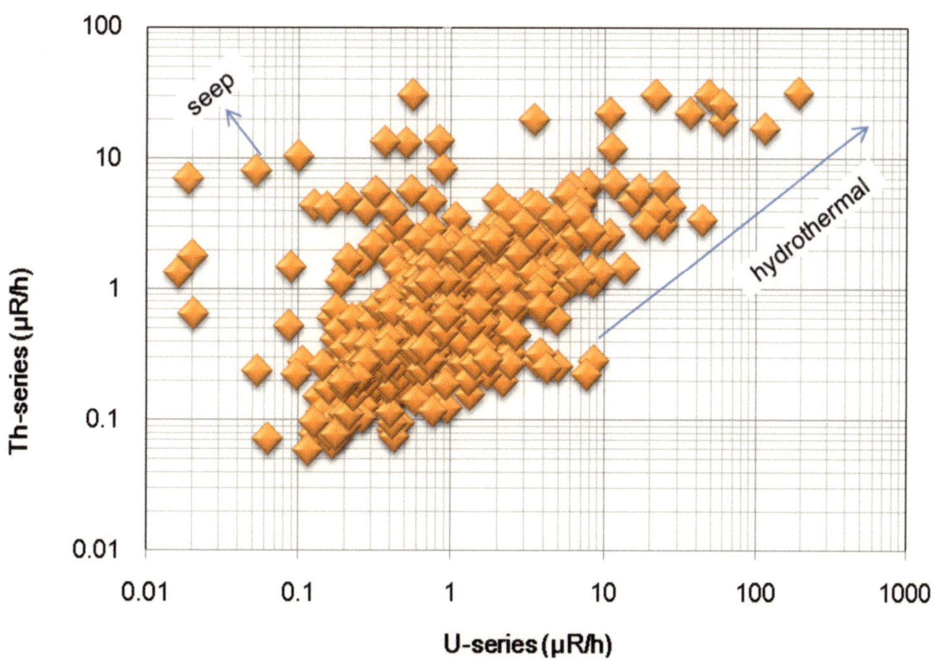


Fig. 7. The relationship of maximum dose rate from U-series and Th-series nuclei. Above the dense data distribution well correlate with U-series increase, a few tens of relatively high Th-series dose rate data were found.

### 3.1.1 Environmental gamma radiation in Mid-Ocean Ridges

In Figure 7, several points of moderately high Th-series contribution but very little U-series contributions are found: >1 μR/h and < 0.1 μR/h of respective Th-series and U-series dose rate. These data came from Indian Ocean either at the Rodriguez Triple Junction (RTJ) or at Atlantis Bank (AB). Those areas are well away from any continents or large land masses but very tectonic active area, thus very little terrigeneous sediments accumulated. RTJ is a Ridge-Ridge-Ridge type triple junction where very slow spreading Southwest Indian Ridge (SWIR) is propagating toward the junction of Central Indian Ridge and Southeast Indian Ridge. The magmatic activity of SWIR is very low due to its very slow spreading, thus numbers of active faults developed in the area. AB is one of the Oceanic Core Complex[3] that tectonically exposed and consisted of gabbros and peridotites. It locates ~100km away from present-days ridge axis, but, it is regarded to be exposed tectonically near the ridge axis approx. 12 Ma. The gabbros are in the category of mafic plutonic rocks that usually contains very little radioactive nuclei. These relatively old age and the composition of seabed caused such low gamma radiations in the areas. These tendencies are in the context of the Th-series enrichment in tectonic active area.

## 4. Long term monitoring at Hatsushima station, Sagami-bay

JAMSTEC has been carrying out multi-disciplinary real time long term observation on deep seafloor at a depth of 1175 m off Hatsushima Island in Sagami Bay. There, a cabled observatory was deployed with several kinds of sensors e.g. video cameras, a CTD sensor that measures conductivity of seawater, ambient temperature and depth of water, a seismometer, since 1993 (Momma et al., 1998). The main target of the observatory is to investigate the environmental and biological phenomena of the cold seepages around the observatory that feed a large number of chemo-synthetic biological communities which are mainly consisted of Vesicomyid clam.

In March 2000, the original observatory, which had been deployed in September 1993, was retrieved and fully renewed. The renewed observatory and the submarine cable connected were deployed at the position of approximately 40 m northward from the original one (Iwase, 2004). At the renewal, a gamma-ray sensor with NaI(Tl) scintillator was attached to the observatory. Since then, a long term environmental gamma-ray monitoring was started. The renewed observatory was retrieved again in March 2002 to repair, and re-deployed in November 2002 approximately 40 m southward from the previous position had deployed; i.e. the observatory relocated into almost the same position where the original observatory had located. Then, gamma-ray observation has resumed and continued at the same position for almost 9 years to date.

The specification of the gamma-ray sensor equipped to the station is almost the same as those for submersibles or a ROVs of JAMSTEC: three inch spherical NaI(Tl) scintillation counter and 256-channel PHA. It is stored in a titanium container (Plate 2). The output signal with a 9600 baud RS232C interface is transmitted to the shore station in Hatsushima

[3] Oceanic Core Complex (OCC) is a domy exposure of lower lithological units of oceanic plates. It is frequently observed along Mid-Ocean Ridges with slow spreading rate, e.g. Mid Atlantic Ridge or Southwest Indian Ridge.

Island through the electro-optical submarine cable. The energy spectra can be obtained by automated calculation every ten minutes at the shore station, i.e. each dataset of energy spectrum is the summation of ten minute measurement. The gamma-ray sensor unit is installed to touch its scintillator side on seabed in which scintillator is attached downward to maximize its sensitivity (Plate 3).

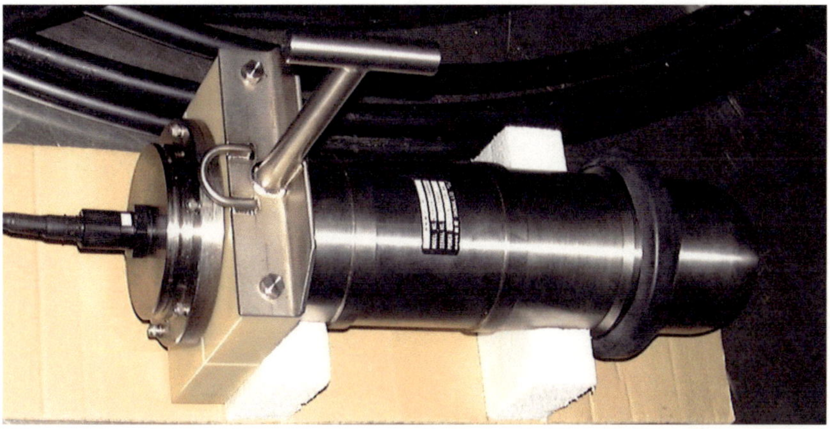

Plate 2. Gamma ray sensor of cabled observatory off Hatsushima Island.

Plate 3. Cabled observatory off Hatsushima Island and gamma ray sensor (denoted by circle) deployed on seafloor.

It is known that output signal of NaI(Tl) scintillator is affected by temperature variation. However, since the water temperature at the observation site on deep seafloor is approx. 3 °C and shows very small perturbation, the influence with temperature is negligible. On the other hand, some kind of signal drift associated with aging could occur. Fig. 8 shows the spectra obtained on January 1st in 2003, 2005, 2007, 2009 and 2011. Each spectrum is accumulated for one day. Prominent peaks of natural radiation $^{214}$Bi, $^{40}$K and $^{208}$Tl are remarked. It is obvious that each peak linearly shifts to lower channel as time passes.

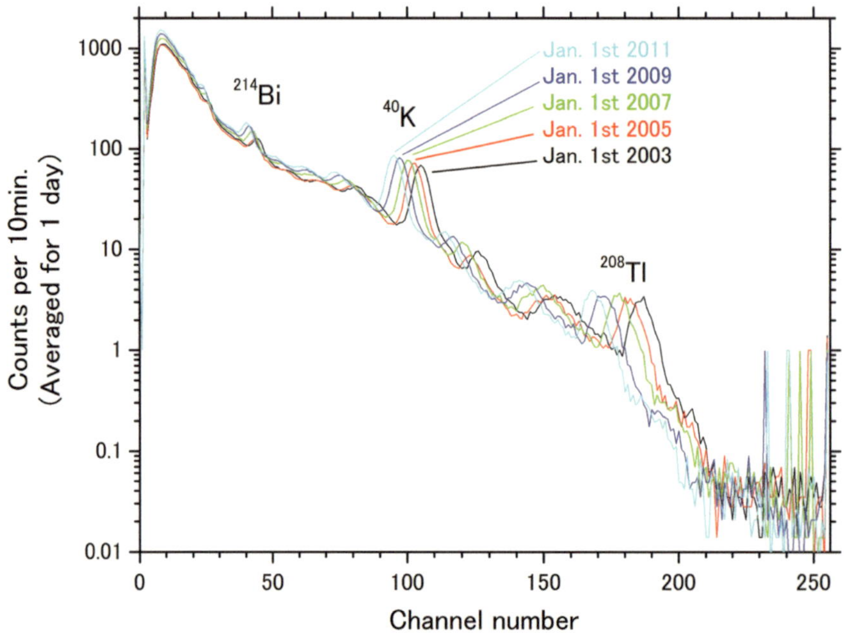


Fig. 8. Gamma ray energy spectra observed with the observatory.

The correspondence between channel number and energy was calibrated by using those three peaks ($^{214}$Bi of U-series, $^{40}$K and $^{208}$Tl of Th-series) in the one day averaged spectrum. The centre position of each peak was calculated by curve fitting. Fitting function is the combination of Gaussian and linear function as follows,

$$y = A\exp\left\{-4ln2\left[\left(x-p\right)/\mathrm{FWHM}\right]^2\right\} + ax + b, \tag{1}$$

where $x$ is channel number, $y$ is the number of counts, $A$ is the peak height, $p$ is the centre position of the peak in units of fractional channel number, FWHM is the full width at half maximum of the peak in units of fractional channel number, $a$ and $b$ are the constant parameters.

As the result, the centre position of each energy peak decreased at roughly constant rate. In case of $^{40}$K (1461 keV), the centre position of the peak decreased as large as 10 channels for the period of 8 years (Fig. 9), while the relation between the channel number and energy stayed linear (Fig. 10).

On the other hand, the full width at half maximum (FWHM) of each peak, which was calculated at the same time by the curve fitting, stayed almost constant as are shown in Fig. 11 (a)-(c) for $^{214}$Bi, $^{40}$K and $^{208}$Tl, respectively.

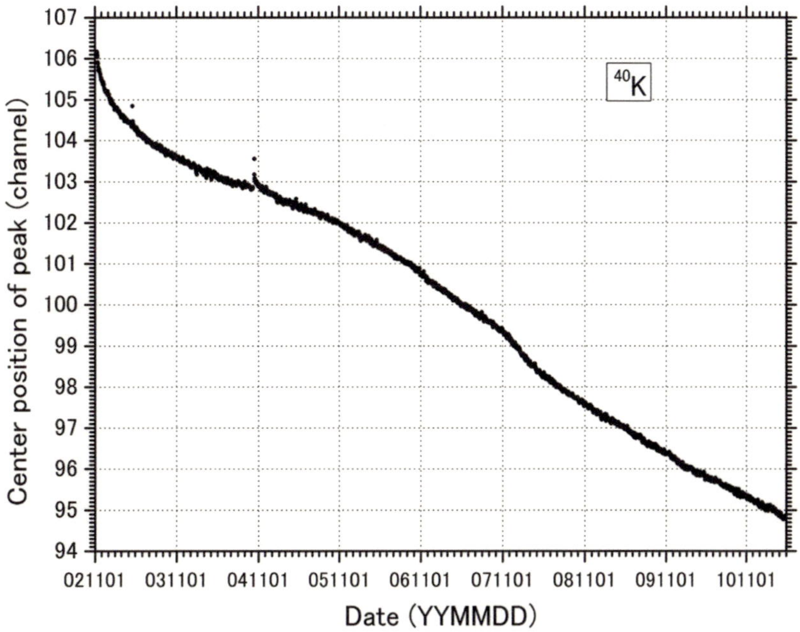


Fig. 9. Temporal change of the center position of $^{40}K$ peak.

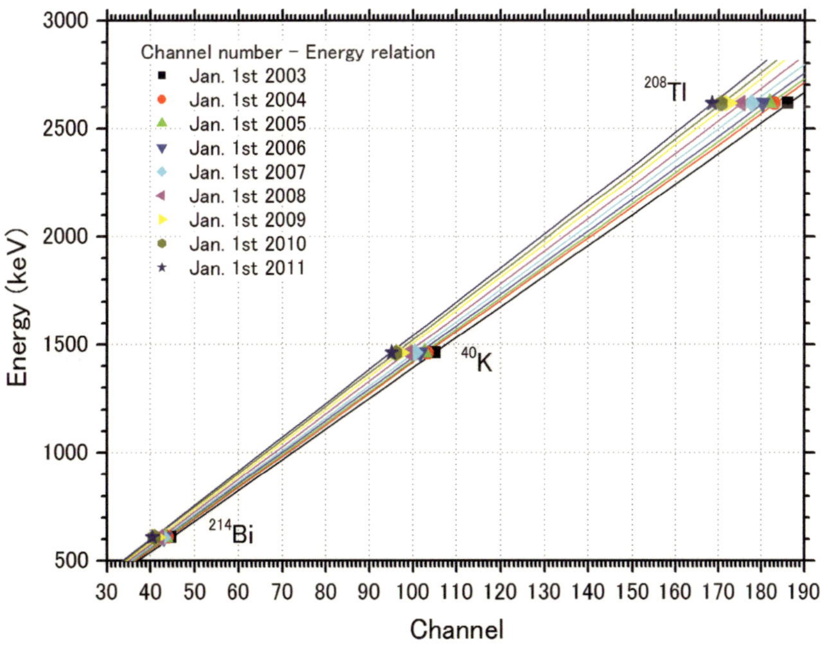


Fig. 10. Temporal change of relation between channel number and energy.

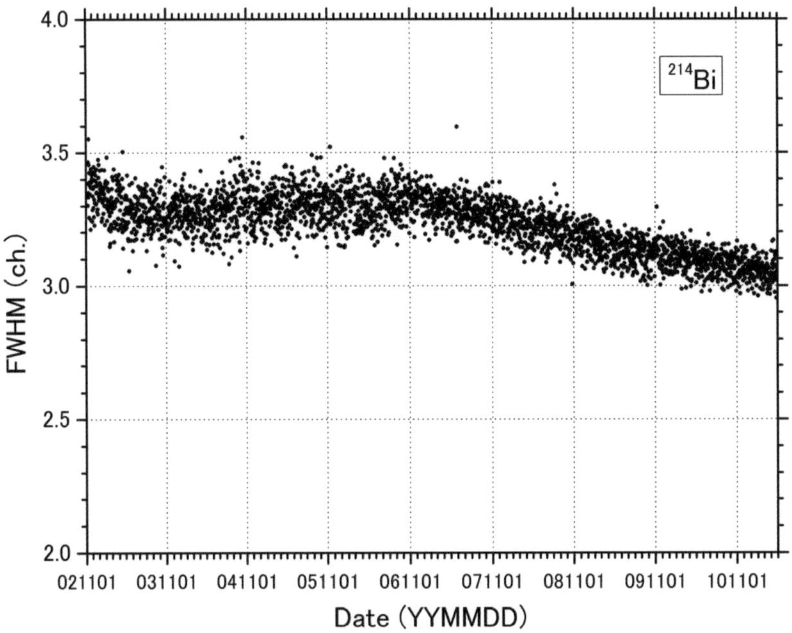

214Bi
FWHM (ch.)
4.0
3.5
3.0
2.5
2.0
021101 031101 041101 051101 061101 071101 081101 091101 101101
Date (YYMMDD)


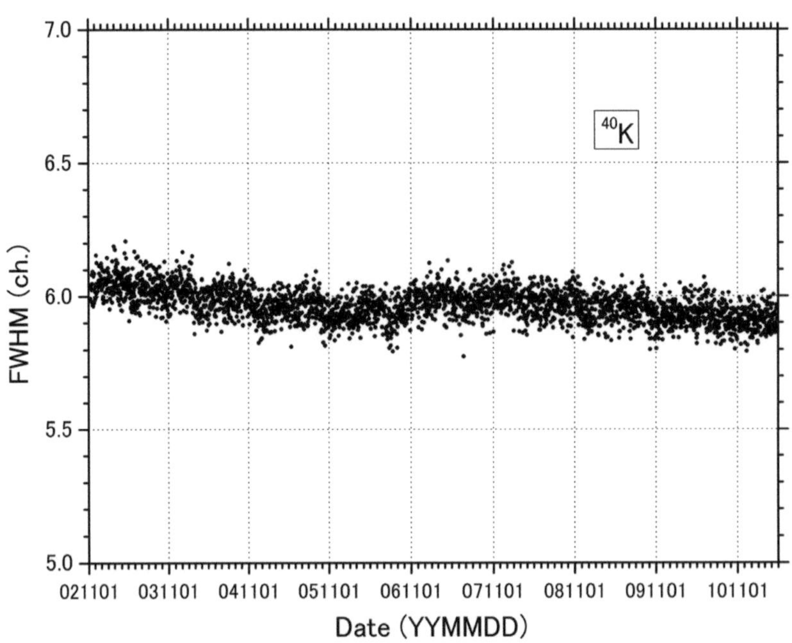

40K
FWHM (ch.)
7.0
6.5
6.0
5.5
5.0
021101 031101 041101 051101 061101 071101 081101 091101 101101
Date (YYMMDD)

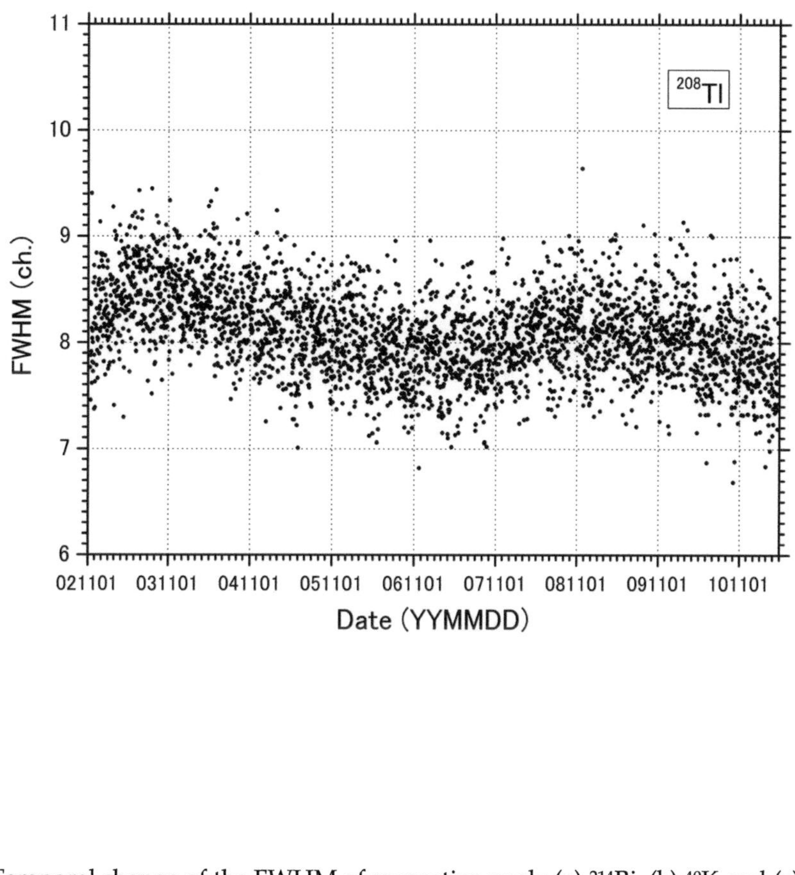


Fig. 11. Temporal change of the FWHM of respective peak: (a) $^{214}$Bi, (b) $^{40}$K and (c) $^{208}$Tl.

By using those results, long term fluctuation of net area of each peak, which corresponds to each radiation does rate, was calculated by using the same fitting function. Here, 30 day simple moving average of the above result on the centre position of the peak was used for $p$ and the average of whole period on the FWHM (3.23, 5.97 and 8.10 for $^{214}$Bi, $^{40}$K and $^{208}$Tl, respectively) was used as known variables in the equation, while $A$, $a$ and $b$ were the unknown parameter. The net area of each peak S is obtained by the following equation.

$$S = A \cdot \mathrm{FWHM} \cdot \sqrt{\pi / \ln 2} / 2 \quad (2)$$

The respective result for $^{214}$Bi, $^{40}$K and $^{208}$Tl is shown in Fig. 12.

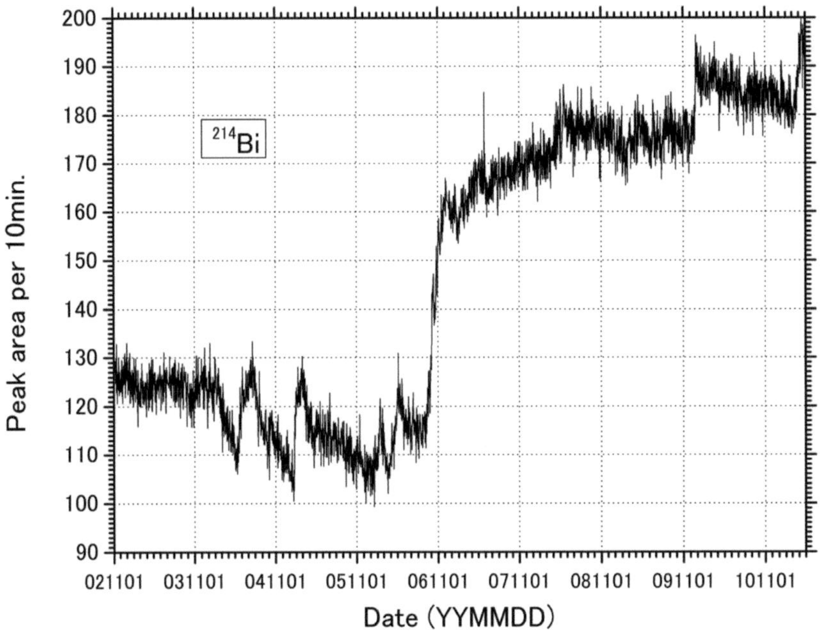

214Bi
Peak area per 10min.
200
190
180
170
160
150
140
130
120
110
100
90
021101 031101 041101 051101 061101 071101 081101 091101 101101
Date (YYMMDD)


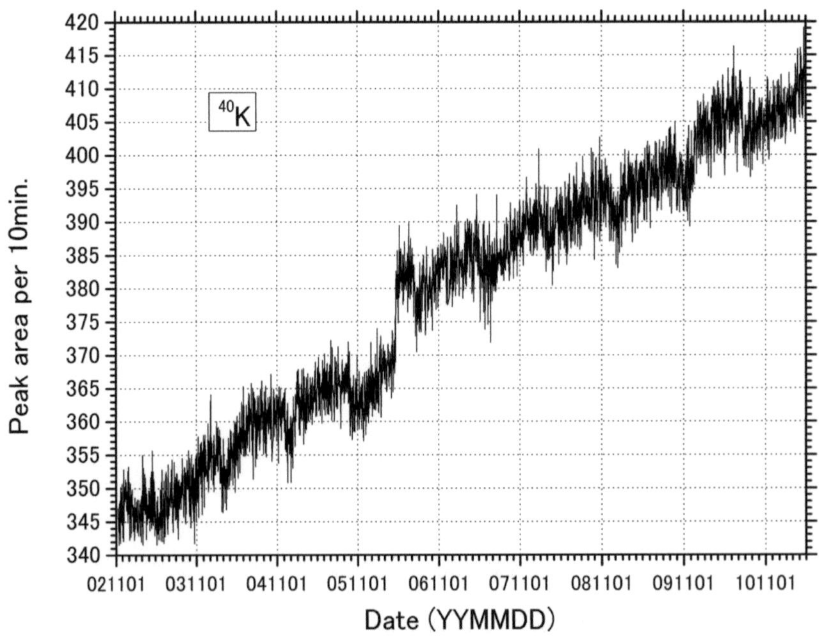

40K
Peak area per 10min.
420
415
410
405
400
395
390
385
380
375
370
365
360
355
350
345
340
021101 031101 041101 051101 061101 071101 081101 091101 101101
Date (YYMMDD)

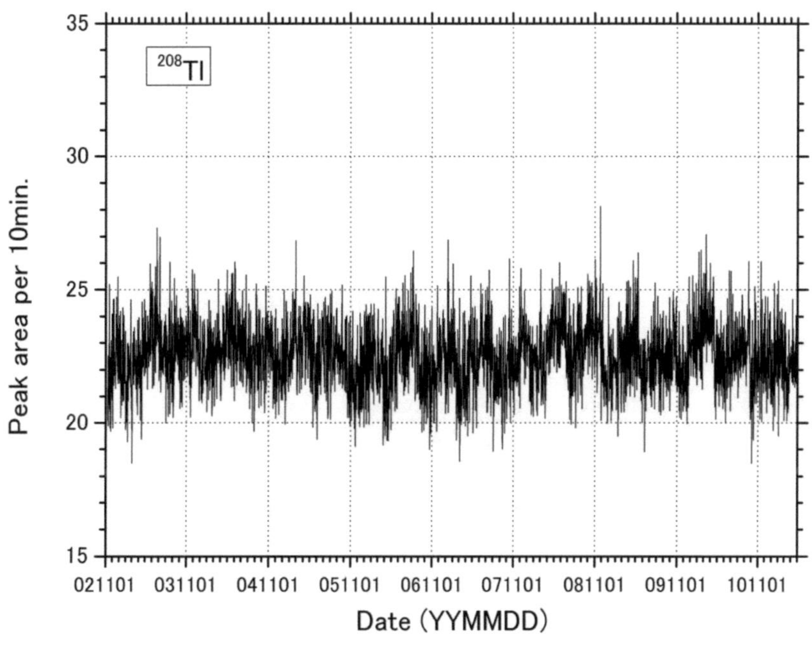


Fig. 12. Temporal change of the respective peak area; (a) $^{214}$Bi, (b) $^{40}$K and (c) $^{208}$Tl.

In Fig. 12 (a), significant increase of $^{214}Bi$ peak area was observed in October 2006. The reason of this fluctuation is under study at present, though it may suggest the fluctuation of seepage or may reflect some tectonic deformation. Another increases which are less significant than that in October 2006 were observed several times, many of which seem to occur in spring. Since the increase of the amount of suspended materials have been also observed in spring, those may have some relation. In Fig. 12 (b), steady increase is observed. Although the reason is yet to be investigated, it may be caused by instrumental reason, e,g drift. Episodic increase of $^{40}K$ peak area was observed when M5.8 earthquake occurred east off Izu Peninsula on April 21st in 2006 which caused mudflow (Iwase et al, 2007). When M5 class earthquakes occurred on December 17th and 18th in 2009, increase of $^{40}K$ peak area seems to be less significant. This difference suggests the difference of mudflow composition as their suspended materials. The peak area of $^{208}Tl$ seems to be constant, though it contains somewhat periodical fluctuation which is caused by some error in calculation. Main reason is the selection of ROI (Range of Interest) channel for curve fitting calculation. While the variable $p$ (the centre position of peak) in fitting function is fractional channel number, ROI is not, and then some discontinuity in the result occurs. Those preliminary results need much detailed evaluation in both technical and other environmental aspects.

## 5. Prospect and concluding remarks

One of the ways to improve the current model as for the proto-type apparatus is to develop stand-alone type of detector with battery and data logger system likely to an on-land system; it could deploy on the seabed for a year or more to accumulate the signals. Such a trial has already done by Ashi et al. (2003). Alternatively, once 10-times higher sensitivity of sensor achieved, such an apparatus may become quite powerful tool to monitor the rapid change of gamma ray intensities in deep sea environment, e.g. vibration of hydrothermal venting, tidal change of seepages. An application of plastic scintillator partly replacing the pressure hull is one of the available solutions (Shitashima et al., 2009).

## 6. Acknowledgments

All operations at sea had been supported numerous supports of captains, crews of research vessels, operation teams of submersibles or ROVs and technicians. Throughout the development, the authors express their thanks to colleagues of Deep Sea Research Department of JAMSTEC.

## 7. References

Ashi, J., Kinoshita, M., Kuramoto, S.'i., Morita, S. & Saito, S. (2003). Seafloor gamma ray measurements around active faults by standalone system, *JAMSTEC J. Deep Sea Research,* vol.22, pp.179-187 (in Japanese w/ English abstract)

Bristow, Q. (1983). Airborne γ-ray spectrometry in uranium exploration- Principles and current practice, *The International Journal of Applied Radiation and Isotopes*, vol.34, No.1, pp.199-229

Hattori, M., Kobayashi, Y. & Okano, M. (1997). Sea bottom radioactivity measurement systems in Japan, *Proc. 7th Internatl. Offshore and Polar Eng. Conf.*, vol.1, pp.116-119

Hattori, M., Okano, M., & Togawa, O. (2000). Sea Bottom gamma ray measurement by NaI(Tl) Scintillation spectrometers installed on manned submersibles, ROV and sea bottom long term observatory, *Proc. 2000 International Symposium on Underwater Technology*, IEEE Catalog Number 00EX418, pp. 212-217.

Hattori, M. & Okano, M. (2001). New results of sea bottom radioactivity measurement, *JAMSTEC J. Deep Sea Res.*, vol.18, pp.1-13 (in Japanese w/ English abstract)

Hattori, M. & Okano, M. (2002). Sea bottom gamma ray measurement - Results of study and modeling of sea bottom radioactive environment -, *JAMSTEC J. Deep Sea Res.*, vol.20, pp.37-52 (in Japanese w/English abstract)

Ito, T., Kinoshita, M., Saito, S., Machiyama, H., Shima, S., Gasa, S., Togawa, O. & Okano, M. (2005). Studies on Applications of Detectors for Marine Radioactivity and Methodologies for Data Analysis (Joint Research), *JAERI-research 2005-028*, pp. 127, Japan Atomic Energy Research Institute, Ibaraki-ken, Japan.

Iwase, R., Mitsuzawa, K., Hirata, K.; Kaiho, Y., Kawaguchi, K., Fujie, G. & Mikada, H. (2001). Renewal of "Real-time Deep Seafloor Observatory off Hatsushima Island in Sagami Bay", *JAMSTEC J. Deep Sea Research*, vol.18, pp.185-192 (in Japanese w/ English abstract)

Iwase, R. (2004). 10 Year Video Observation on Deep Seafloor at Cold Seepage Site in Sagami Bay, Central Japan, *Proc. OCEANS'04/TECHNO-OCEAN'04: 2200-2205*

Iwase, R., Goto, T., Kikuchi, T. & Mizutani, K. (2007). Earthquake Accompanied by Mudflow Observed by a Cabled Observatory off Hatsushima Island in Sagami Bay in April 2006, *Proc. 2007 Symposium on Underwater Technology and Workshop on Scientific Use of Submarine Cables and Related Technologies*, pp.472-475

Jones, D.G., Roberts, P.D. & Miller, J.M. (1988). The distribution of gamma-emitting radionuclides in surface subtidal sediments near the Sellafield plant, *Estuarine, Coastal and Shelf Science*, vol.27, pp.143-161

Kumagai, H. & Okano, M. (1982). A portable scintillation spectrometer for environmental radiation measurements, *Reports of the Institute of Physical and Chemical Research*, vol. 58, No.1, pp.1-10

Momma, H., Iwase, R., Mitsuzawa, K., Kaiho, Y. & Fujiwara, Y. (1998). Preliminary results of a three-year continuous observation by a deep seafloor observatory in Sagami Bay, central Japan, *Phys. of Earth and Planet. Inter.*, vol.108, pp.263-274

Okano, M., Izumo, K., Kumagai, H., Katou, T., Nishida, M., Hamada, T. & Kodama, M. (1980). Mearurement of environmental radiations with a scintillation spectrometer equipped with a spherical NaI(Tl) Scintillator, *Natural Radiation Environment III, Symposium Series DOE51 (CONF-780422)*, pp.896-911

Shitashima, K. (2009) Development of in-situ radon sensor using plastic scintillator, CRIEPI Research Report, V08054, pp. 17

Yoshida, N. & H. Tsukahara (1987) A γ-ray spectral survey on giant clam colonies using the submersible Shinkai2000. *JAMSTEC J. Deep Sea Res.,* Vol.3, pp.105-112 (in Japanese w/English abstract).

# 5

# Gamma Dose Rates of Natural Radioactivity in Adana Region in Turkey

Meltem Degerlier
*Nevsehir University, Science and Art Faculty, Physics Department, Nevsehir Turkey*

## 1. Introduction

We are all exposed to ionizing radiation from natural sources at all times. This radiation is called natural background radiation. Background radiation is the radiation constantly present in the natural environment of the Earth, which is emitted by natural and artificial sources. Natural radioactivity is wide spread in the earth's environment; it exists in soil, plants, water and air. Exposure of radiation mainly come from natural radiation (85 %). The assessment of gamma radiation doses from natural sources is of particular importance because natural radiation is the largest contributor of external dose to the world population (UNSCEAR,2000; Narayana N. et al.,2007) The exposure of human beings to ionizing radiation from natural sources is a continuing and feature of life on earth inescapable (UNSCEAR Report 2000). Throughout the history of life on earth, organisms have been continuosly exposed to radiations from radionuclides produced by cosmic ray interaction in the atmosphere and radiations from naturally occuring substances that are spatially distributed in all living and non-living components of the biosphere.(Whicker F.W. And Schultz, 1982)

Environmental natural gamma radiation is formed from terrestrial and cosmic sources (Merdanoglu and Altinsoy, 2006, M.Degerlier et al., 2008) It comes from two primary sources: cosmic radiation and terrestrial sources. The worldwide average background dose for a human being is about 2.4 millisievert (mSv) per year. This exposure is mostly from cosmic radiation and natural radionuclides in the environment (including those within the body).

The main sources of natural background radiation are radioactive substances in the earth's crust, emanation of radioactive gas from the earth ,cosmic rays from outer space which bombard the earth, trace amounts of radioactivity in the body.

Sources in the Earth include sources in water, soil and food which are incorporated to the human body, to building materials, and to products that incorporate radioactive sources from nature, sources from outer space are the radiation produced by the atomic bombardment of the upper atmosphere by high-energy cosmic rays and sources in the atmosphere, such as the radon gas released from the Earth's crust, which then decays into radioactive atoms that attach to airborne dust, and other particulate (granular, powder) materials.

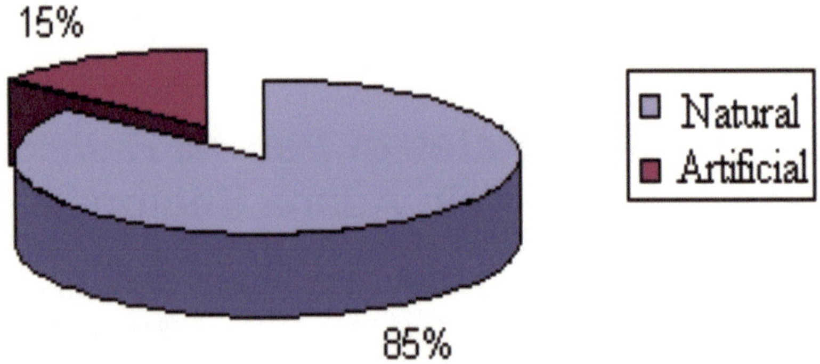


Fig. 1. Exposure radiation dose rates from natural and artificial sources

| Radiation | | UNSCEAR | |
|---|---|---|---|
| Type | Source | World Average (mSv) | Typical Range (mSv) |
| Natural | Air | 1.26 | 0.2-10.0 |
| | Internal | 0.29 | 0.2-1.0 |
| | Terrestrial | 0.48 | 0.3-1.0 |
| | Cosmic | 0.39 | 0.3-1.0 |
| | Total | 2.40 | 1.0-13.0 |
| Man Made | Medical | 0.60 | 0.03-2.0 |
| | Fallout | 0.007 | 0-1+ |
| | others | 0.0052 | 0-20 |

Table 1. Exposure dose rates as mSv from natural and artificial sources in the World

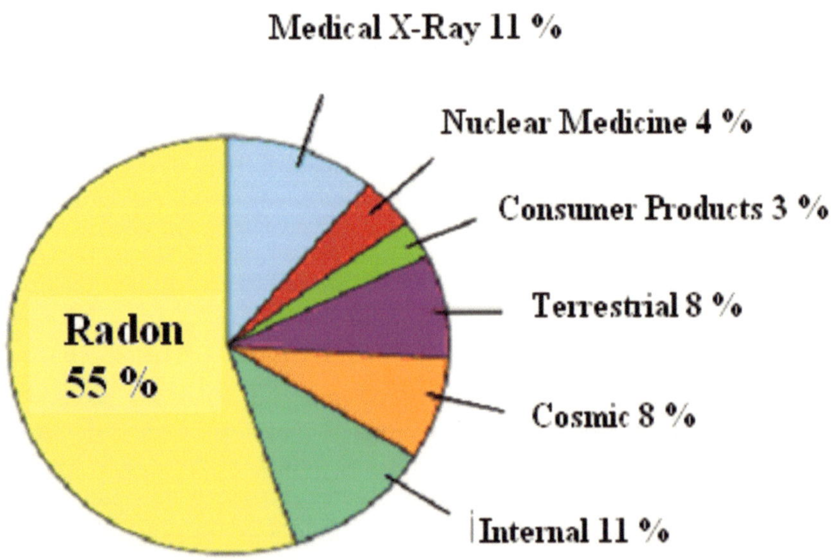


Fig. 2. Exposure dose percentage from natural and artificial radiation sources

On the ground, cosmic radiation makes up on average about 17 % of the natural background radiation to which we are all exposed. The rest consists of radon gas (50%), radiation from minerals in the soil (20 %), and radiation in our bodies from food and water (13 %).

Some radioactive materials - most of which are naturally occurring elements - are actually air pollutants. All of them, as a whole, are a relatively small proportion of the many elements and chemicals that are considered air pollution. Radon is the most significant of these elements, but most radon exposure stems from the indoor environment. Improving technology continues to minimize man-made radioactive air pollutants and monitor air quality.

## 2. Terrestrial

Naturally occurring radionuclides of terrestrial origin (also called primordial radionuclides) are present in various degrees in all media in the environment, including the human body itself. Only those radionuclides with half-lives comparable to the age of the earth, and their decay products, exist in significant quantities in these materials. Irradiation of the human body from external sources is mainly by gamma radiation from radionuclides in the $^{238}$U and $^{232}$Th series and from $^{40}$K. These radionuclides are also present in the body and irradiate the various organs with alpha and beta particles, as well as gamma rays. Some other terrestrial radionuclides, including those of the $^{235}$U series, $^{87}$Rb, $^{138}$La, $^{147}$Sm and $^{176}$Lu exist in nature but at such low levels that their contributions to the dose in humans are small.

Natural radionuclides in soil generate a significant componed of the background radiation exposure of the population (Karahan and Bayulken, 2000) Gamma radiation intensity in a region depends on soil and geographic structure.

External exposures outdoors arise from terrestrial radionuclides present at trace levels in all soils. The specific levels are related to the types of rock from which the soils originate. Higher radiation levels are associated with igneous rocks, such as granite and lower levels with sedimentary rocks. There are exceptions however as some shales and phosphate rocks have relatively high content of radionuclides. There have been many surveys to determine the background levels of radionuclides in soils, which can in turn be related to the absorbed dose rates in air. The latter can easily be measured directly, and these results provide an even more extensive evaluation of the background exposure levels in different countries. All of these spectrometric measurements indicate that the three components of the external radiation field, namely from the gamma emitting radionuclides in the $^{238}$U and $^{232}$Th series and $^{40}$K, make approximately equal contributions to the externally incident gamma radiation dose to individuals in typical situations both outdoors and indoors.

The radionuclides in the uranium and thorium decay chains cannot be assumed to be in radioactive equilibrium. The isotopes $^{238}$U and $^{234}$U are in approximate equilibrium as they are separated by two much shorter lived nuclides, $^{234}$Th and $^{234}$Pa. The decay process itself may however allow some dissociation of the decay radionuclide from the source material, facilitating subsequent environmental transfer. Thus, $^{234}$U may be somewhat deficient relative to $^{238}$U in soils and enhanced in rivers and the sea. The radionuclide $^{226}$Ra in this chain may have slightly different concentrations than $^{238}$U because separation may occur between its parent $^{230}$Th and uranium and because radium has greater mobility in the environment. The decay products of $^{226}$Ra include the gaseous element radon, which diffuses out of the soil, reducing the exposure rate from the $^{238}$U series. The radon radionuclide in this series $^{222}$Rn has a half life of only a few days but it has two longer lived decay products, $^{210}$Pb and $^{210}$Po, which are important in dose evaluations. For the $^{232}$Th series, similar considerations apply. The radionuclide $^{228}$Ra has a sufficiently long half-life that may allow some separation from its parent $^{232}$Th. The gaseous element of the chain $^{220}$Rn has a very short half life and long lived decay products.

The activity concentration of $^{40}$K in soil is an order of magnitude higher than that of $^{238}$U or $^{232}$Th.

Terrestrial radiation is due to various radioactive nuclides that are present in soil, water, air and their abundance changes depending on the geological and geographical features of region (UNSCEAR Report 2000) The intensity of the terrestrial natural radioactivity varies by an order of magnitude for different regions of the world due to geological and environmental factors (Patra A.K.et al, 2006)

The variations in the abundance and distribution of the primordial radionuclides in the environment account for the spatial variations in the natural gamma radioactivity of such environments (Isinkaye M.O, et al., 2008) The terrestrial component is due to the radioactive nuclides that are present in air, soils, rocks, water and building materials in amounts that vary significantly depending on the geological and geografical features of a region.

Radionuclides when released to the atmosphere, undergo decay in transit or are deposited on Earth's surface by wet or dry deposition within relatively short periods. They are initially deposited on the upper surface of the soil, but are quickly weathered into the first few centimeters of the soil (UNSCEAR Report,2000; Isinkaye M.O et al., 2008)

In its first assessment of representative concentrations of these radionuclides in soil, in the UNSCEAR 1982. Committee suggested the values of 370,25 and 25 Bq kg$^{-1}$ for $^{40}$K, $^{238}$U and $^{232}$Th respectively.

Direct measurements of absorbed dose rates in air have been carried out in the last few decades in many countries of the world.

## 3. Cosmic

The main contributors to natural radiation are high energy cosmic ray particles incident on the Earth's atmosphere and radioactive nuclides that originated in the Earth's crust. Humans are affected by both external and internal exposures (UNSCEAR Report 2000)

Cosmic radiation is formed due to high energy particles that come from outer space and continually bombard the Earth. These cosmic rays interact with the nuclei in atmosphere, producing a cascade of interactions and secondary reaction products that contribute to cosmic ray exposures. The cosmic ray interactions also produce radioactive nuclei known as cosmogenic radionuclides (UNSCEAR 2000)

Although cosmic radiation increases with increasing altitude, it could be expected that people living at high altitudes suffer more from cosmic rays than those at sea level. Because of the Earth's magnetic field, the cosmic ray intensity varies with latitude, the lowest value being at the geomagnetic equator.

Cosmic radiation observed at a high elevation would be expected to have higher counts as a result of less atmosphere above the flight line.

The cosmic radiation originates from space as cosmic rays whose contribution to background changes mainly with elevation and latitude. Cosmic radiation consist of energetic charged particles, such as protons and helium ions, moving through space. They originate from events beyond our solar system and from the sun. When these particles enter the Earth's atmosphere they collide with, and disrupt, atoms in our atmosphere, producing secondary, less intense, radiation. By the time cosmic radiation reaches the ground its intensity has been considerably reduced.

The amount, or intensity, of cosmic radiation depends on altitude and latitude, as well as the stage of the solar cycle. The Earth's atmosphere provides considerable protection from cosmic radiation. At commercial aircraft altitudes the protective layer of the Earth's atmosphere is much thinner than it is on the ground and the intensity of cosmic radiation is approximately 100 times greater at these altitudes than it is on the ground.

The Earth's magnetic field can deflect some of the cosmic radiation away from the Earth. The shielding ability of the magnetic field is most effective over the equator and least effective over the poles. The intensity of cosmic radiation at aircraft altitudes around the equator is about three times less than at the poles.

The sun's magnetic field can also deflect cosmic radiation away from the Earth. The strength of the sun's magnetic field varies with the approximate 11 year cycle of rise and decline of solar activity (solar cycle). When solar activity is low (solar minimum), the magnetic field is less effective in deflecting cosmic radiation; cosmic radiation reaching the Earth will be more intense during solar minimum. The effect of solar activity on intensity of cosmic radiation is much smaller than that caused by altitude or latitude. The sun ejects energetic particles, such as protons (solar flares), which may also contribute to the intensity of cosmic radiation. However, only on very infrequent occasions would solar flares have sufficient energy to increase the intensity of cosmic radiation at commercial aircraft altitudes.

## 4. Materials and methods

In order to determine the outdoor gamma dose rates region and activity concentrations in soil samples is divided to 6 basic geographic areas in Adana region in Turkey. Each geographic area called as a sampling station. This region is located in the southern part of Turkey.

The outdoor gamma dose rates were measured by Eberline smart portable device (ESP-2) connected with and SPA-6 model plastic scintillation detector. Measurements were taken in air for two minutes at 1 m above the ground and the gamma dose rates were recorded as $\mu Rh^{-1}$ . The gamma absorbed doses in nGy $h^{-1}$ were also converted to annual effective dose in mSv $y^{-1}$ as proposed by UNSCEAR .

SPA-6's calibration was done using $^{137}Cs$ with an electrometer device for certain distances in the laboratory.

Sampling stations were chosen uncultivated and near to populated areas to understand the amount of dose received by the population because of absorbed gamma dose rate in air. At each measurement a reading was taken in air for 1 h at 1 m above the ground level. The instrument calculates an average 1 h exposure rate based on the multiple measurement results. The results include both terrestrial and cosmic ray components of gamma radiation level that was recorded in units of $\mu Rh^{-1}$.

Soil samples were taken with 25 cm diameter cores collected at different locations and different depths ranging from 0 to 30 cm. These samples were taken from uncultivated fields and sampling stations were chosen close to populated areas.

The soil samples were dried, pulverized, homogenized and sieved through 2 mm mesh. The meshed soil samples were transferred to Marinelli beakers of 1000 ml capacity. The soil samples were weighed, carefully sealed and stored for 30 days to allow secular equilibrium between thorium and radium and their products (Mollah et al., 1987)

Gamma spectropic measurements were performed using a coaxial HPGe detector having a 16 % relative efficiency. A detection system containing a Canberra Model 2020 Amplifier and a Canberra S-85 Multi Channel Analyzer with Model 8087 4K ADC was used for the measurements. The detector was shielded in a 10 cm thick lead well, internally lined with 2 mm thick copper and 2 mm thick cadmium foils. The overall detector resolution (FWHM) of 1,9 keV was obtained for the 1332 keV gamma line of $^{60}Co$. Energy calibration and relative efficiency calibration of the gamma spectrometer were carried out using $^{109}Cd$, $^{57}Co$, $^{113}Sn$, $^{134}Cs$, $^{137}Cs$, $^{188}Y$ and $^{60}Co$ calibration sources in 1000 ml Marinelli beaker covering the energy range from 80 to 2500 keV. The counting time for each sample, as well as for background was 50,000 s.

Gamma spectroscopy was used to determine the activities of $^{238}U$, $^{232}Th$, $^{40}K$ and $^{137}Cs$. For concentrations of $^{232}Th$ and $^{238}U$ the following gamma transition lines were used; $^{232}Th$ series: $^{228}Ac$ (911 keV), $^{208}Tl$ (583.1keV); $^{238}U$ series : $^{214}Pb$(351.9 keV) and $^{214}Bi$ (609.2 keV).

The contribution of natural radionuclides to the absorbed dose rate in air depends on the concentration of the radionuclides in soil. The largest part of the gamma radiation comes from terrestrial radionuclides. There is a direct correlation between terrestrial gamma radiation and radionuclide concentration in soil.

## 5. Results and discussions

The radionuclide activity concentrations in the soil samples taken from 6 different locations are reported in Table2. The mean activity concentrations of $^{214}Pb$ and $^{214}Bi$ of $^{238}U$ series are 16,81 and 14,65 Bq $kg^{-1}$ , respectively. The mean activity concentrations of $^{228}Ac$ and $^{208}Tl$ of $^{232}Th$ series are 24,34 and 26,21 Bq $kg^{-1}$, respectively. The worldwide average concentrations of $^{238}U$, $^{232}Th$ are reported by UNSCEAR 2000 as 35, 30 Bq $kg^{-1}$. The average concentrations of $^{214}Pb$ and $^{214}Bi$ of $^{238}U$ and $^{228}Ac$ and $^{208}Tl$ of $^{232}Th$ series are lower than world average. The radioanuclide activity concentrations are shown in Figure 4 for $^{214}Pb$, in Figure 5 for $^{214}Bi$, in Figure 6 for $^{228}Ac$, in Figure 7 for $^{208}Tl$. The highest value was measured in Saimbeyli town where is on a high elevated place as 28.33 Bq $kg^{-1}$ for $^{214}Pb$, 23.06 Bq $kg^{-1}$ for $^{214}Bi$ of $^{238}U$ series and 49.87 Bq $kg^{-1}$ for $^{228}Ac$, 54.89 Bq $kg^{-1}$ for $^{208}Tl$ of $^{232}Th$ series. The lowest value was measured in Yeniyayla (Yuregir) town as as 5.80 Bq $kg^{-1}$ for $^{214}Pb$, 4.83 Bq $kg^{-1}$ for $^{214}Bi$ of $^{238}U$ series and 8.38 Bq $kg^{-1}$ for $^{228}Ac$, 8.36 Bq $kg^{-1}$ for $^{208}Tl$ of $^{232}Th$ series.

The outdoor gamma dose rates was shown in Table 2 and Figure 8.The average outdoor gamma dose rates in air is 76.2 nGy $h^{-1}$. The highest value was measured as 134 nGy $h^{-1}$ in Feke town. The lowest outdoor gamma dose rate was measured as 49.5 nGy $h^{-1}$ in Karatas town.

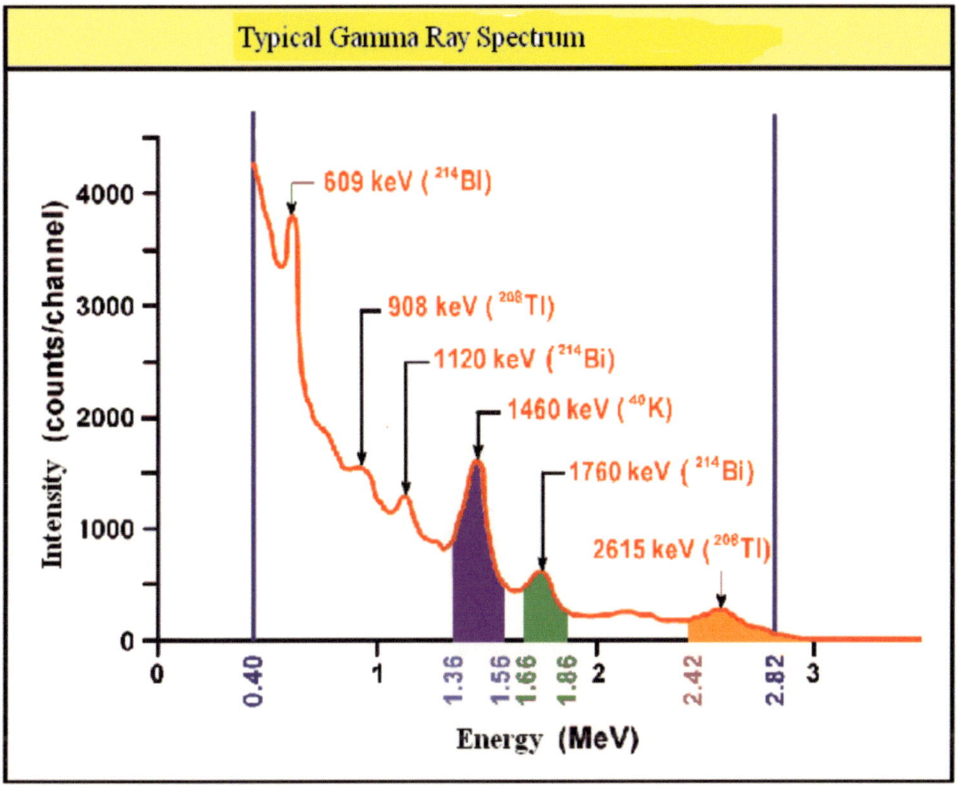


Fig. 3. Typical Gamma Ray Spectrum

| | U-238 series | | Th-232 Series | |
|---|---|---|---|---|
| | Pb-214 | Bi-214 | Ac-228 | Tl-208 |
| Balcali Fen Edebiyat Fakultesi | 17.66 | 16.17 | 23.63 | 27.19 |
| Yeniyayla (Yuregir) | 5.80 | 4.83 | 8.38 | 8.36 |
| Feke (Center) | 22.59 | 21.34 | 39.38 | 43.47 |
| Ceyhan (Center) | 17.38 | 16.77 | 16.15 | 14.98 |
| Karatas (Center) | 9.10 | 9.75 | 8.65 | 8.37 |
| Saimbeyli (Center) | 28.33 | 23.06 | 49.87 | 54.89 |

Table 2. Activity concentrations in soil samples as Bq $kg^{-1}$

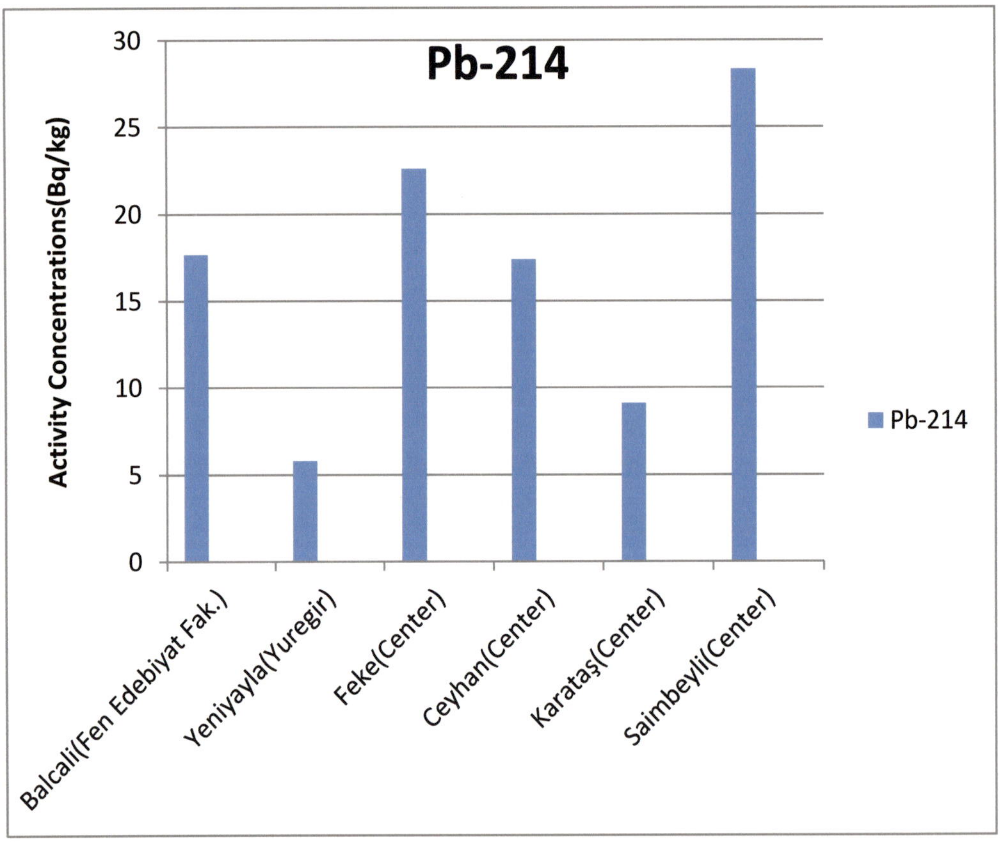


Fig. 4. Pb-214 activity concentrations (Bq kg$^{-1}$) in soil samplesin Adana Region

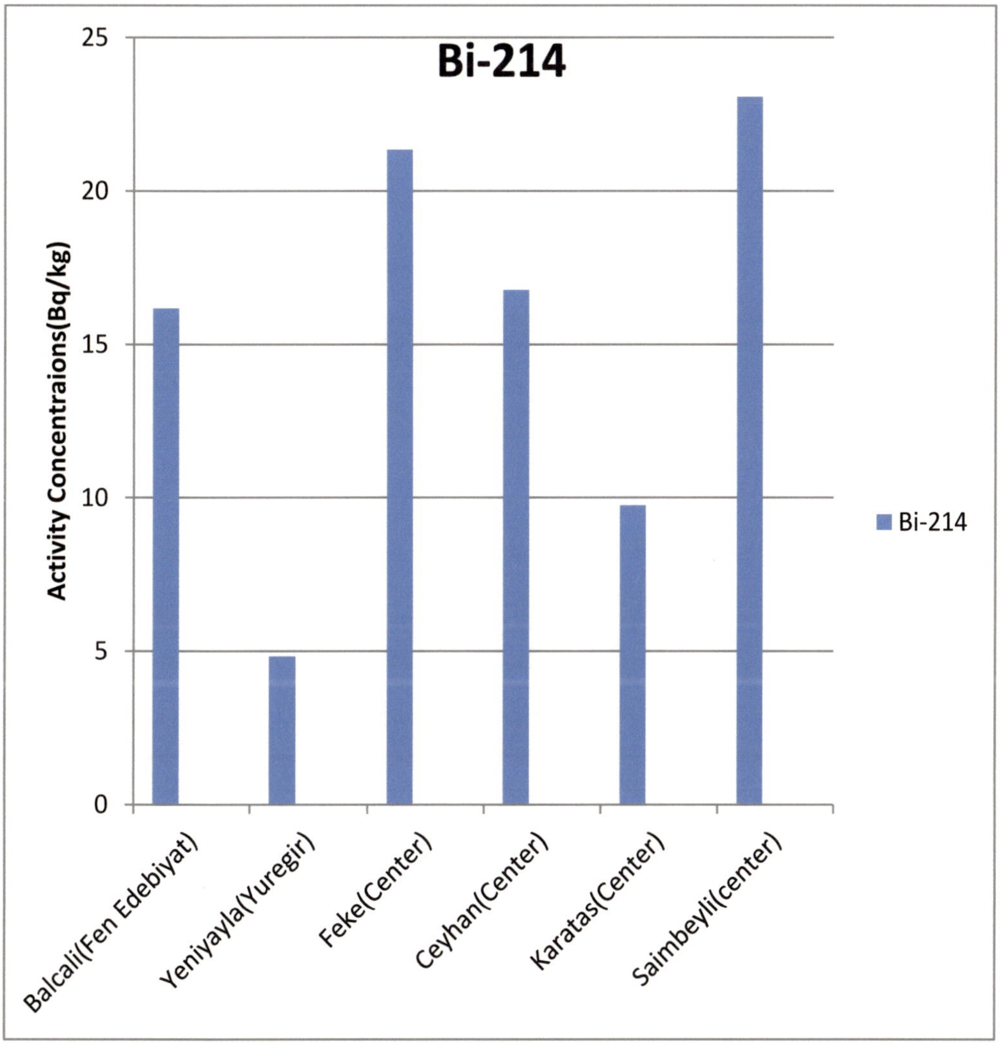


Fig. 5. Bi-214 activity concentrations (Bq $kg^{-1}$) in soil samples in Adana region

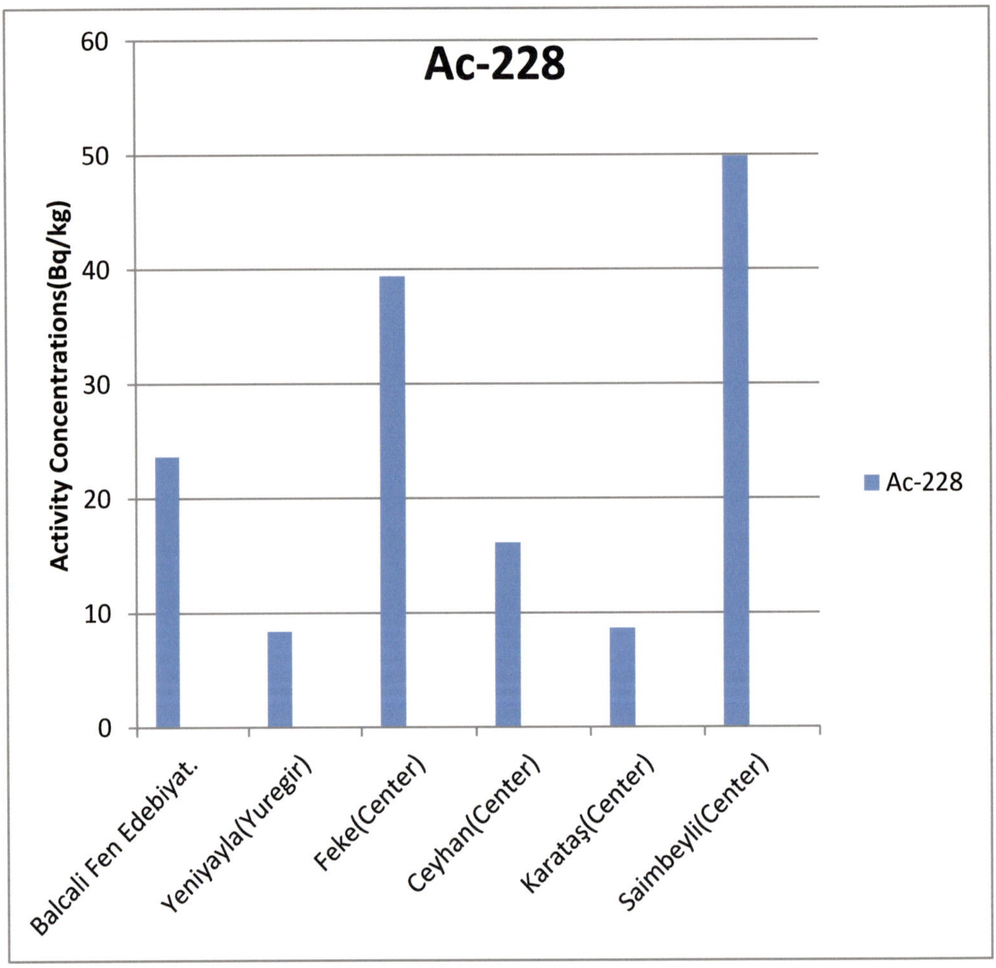


Fig. 6. Ac-228 activity concentrations (Bq $kg^{-1}$) in soil samples in Adana Region.

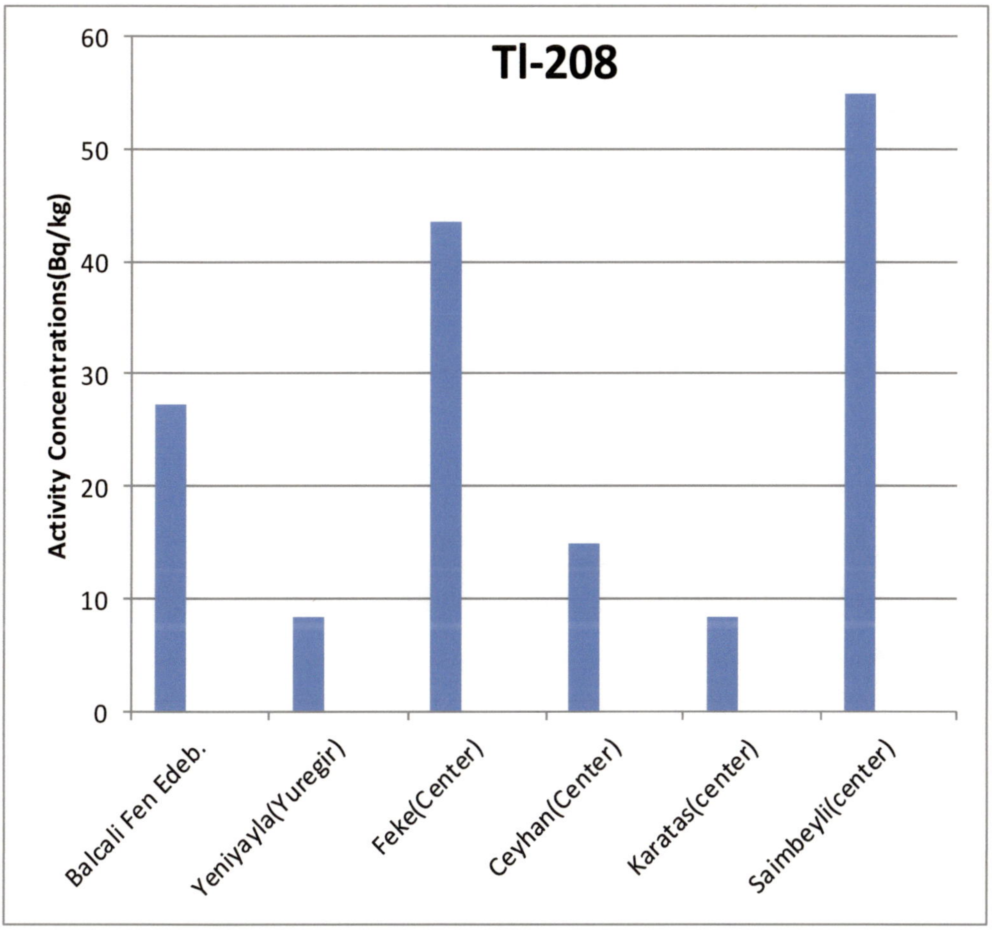


Fig. 7. Tl -208 Activity concentrations (Bq kg$^{-1}$) in soil samples in Adana Region

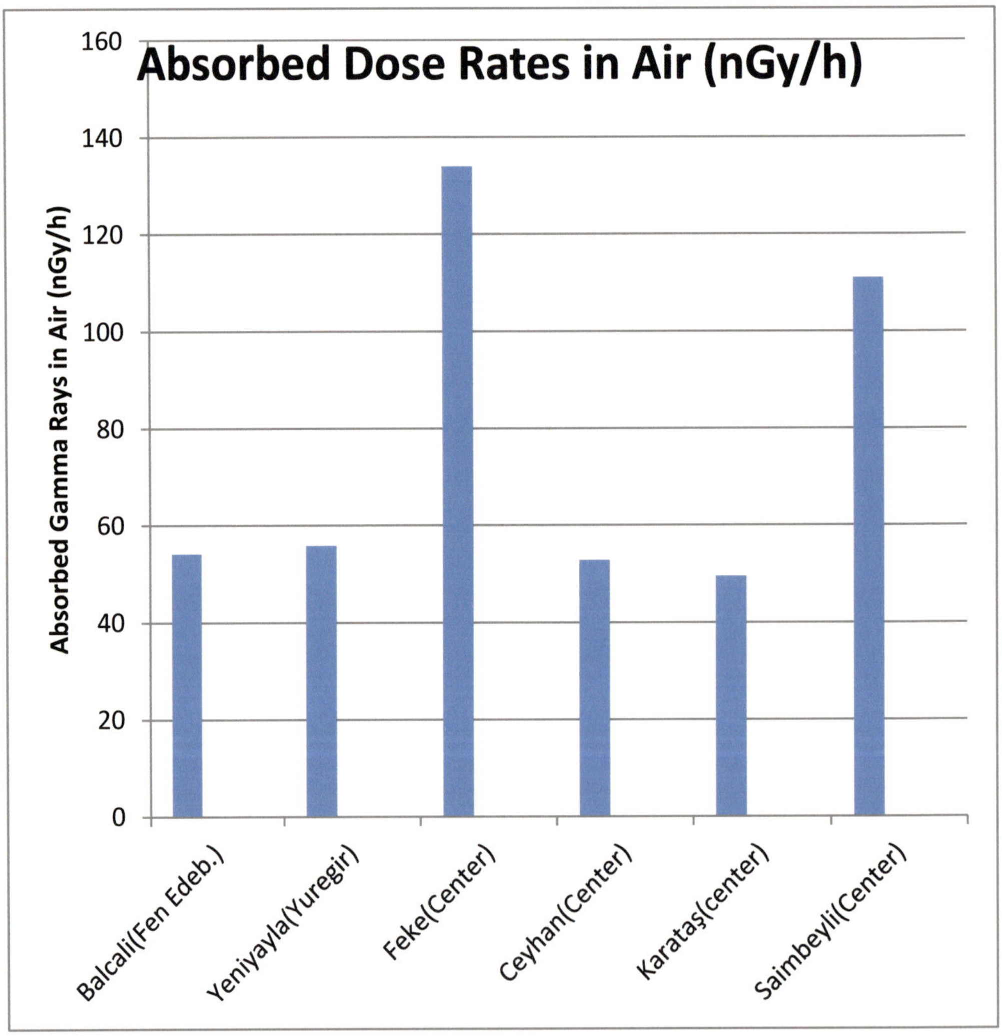


Fig. 8. Absorbed Dose Rates in Air (nGy $h^{-1}$)

| Places | Measured Absorbed Equivalent (nGy/h) |
|---|---|
| | |
| Balcali (Fen Edebiyat Fak.) | 54,1 |
| Yeniyayla (Yuregir) | 55,8 |
| Feke(Center) | 134 |
| Ceyhan(Center) | 52,8 |
| Karataş(Center) | 49,5 |
| Saimbeyli (Center) | 111 |
| | |

Table 3. Outdoor gamma dose rates in air as nGy $h^{-1}$

## 6. References


Blundy, J. and Wood, B. (1991) Crystal-chemical controls on the partitioning of Sr and Ba between plagioclase feldspar, silicate melts, and hydrothermal solutions.Geochimica et Cosmochimica Acta, 55, 193–209.

Degerlier M. And Ozger G., Assessment of Gamma Dose Rates In Air In Adana/Turkey, Radiation Protection Dosimetry,132, No.3,350-356,2008.

Isınkaye, M.O.. Radiometric Assessment of Natural Radioactivity Levels of Bituminous Soil In Agbabu, Southwest Nigeria. Radiat. Meas. 43, 125-128, 2008.

Karahan G. and Bayulken A. Assessment of Gamma Dose rates Around Istanbul(Turkey), J.Environ.Radioact.47,213-221,2000.

Merdano˘glu, B., Altınsoy, N., 2006. Radioactivity concentrations and dose assessment for soil samples from Kestanbol granite area, Turkey. Radiation Protection Dosimetry 121 (No. 4), 399-405

Mollah, S., Rahman, N.M., Kodlus, M.A., Husain, S.R., 1987. Measurement of high natural background radiation levels by TLD at Cox and Bazar coastal areas in Bangladesh. Radiation Protection Dosimetry 18 (1), 39-41.

Narayana, Y., Rajashekara, K.M. and Siddoppa, K. Natural Radioactivity in Some Major Rivers of Coastal Karnataka on the Southwest Coast of India. 95(2-3), 98-106,2007.

Patra, A. K., Sudhakar, J., Ravi, P.M., James, J.P. and Hedge, A.G. Natural Radioactivity Distribution in Geological Matrices Around Kaiga Environment. J.Radioanal.Nucl. Chem. 270(2), 307-312, 2006.

UNSCEAR REPORT. United Nations Scientific Committee on The Effects of Atomic Radiation Sources, Effects and Risks of Ionizing Radiations. New Tork: United Nations Publication,2000.

Whicker, F.W. and Schultz, V. Radioecology Nuclear Energy and the Environment(Boca Raton, FL:CRC Press, Inc.),1982.

# Part 3

## Materials Science

# 6

# Gamma Radiation as a Novel Technology for Development of New Generation Concrete

Gonzalo Martínez-Barrera[1], Carmina Menchaca Campos[2] and Fernando Ureña-Nuñez[3]

[1]*Laboratorio de Investigación y Desarrollo de Materiales Avanzados (LIDMA), Facultad de Química, Universidad Autónoma del Estado de México, Km.12 de la Carretera Toluca-Atlacomulco, San Cayetano,*

[2]*Centro de Investigación en Ingeniería y Ciencias Aplicadas (CIICAp), Universidad Autónoma del Estado de Morelos, Cuernavaca Morelos*

[3]*Instituto Nacional de Investigaciones Nucleares, Carretera México-Toluca S/N, La Marquesa Ocoyoacac,*

*Mexico*

## 1. Introduction

Global development demand new technological creations at less costs. Inside the specialized industry, composite materials usage is well known. That is why it is of major importance research and development of new composite materials for technological applications, as in the construction industry.

The composite materials usage has increased rapidly during the last few decades in building construction; showing different properties as reduction in weight, reduction on the fatigue and corrosion. As we know, a key issue with composite technology is that the final characteristics of a material are established at the time of fabrication of the goods. Therefore, part design, fabrication development and material characterization must proceed concurrently.

The production of polymer concrete can be developed for semi-industrial and industrial scales for its economical advantages, as well as environmental benefits if its main raw materials are wastes. In such composite materials the polymeric materials are able to compete and contribute substantially to the development of better, cheaper and more functional products.

In polymer concrete some engineering polymers are widely used due to their excellent mechanical, thermal and chemical properties. These properties are direct consequence of their composition as well as their molecular structure. Nowadays, mechanical properties of polymeric materials are of great interest and can be improved by composition and/or morphological modifications, in order to change the softening temperature and get their elastic solid state back [Menchaca et al., 2011].

The developments in the study and applications of radiation effects have been rapidly increasing research activity towards the development and understanding of novel synthetic materials with particular emphasis on their properties, synthesize, analyze and modify such new materials.

Radiation chemistry is defined as the study of chemical effects caused by the passage of ionizing radiation through matter. Ionizing radiation comes from substances undergoing nuclear transformations, from outer space in the form of cosmic rays and from particles accelerators. It includes α, β, and γ rays from radioactive nuclei, charged particles such as protons and deuterons and X-rays of wavelength less than approximately 250 Å [Wilson, 1974 ].

Each particle of ray of ionizing radiation produces a large number of ionized and excited molecules along its track. The ionizing radiation is no selective and may interact with any molecule in its path and raise it to any of its possible ionized and/or excited states (Figure 1). The heterogeneity of the latter type of reaction is especially marked in the liquid and solid state [Wilson, 1974].

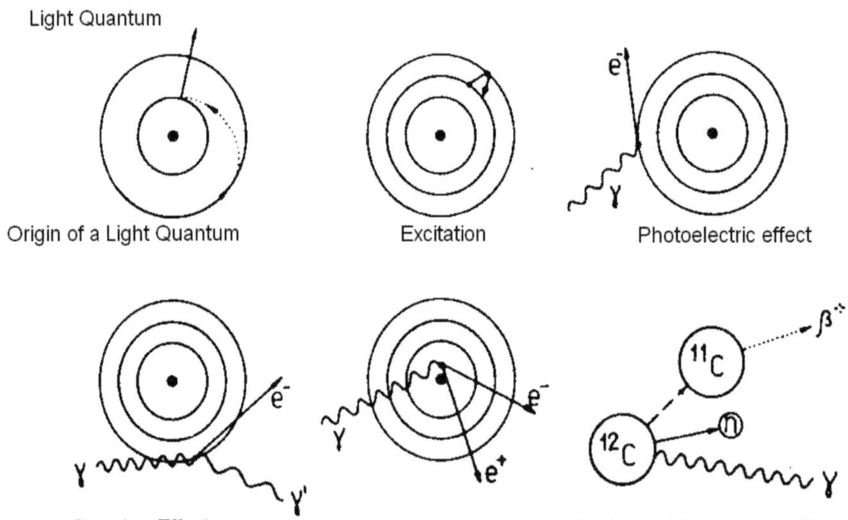


Fig. 1. Effects produced by radiation throughout matter.

There are two main types of radiation sources: a) radio isotopes, and b) devices such as X-rays tubes and electron accelerators. The isotope most frequently used as radiation source is $^{60}Co$, mainly because of its advantageous properties: availability, high energy gamma-rays, 5.3-year half live [Wilson, 1974].

Once that high energy or ionizing radiation penetrates into the matter, its energy is lost due to the interaction with the molecular valence orbital electrons that are found on its path. As a result, these electrons are promoted to higher levels of energy (excitation) or pulled out from their orbital (ionization). The basic chemical transformations that happen in a molecule subject to irradiations can be summarized as follow in Figure 2 [Mykiake, 1960]:

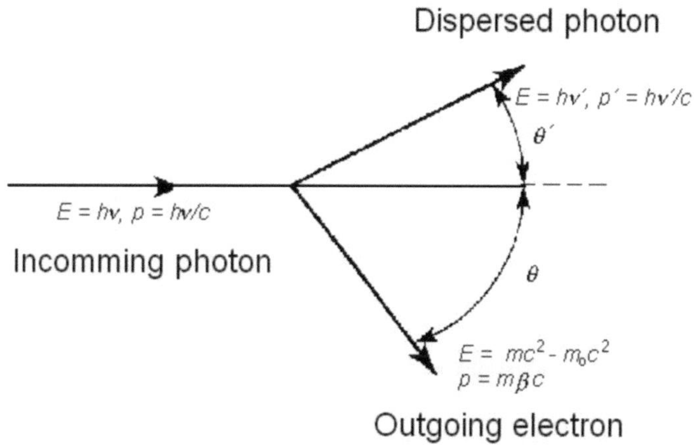


Fig. 2. Basic chemical transformations in a molecule subject to irradiation.

Ionizing radiation can be used for understanding mechanism of polymerization reaction as well as for initiation of the polymerization process. Some of the advantages of the radiation initiated polymerization over the conventional methods are: a) Curing at ambient temperature; b) absence of foreign matter, like initiator, catalyst, additives; c) polymerization at low temperature, or in solid state, d) rate of the initiation step can easily be controlled by varying dose rate, e) better solvent resistance of the polymer and its improved shape stability with respect to aging and to high temperatures [Chapiro, 2002]; f) better control of part dimensions and elimination of internal stresses which reduce material strength; shorter curing times; and no emission of volatiles to the environment.

The radiation chemistry of polymers provides a research field that is full of fresh and stimulating discoveries. Polymers are most sensitive to slight variations of the chemical bond, in this way, initial features and properties may be varied and new materials may eventually be tailored [Charlesby, 1952; Dole, 1950].

When polymers are irradiated their extremely long molecular chains can be broken easily by the absorption of a quantum of energy above the energy of the covalent bond of the main carbon chain, which typically is in the range of 5 - 10 eV. The energy of gamma photons of some MeV surpasses by many orders of magnitude this minimum value, representing a high risk of degradation to all kind of polymers, naturals and synthetics alike. However, by controlling the applied doses, degradation of polymers of large molecular mass - or even of cross-linked molecular structures - has been a field of radiation application.

As it is known one of the main effects of the ionizing radiation over polymeric materials is the formation of *cross-linked* molecular structures, and the degree of the cross-linked effect depends on the applied doses. In polymer chemistry, when a synthetic polymer is said to be "cross-linked", it usually means that the entire bulk of the polymer has been exposed to the cross-linking method. The resulting modification of mechanical properties depends strongly on the cross-link density. Low cross-link densities raise the viscosities of polymer melts. Intermediate cross-link densities transform gummy polymers into materials that have elastomeric properties and potentially high strengths. Very high cross-link densities can cause materials to become very rigid or glassy.

Cross-links can be formed by chemical reactions that are initiated by heat, pressure, change in pH, or radiation. For example, mixing of a unpolymerized or partially polymerized resin with specific chemicals called crosslinking reagents results in a chemical reaction that forms cross-links. Cross-linking can also be induced in materials that are normally thermoplastic through exposure to a radiation source, such as electron beam, gamma radiation, or UV light.

Cross-links are the characteristic property of thermosetting plastic materials. In most cases, them is irreversible, and the resulting thermosetting material will degrade or burn if heated, without melting. Especially in the case of commercially used plastics, once a substance is cross-linked, the product is very hard or impossible to recycle. In some cases, though, if the cross-link bonds are sufficiently different, chemically, from the bonds forming the polymers, the process can be reversed.

Another result of polymers irradiation is that smaller hydrocarbon chains will be formed (lighter hydrocarbons and gases) as well as heavier hydrocarbons by recombination of broken chains into larger ones. This recombination of broken hydrocarbon chains into longer ones is called *polymerization*. Polymerization is one of the chemical reactions that takes place in organic compounds during irradiation and is responsible for changes in the properties of this material. Some other chemical reactions in organic compounds that can be caused by radiation are oxidation, halogenation, and changes in isomerism.

The polymerization mechanism is used in some industrial applications to change the character of plastics after they are in place; for example, wood is impregnated with a light plastic and then cross-bonded (polymerized) by irradiating it to make it more sturdy. This change in properties, whether it be a lubricant, electrical insulation, or gaskets, is of concern when choosing materials for use near nuclear reactors.

When the geometry of the bond structure is modified using gamma-irradiation, the characteristics of the long chains of polymers vary, thus some changes in polymer properties can be explained through induced chain strength, chain re-orientation and crystallinity. On the other hand, depending on the dose *cross-linking* or *chain scissions* may be present in irradiated polymers.

It has been claimed that chain scission occurs either in the amorphous region [Pattel & Keller, 1975; Jenkins & Keller, 1975; Ungar & Keller, 1980] or inside the crystals [Hoseman et al., 1972; Loboda-Cackovic et al., 1974]. Also it was reported that both process begin with the formation of free radicals [Timus et al., 2000; Valenza et al., 1999; Bittner et al., 1999] followed by the Compton Effect [Bittner et al., 1999; Yu & Li, 1998]. Some researchers establish that the main process in polymers, due to high radiation energy, is that of cross-linking [Balabanovich et al., 1999; Charlesby, 1960]. Others propose the chain scission as the main effect [Timus et al., 2000; Bittner et al., 1999] and even some others show that both processes can happen [Timus et al., 2000; Valenza et al., 1999; Balabanovich et al., 1999; Charlesby, 1960; Barkhudaryan, 2000a; Barkhudaryan, 2000b;, Delley et al., 1957; Gupta & Deshmukh, 1983; Li & Zhang, 1997; Zhang et al., 2000] all of them as a function of the experimental conditions and the type of polymer under study. Also it was reported that both processes begin with the formation of free radicals [Timus et al., 2000; Valenza et al., 1999; Bittner et al., 1999].

A mechanism for chain scission occurring in the amorphous zone of the nylon 6,12 at high radiation dose which takes into account the Compton Effect has been proposed [Menchaca et al., 2003]. In the same work, cross-link is under consideration in the low dose region. An study of the low gamma irradiation dose (up to 50 kGy) has shown that the cross-link process is taking place in the amorphous zone of the nylon 6,12 [Menchaca et al., 2003; Menchaca et al., 2008]. When gamma radiation at low dose is applied both, the fusion temperature and the crystallinity degree show evidence of increments [Thanki et al., 2001], as well as a partial and repairable damage in the amorphous zone, is reported [Malek et al., 2001]. The last phenomenon produces the so called re-polymerization process involving chain reorganization.

## 2. Effects of gamma radiation on components of concrete

### 2.1 Polymeric materials: Resin and fibers

The unsaturated polyester resins (UP) are most widely used thermosetting resins and are being increasingly applied for various purposes because of their easy handling, balanced mechanical and chemical characteristics and a cheap price.

The cross-linking reaction of UP resins is usually initiated by a thermal or redox initiator. The cross-linking reaction occurs by heterogeneous free radical mechanism and it follows different periods: a) The induction period during which there is no cross-linking until the inhibitor is used up; b) The propagation period: the reaction starts and its rate depends on the mass law. As the 3-D network appears, it reduces the availability of reactants; diffusion-controlled part of propagation period begins. When, because of restrictions imposed by the network, termination of macro-radicals ceases, the reaction rate significantly increases and so called "gel effect" occurs; c) In the final reaction period, vitrification of the system takes place and the cross-linking stops; the propagation period of the cross-linking reaction should be distinguished from the free radical reaction step of the same name.

The micro-gels are caused by intra-molecular reaction between polyester insaturations and some styrene molecules present inside the polyester coil because the concentration of styrene inside the coil is lower [Jurkin & Pucic, 2006]. Further in the reaction, vinyl monomers interconnect micro-gels to produce a 3-D network, and the resin system abruptly changes from a viscous liquid into a hard thermo-set solid. Still, a part of un-reacted polyester double bonds remain mostly buried inside micro-gels.

As we know the effects of the passage of electromagnetic radiations through matter produces three main type processes: a) Photoelectric effect, b) Scattering of free electrons as Thompson, Rayleigh and Compton Effect, and c) Electron-positron pair production [Menchaca et al., 2011]. These effects are permitted by the energy range that the particle or photon radiation can give to the molecules, atoms or ions in the matter structure. However in gamma irradiated polymeric materials, for instance, the Compton Effect is the most important due to the energy of the gamma photons (1.17 MeV and 1.33 MeV) and the low density of the polymers.

The effects of ionizing radiation in polymers depend on the structure and density of each polymer. These effects can be: cross-link of the molecular chain of the polymer, damage in crystalline regions, degradation of the polymer, and the possibility of the molecular weight changes of some polymers for changing physical and chemical properties [Martinez-Barrera et al., 2004].

The alternative route of curing UP resins is radiation processing that has many advantages over the conventional methods: no catalyst or additives are needed to initiate the reaction and it can be performed at low temperatures. The initiation is homogeneous throughout the system and the rate of cross-linking is easily controlled by varying the dose rate.

Curing polyester resins involves chain scissions which result in the formation of free radicals. The radicals react with the double-bonds and release strain energy resulting in polymerization. The recovery probability of the radicals decreases according to the chain stress and the scission of the chemical bond increase. A dependence among the chain lengths, the strain and their rupture is done; the shortest chains have the highest strain energy and they break first [Nishiura et al., 1999].

By contrast, in gamma irradiated resins the reaction runs smoothly and the product is flawless - unlike badly foamed products obtained when using catalysts. As we know thermoset resin containing double bonds (C=C) in the presence of a monomer, when exposed to a limited dose, will partially cross-link to form a stable 3-D gel. As a result, the mixture is no longer a viscous liquid. Instead, it becomes a viscoelastic gel. When irradiation impacts resin that contains an initiator, active species are created along the path of the incident electrons. These active species then react to create cross-links in the material. As a result, clusters of cross-linked material have formed, giving structure to the material. This in turn, restricts the movement of the large polyester oligomers and the material develops viscoelastic properties.

Polyester isophthalic resins dissolved in 36% by weight styrene were irradiated at six different doses (from 5.5 to 33.3 kGy) with a ratio dose of 4.1 kGy [Woods & Pikaev, 1994]. Several results were found. By measuring the deformation due to a compressive load on the sample resin, a relationship was developed between the compressive load and the degree of cross-linking. The maximum compressive loads for sample resins without initiator varying from 20 to 200 N, these values are lower than those for sample resins with initiator (from 30 to 320 N), which means a maximum difference of 116% [Czayka et al., 2007]. This would indicate that under irradiation, some of the initiator is contributing to cross-linking.

More interesting is the comparison between modifications with irradiation vs with chemicals. The values for irradiated resins show 151 MPa as maximum tensile, 10.9 GPa for tensile modulus and 1.8% of the strain at break. These values are 18, 14 and 20% bigger, respectively, than those for modificated resins by chemicals [Czayka et al., 2007]. Moreover, for the total enthalpy, which is represented by the area under the heat flow vs. time curve; The enthalpy transitions became broader as the cross-links form with dose increase, there is less mobility and hence, the rate of cross-linking decrease. The enthalpy change, decrease from 242 to 185 (J/g) and the fraction cross-linked increase from 0.19 to 0.38 when increase the irradiation dose.

The degree of polymerization can be follow by selected wavelengths. The wavelengths for styrene monomer and for polystyrene that may be polymerized during irradiation include the carbon–carbon double bonds at 982 $cm^{-1}$. If radiation is cross-linking the material, then the magnitude of this peak should decrease with increasing dose levels. When the dose increases the peak height for the styrene decreases, in the region where the polystyrene peak should occur, there is none. This means that there is no measurable indication that homo-polymerization of the styrene is occurring during irradiation.

A total and fast cure for certain polymers is achieved by gamma radiation in the cases when the catalyst does not complete its function, as total polymerization process of the polymer resin [Delahaye et al., 1998; Martínez-Barrera et al., 2006]. This eliminates the need for further additives or monomers. Nevertheless, there are limitations, such as excessive rise in the temperature of the polymer due to the high exothermic nature of polymerization. Moreover, the required doses for total cure strongly depend on the composition used; it is necessary to evaluate the rate of cure progress.

During UP resin cross-linking, full conversion is never reached. Post irradiation reaction polymerization continues after the irradiation was ended due to free radicals formed trapped and stabilized in macromolecules or in the network. In gamma irradiated UP resins (from 1 to 6 kGy at dose rate of 39.13 kGy/h), depending on the dose at which the irradiation was terminated, the samples are in form of viscous liquid, gelled material or, at the highest dose, glassy solid. At doses below 3.5 kGy the samples remained liquid, while those irradiated to higher doses formed measurable quantity of insoluble gel.

The extraction analysis offers the possibility to analyze the post-irradiation changes of the free styrene content, separately of those of the gel content. During the 15 days post-irradiation period monitoring, the gel content increased while the corresponding free styrene content decreased [Jurkin & Pucic, 2006]. Sharp rise in viscosity at 4 kGy, caused by the more dense network formed during irradiation, greatly reduced the post-irradiation cross-linking.

The extent of post-reaction increased again as the irradiation was terminated in the gel-effect dose range, 4.5 to 5 kGy. At the highest dose, 6 kGy, radiation reaction approached maximum conversion and the system vitrified, thus impeding the post-effect at room temperature, so the fraction of post-irradiation formed gel decreased again.

Still the shape of DSC traces and corresponding heats of the residual reaction offer plenty information on the post-irradiation cross-linking. At all doses above the induction period threshold (3 kGy), on the day of the irradiation, two exothermic processes were seen [Jurkin & Pucic, 2006]. The lower temperature process had a maximum at about 120°C and the broad higher temperature exotherm had a maximum between 160 and 200°C. The lower temperature process was attributed to both the styrene-polyester copolymerization and the styrene homopolymerization.

In the case of polymeric fibers, the re-polymerization and reorientation processes are favoured when they are irradiated, producing longer but oriented chains, compared to the original ones. The scission chain mechanism produced by gamma irradiation, primordially located in the amorphous zone of the fiber, is not enough to break down the carbon-heteroatom or carbon-carbon bonds continuously and produce free radicals. In fact, the few free radicals produced react immediately to form long chains. Nevertheless, such kind of energy is not enough to break the bonds repeatedly and produce smaller species, as it happens in irradiated-fiber at higher dose [Menchaca et al., 2009]. Such mechanism predicts an increase in the fusion temperature, as well as in its crystallinity, since the chains are broken down in the amorphous zone and reoriented, yielding new crystalline zones [Menchaca et al., 2011].

The structural changes caused on irradiated-nylon fibers are reflected in their morphology. It can be observed different kind of modifications on the surface depending on the radiation dose applied. Figure 3a shows the morphology of the non-irradiated nylon fiber, which includes a surface with protuberances, strips and scratches. When the radiation dose increases to 5 kGy small particles are scrapped of these protuberances and better strips definitions are noticed. For higher irradiation dose more roughness and scratches are formed (Figures 3b), therefore superficial defects and scratches are more evident, as the dose is augmented (Figure 3c).

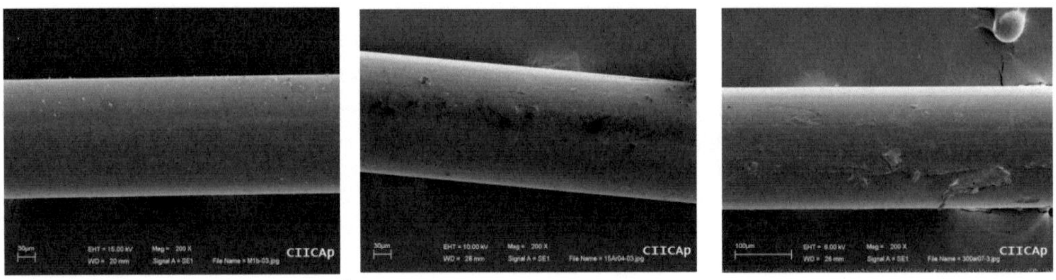


3a. Non-irradiated sample. 3b. Irradiate at 50 kGy. 3c. Irradiated at 200 kGy.

Fig. 3. SEM images of non-irradiated and irradiated nylon 6,12 fibers.

The damages can be related to crystallinity changes, because the formation of oligomers during gamma irradiation exposure provokes changes in the density, visco-elasticity, rheological and mechanical properties [Menchaca et al., 2011]. At the same time, these "damages" due to the radiation exposure increase the surface roughness [Menchaca et al., 2010] helping to grip some other kind of materials to the polymer in order to get materials with enhanced properties, e.g. concrete.

A post-irradiation study on nylon fibers show changes in the morphology with time. For the freshly irradiated fibers, mean crystal size tends to diminish (Figure 4a) probably because chain scission is generating more crystalline areas but with less size. Almost the same behavior is observed in the 3 years irradiated fibers, where mean crystal size tends to diminish from 15 to 50 kGy. Above 100 kGy there are not observed changes. The differences between freshly irradiated and 3 years irradiated fibers are the crystallites sizes. With time, crystal size increase meaning that whatever the reaction mechanism is taking place, after three years generation of new crystalline areas still goes on (Figure 4b).

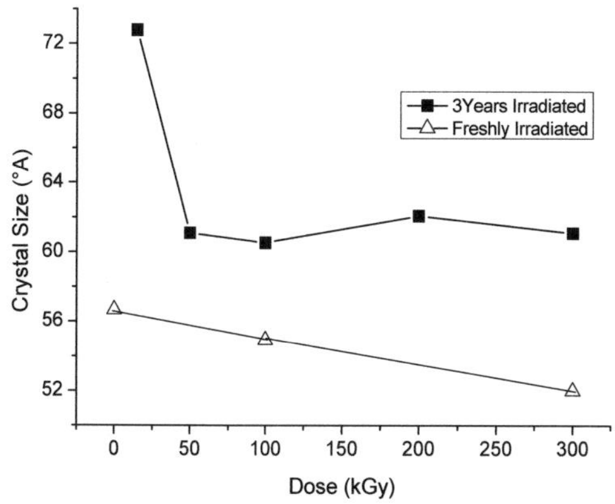


4a. Mean crystal size of gamma irradiated fibers.

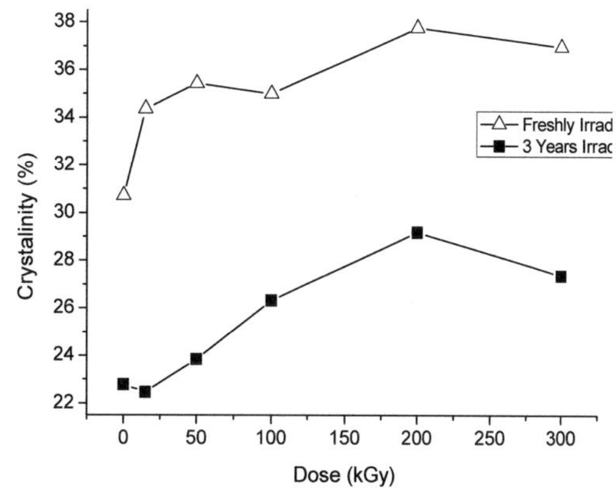


4b. Crystallinity of gamma irradiated fibers.

Fig. 4. Crystal size and crystallinity behavior of gamma irradiated fibers.

On the other hand, the melting point has a general trend to decrease with no significant differences in temperature, just a slight increase as a function of time for the 3 years irradiated condition (Figure 5). But, the peculiar low dose response can be observed again. At 15 kGy and 50 kGy the melting point for gamma irradiated fibers is above or at least at the same temperature compared to the non-irradiated fibers. This phenomenon is attributed to the cross-link or partial damage [Wilson, 1974; Menchaca et al., 2003], and the decrease in the fusion temperature is ascribed to the chain scission or permanent damage that yields oligomers [Ramesh, 1999].

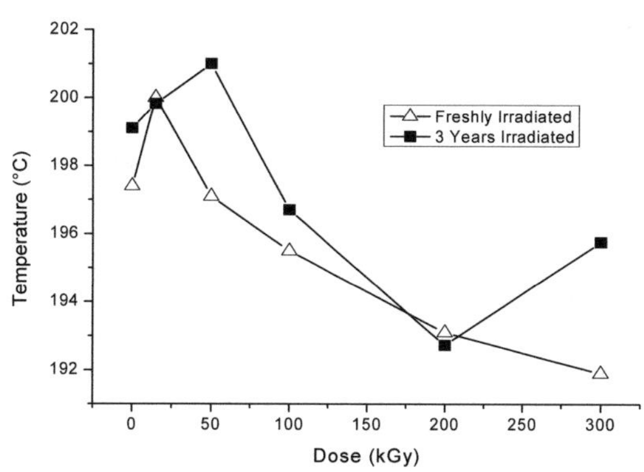


Fig. 5. Post-irradiation behavior of melting point.

Gamma irradiation causes both immediate and time dependent changes in the mechanical properties and there is considerable experimental evidence that the time dependent effects arise from the presence of long lived free radicals. Chain scission processes occur both during and after irradiation, leading to release of inter-lamella tie chain material which then causes an increase in crystallinity. Mechanical changes can be closely related to the crystallinity increase and are of considerable importance in property critical applications.

Research concerning to the micro-mechanical deformation mechanisms of irradiated and non-irradiated isotactic polypropylene (iPP), studied as a function of temperature above the glass transition has been reported [Zhang & Cameron, 1999]. Several deformation mechanisms were identified and included lamellar separation, shear, stable and unstable fibrillated deformation and cavitation. The ductile-brittle transition rises dramatically with irradiation, while the glass transition shows only a small increase. This observation is explained by irradiation, through chain scission and cross-linking, having a dominant effect on large-scale plastic deformation and a lesser effect on the deformation which relies on the amorphous phase alone [Zhang & Cameron, 1999].

Polypropylene is an important structural material. In some circumstances, for example during sterilization for medical applications, it is subjected to significant doses of gamma irradiation, inducing significant change in mechanical properties. The response of the semi-crystalline structure to mechanical stress, before and after irradiation, is therefore of considerable interest. Isotactic polypropylene is a semi-crystalline polymer, which may crystallize into one of the three isomorphs, termed $\alpha$, $\beta$ and $\gamma$ [Varga, 1995]. Conventional thermal processes result in spherulites with crystals of the monoclinic $\alpha$ isomorph [Varga, 1995]. All such spherulites possess radial lamellae, but under certain crystallization conditions, tangential lamellae may also be present [Padden, 1995; Norton & Keller, 1985; Lotz & Wittman, 1986; Olley & Bassett, 1989; Padden & Keith, 1959; Idrissi et al., 1985]. It is reported that most bulk crystallised samples contain spherulites of mixed character [Norton & Keller, 1985] in which some areas of the spherulite are richer in tangential lamellae than others [Padden & Keith, 1959]. A range of spherulite types will be present if the crystallization temperature is not constant.

Polypropylene subjected to gamma irradiation undergoes cross-linking and scission [Nishimoto & Kagiya, 1992; Carlsson & Chmela, 1990]. In the presence of air, oxidation will enhance these effects [Carlsson et al., 1985]. Post irradiation ageing may occur as free radicals formed during irradiation react after the irradiation has ceased [Carlsson & Chmela, 1990; Carlsson et al., 1985]. Studies on isotactic polypropylene indicate that irradiation at 50 kGy in air causes slight increment of the crystallinity and the glass transition temperature to rise by a few degrees [Zhang & Cameron, 1999]. The changes to the dimensions of the lamellar architecture are small. It is widely reported however, that radiation does introduce major deterioration in mechanical properties of polypropylene as a consequence of chain scission and cross-linking [Kagiya et al., 1985; Nishimoto et al., 1991; Rolando, 1993; Martakis et al., 1994; Nishimoto et al., 1986; Kholyou & Katbab, 1993].

### 2.2 Mineral aggregates

Different mechanisms come into play when gamma radiation is applied on mineral aggregates. A few studies have been carried out. Some results have been reported for calcium bentonite which consists of a coarse fraction and a clay fraction. The concentration of radiation-induced defects increases with increasing dose. The coarse fraction has a higher concentration of defects (more than one order of magnitude) than the clay fraction. These results are consistent with the fact the coarse fraction contains minerals (silica and plagioclase groups), that are very sensitive to radiation. In the case of the clay fraction, the gamma radiation promotes defects in its crystalline lattice, mainly affecting the stability of the Al-O and Si-O bonds. The defects are holes trapped in the former positions of O atoms in the structure [Dies et al., 1999].

One of the most important applications of calcium bentonite is as engineering barrier for long-living radioactive waste materials from the nuclear industry such as soluble salts, aqueous solutions of nitrates, oxides and glasses. The requirements for acting as an engineering barrier include radiation and thermal stability - and also structural integrity.

## 3. Polymer matrix + polymeric fibers + mineral aggregates: Effects of gamma radiation

Very little information concerning the effects of gamma radiation in composites of the type polymer matrix + mineral aggregates + polymeric fibers has been developed. Nevertheless, in the last decade studies on the effects in the bonding interaction at the interface, as well as modifications of the polymer phase and mineral aggregates (fillers) are of potential interest. Moreover areas involving predictions of the useful service lifetime in different service environments are also important to consider.

Hydraulic concrete surface coated with a solution of polymethyl methacrylate, loading from 4.7 to 5.1 wt%. The methyl methacrylate (MMA) forms a hard glassy polymer, strongly bonded to the cement matrix, which substantially improves the properties of the original concrete [Levitt et al., 1973]. The presence of moisture can reduced the polymer loadings; the initial surface absorption results reached their best values when water is present in the concrete at the time of impregnation. The values in excess indicate that in polymer concrete composites, the strength of the impregnated concrete matrix may exceed that of the added flint gravel aggregate (gravel+sand). The fracture occurs through shear failure of the aggregate, and this behavior is typical for all the impregnated samples.

Hydraulic concretes were soaked in the unsaturated polyester resin at different impregnation times ranged from 1 to 15 hours. The addition of polymer to the hardened concrete causes healing of micro-fractures and produces improved bonding between the cement paste and aggregate [Ismail et al., 1998]. The main factor which influences the unsaturated liquid absorption is the accessibility (i.e., permeability) of polyester to the pores of the samples. The degree of polymer impregnation increases with the increase in impregnation time reaching a saturation state at 5 hours, after which the degree of impregnation is relatively constant up to 15 hours. So, the degree of incorporation is namely dependent upon the amount of monomer introduced into the porous samples.

The final strength of the composite is dependent on a number of factors: namely the extent of the impregnation and filling of pores, the type and content of resin, the size of aggregate. The type of polymer and its ability to carry stress the degree of conversion of monomers to polymer during the polymerization, the formation of a continuous polymer phase and the mechanical properties of polymer.

The mechanical properties for polyester-filler composites depended on the type and amount of filler and also on the particle size of the filler used. Nevertheless, high filler content is important from an economical point of view and if this is a recycled material comes from waste materials, as recycled PET form soft drink bottles and marble waste materials, the PC has acceptable physical properties, good mechanical integrity, enhanced chemical characterization, and providing better heat and flame resistance [Tawfik & Eskander, 2006].

In the case of polymer concrete, the physical and mechanical properties of cross-linked polyesters depend mainly on the type and ratio of the copolymerizable monomer used, for example of the styrene content.

Different concentrations for PC are suggested: for example 88 % of minerals and 12 % of styrenated polyester resin (SP); the minerals include 30% basalt (diameters: 0.5-1.0 cm), 40% marble (0.1 - 0.5 cm), and 30% marble (> 0.5 cm). It is recommended to dried the filler at elevated temperatures (for example at 273°C). For these PCs the compressive strength values increase as the styrene content increase, until a maximum of 123 MPa, at a polyester/styrene ratio of 60:40 wt %. Moreover, lower percentages of basalt reduce 36 % the compressive strength. Other studied on hydraulic concrete whose surface was coated with a solution of polymethyl methacrylate, the compressive strength increases according to the aging, after one day is 45 MPa and at 28 days 60 MPa.

Physicochemical modifications by using chemical attack or thermal processes are consuming time and money. An alternative for the preparation of composites is to use *ionizing energy*; the high energy radiation has special advantages, for example the controlled polymerization can be initiated uniformly within substantial thicknesses of material. Moreover, it is possible to improve compatibility between polymer matrix and the mineral aggregates - by means of structural and surface modification of both resin and the aggregates. Thus improvement of the mechanical properties of polymer concrete can be obtained.

Some results have been reported on formation of chemical links between aggregate minerals and polymer chains. For example, in silica + polysiloxane-rubber composites the induced cross-linking enhances crystallization rates and thus enhances mechanical properties at high strains. At the same time, a reduction in polymer-filler interactions at interfaces in silica + siloxanes composites is seen; the silica is modified by irradiation and a high surface area is obtained. Gamma radiation excites electrons sufficiently so that they leave their normal positions (valence to conduction band) producing positive holes and free electrons. Positive holes are electronic defects in the silica $O^{2-}$ matrix created as a result of removal of an electron from the $O^{2-}$ sites, which then become $O^{-}$ sites [Patel et al., 2006].

Mortar with impregnated polymer was subject to gamma irradiation for polymerization. The impregnated samples were subjected to 50 kGy with a dose rate of 10 kGy/h. The physico-mechanical properties were studied. The results show that the polymer loading, compressive strength, and bulk density increase with the increase in the percentage of cross-linking agent as well as the gamma irradiation doses [Ismail et al., 1998]. This behavior is attributed to the amount of polymer deposited in the pores of the specimens.

For a given gamma irradiation dose, the degree of polymer impregnation of the hardened cement mortar samples increases with the increase of the immersion time in the unsaturated polyester resin up to 4-5 hours. Moreover, the compressive strength also increases with the increasing impregnation up to 4 hours reaching an improvement of 59% when comparing to non-irradiated sample. This is attributed to the interaction between calcium silicate hydrates and polyester formed in the pores of the hardened mortar during the polymerization process under the effect of gamma irradiation. A continuous polymer phase is formed within the cement matrix inhibiting the formation and propagation of microcracks in the matrix when the cement gels shrink upon drying.

The gamma radiation polymerization of hydraulic concretes impregnated with methyl methacrylate show substantial improvements: 208% for the compressive strength (respect to non-irradiated concrete =35 MPa), 247 % for the bending strength (non-irradiated =4.4 MPa), and 46% for the dynamic modulus of elasticity (non-irradiated = 27.7 GPa); such elasticity improvements are not sufficiently high to indicate the development of brittleness. The concrete samples were irradiated with gamma radiation from a $^{60}$Cobalt source, at 35 kGy with a dose rate of 0.75 kGy/h [Levitt et al., 1973].

The influence of irradiation dose on the thermal degradation reaction of polyester-styrene resin and polyester-styrene resin/gypsum composites in presence of nitrogen has been investigated. The composites were irradiated at ambient temperature and a dose rate of 6.0 kGy/h. The decomposition temperatures for both systems are determined for irradiation doses ranged between 10 and 320 kGy [Ajji, 2005]. Dose at 20 kGy is more than enough to harden the composites. The TMA thermograms with alternated force show significant increase on the elongation at the glass transition temperature. The Tg for polyester-styrene resin vary from 67 to 78°C, and for polyester-styrene/gypsum from 77 to 86°C, when irradiating from 20 to 320 kGy. This means a 14% of difference when gypsum is aggregated. In both cases, there is a slight increase of the glass transition temperature for low irradiation doses and then the temperature becomes constant. This can be explained that cross-links built between the polymer chains via irradiation reduce the segmental mobility of the chains.

Moreover, the Tg of the polyester-styrene resin/gypsum composites is higher than the glass transition temperature of polyester-styrene resin irradiated at the same dose. This difference is most probably due to the interaction between the polymer matrix and filler particulates. Since the segmental mobility of the chains near the filler particles is reduced, the Tg of the composites increases.

Thermograms of the gypsum powder shows only one step (onset = 107.55oC) related to hydration water, and there is no interference with the other steps related to the polyester or the composite decomposition. As it was mentioned at 20 kGy the polyester-styrene resin is solidificated [Ajji, 2005]. After hardening the polyester resin, a slight change in the decomposition temperature could be observed, because the polyester-styrene/gypsum composite show a loss of 5 and 20% at 200 and 340°C, respectively, which are comparable with polyester-styrene resin (loss of 2 and 20% at 200 and 360°C, respectively).

The decomposition temperature of the polyester-styrene/gypsum composites decrease in presence of the inorganic fillers. Furthermore, the filler seems to have influence on the mechanism of thermal degradation of the polymer. This is a generally observed behavior, that irrespective of the amount and type of filler, the inorganic particles decrease the thermal stability of the polymer composites. The main reason for inducing thermal instability is believed to be an indirect one, improved and effective heat transfer to the polymer phase through the dispersed inorganic phase.

In Poly(methyl acrylate)/phosphate composites the total polymerization conversion is achieved with a dose of 10 kGy of $^{60}Co$ gamma radiation at room temperature. The composites were prepared by converting the liquid monomer/filler systems into polymer by gamma irradiation [Ajji & Alhassanieh, 2011]. A 10 kGy dose is a little higher than the necessary dose for achieving a total polymerization conversion in order to avoid any uncertainty in the bulk of the samples.

The glass transition temperature, Tg of Poly(methyl acrylate)/phosphate composite is higher than the Tg of the pure poly(methyl acrylate); 13 ± 3°C versus 8 ± 3°C, respectively, when both are irradiated at 15 kGy. This difference can be explained with the interaction between the polymer matrix and filler particulates (phosphate particles). Since the segmental mobility of the chains near the filler particles is reduced, and thus the Tg of the composites increases. This behavior has also been observed and documented in other composite systems; the increase in the Tg has been explained on the basis of reduced mobility of molecular segments in the vicinity of the filler particulates.

The presence of polymer chains surrounding phosphate particulates leads to a shielding of the phosphate particles, and consequently to a higher element ratio incorporated in the solid phase. Thus such polymer/composite systems could be considered for using as storage medium for radioactive waste of the studied radionuclides.

For non-irradiated polyester samples a 5% is loss of its initial weight at 260°C, and for irradiated samples at 295°C, while for the polyester-cement composite samples at 395°C. So, the polyester-cement composite has higher thermal stability than the irradiated polyester. This is attributed to chemical reaction of the polyester and cement constituents under the effect of gamma irradiation [Ismail et al., 1998]. At higher decomposition temperatures (500°C) there is no significant difference in thermal decomposition of non-irradiated and irradiated polyester (weight loss about 90%), while for polyester-cement composite is lower.

The break down stress of the polyester-styrene/gypsum composite is lower than the pure polyester–styrene resin when irradiating from 10 to 320 kGy. The values in both cases increase when the radiation dose augment too. The pure polyester-resin varies from 6.5 to 9.0 MPa while the polyester-styrene/gypsum composite varies from 2.0 to 3.6 MPa, it means a maximum difference of 225% [Ajji, 2005].

Such behavior is due that the gypsum (filler) has lower tensile strength than that of the polymer; and the increase of the cross-link density, which increases with increasing the irradiation dose and thus the tensile strength. This difference can be explained considering that only the polymer chains (but not the inorganic filler particles) can build cross-links between the chains.

The hardness percentage is not affected significantly by gamma irradiation; for the pure polyester-resin the percentages vary from 88 to 89.5 MPa and for polyester-styrene/gypsum composite from 91 to 92 MPa.

By solidifying concrete, water is evaporated and cavities are formed in concrete. These cavities are gas pores that cannot rise to the surface of concrete and will be caught on it. More pore formation in the concrete implies diminution on its strength. It is possible to minimize the micro pores by radiation of concrete during solidification. Because as the first micro pores have been filled by water. So, when gamma-ray interacts with water molecules, the hydrogen moves to surface of concrete and the OH participates on Alkali-Silica Reaction (ASR) [Rezaei-Ochbelagh et al., 2010]. By this way, micro pores can be deleted. Therefore, if concrete is radiated during drying process, its strength will increase. Moreover, in the case of a concrete structure with microscopic bubbles, there is a difference in radiation intensity when the ray passes through concrete with air-filled bubbles and without that. SEM micrographs show that concretes without sand are less dense than those with sand. Moreover, the irradiated specimens are denser as compared to the non-radiated ones.

Alkali-silica reaction in concrete is one of the slow chemical reactions. This slow reaction causes severe deterioration of concrete. Nuclear radiations make the aggregate ASR- sensitive and the deterioration of concrete can emerge long after the irradiation. It is therefore important to know the effect of nuclear radiation on the reactivity of aggregates to alkaline solution.

The compressive strength of concrete was measured against gamma radiation for two types of concretes (with and without sand). The specimens were irradiated with $^{137}Cs$ source, with a dose rate of 0.12 Gy/day. The compressive strength of irradiated concrete is more than non-irradiated concrete: 155 vs 241 kg/cm$^2$ for irradiated concrete with sand, and 145 vs 273 kg/cm$^2$ for concrete without sand [Rezaei-Ochbelagh et al., 2010].

## 4. Polymer concrete irradiated by gamma particles

Studies on the effects of gamma ionizing radiation on the curing process and on final properties of polymer concrete are ongoing. Developments include the effects on the mechanical properties. Our developments regarding the influence of fiber reinforcements on polymer concrete and the different behaviors based on the components (polymer resin and mineral aggregates) [Martínez-Barrera et al., 2007; Martínez-Barrera et al., 2008a; Bobadilla-Sánchez et al., 2009; Martínez-Barrera et al., 2009; Martínez-Barrera et al., 2008b; Martínez-Barrera et al., 2010].

In principle one can obtain high compressive and flexural strength, high impact and abrasion resistance, lower weight and lower costs. In general, the compressive strength values increase with the gamma irradiation dose. Moreover, when using $CaCO_3$, the highest compressive strength values are obtained compared to using $SiO_2$ aggregates. Intermediate values are found when using a combination of them ($CaCO_3$ and $SiO_2$).

The influence of polymeric fibers has been established. The Nylon fibers have a rigid shape, which differs from the polypropylene or polyester fibers having a more elastic shape. Thus the compressive strength depends on the material type, that is to say either rigid or elastic. So, it is worth point out that the combination of two minerals and elastic fibers (polyester and polypropylene) and at least 10 kGy of gamma irradiation allows higher values of compressive strain.

The Young's modulus E, can be a defining measure of whether one will obtain a ductile or more brittle concrete. Excepting only polymer concrete with marble and calcium bentonite, the values are higher than the standard value for polyester-based polymer concrete. Moreover, the improvement above that standard is notable: a) 143 % for polymer concrete with $SiO_2$, b) 141 % for polymer concrete with $CaCO_3$, and c) 120 % for polymer concrete with $CaCO_3$+$SiO_2$. Generally the higher the gamma irradiation the higher the Young's modulus and the harder the polymer concrete becomes.

Our studies are summarized in Figures 6 to 11. In Figure 6 is shown the compressive strength of PC compounded with unsaturated polyester resin (UPR) and one or two mineral aggregates, covering gamma doses between 50 and 150 kGy; values for specimens of 100% resin are also shown.

In Figure 6 we see that the compressive strength values increase with the gamma irradiation dose. Moreover, when using $CaCO_3$, the highest compressive strength values are obtained compared when using Marble + Calcium Bentonite (M+CB) aggregates. Intermediate values are found when using Marble aggregates. The standard value of compressive strength for polyester-based PC is 70 - 80 MPa [Martínez-Barrera et al., 2008a]. Thus, considering all PCs the maximal improvement percentage on compressive strength is 68% respect to standard values. Moreover, the values for resin are comparable with those for PC with $CaCO_3$. Such resin could be used for certain applications.

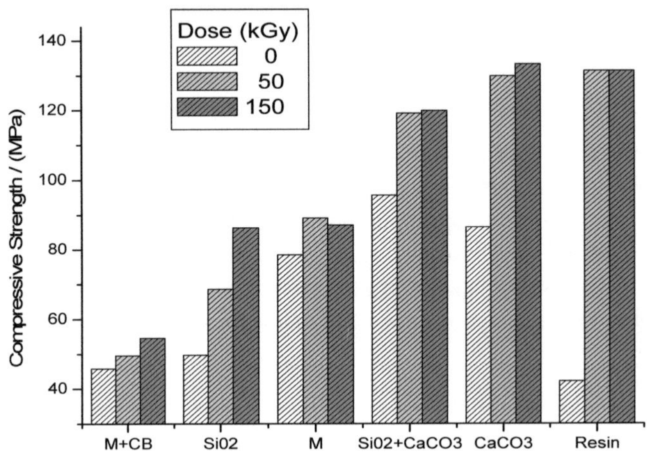


Fig. 6. Compressive strength of polymer concrete compounded with different mineral aggregates.

With respect to fiber-reinforced PCs, Figure 7 shows that the compressive strength values increase when the gamma irradiation dose increases. Nevertheless, lower values are done when comparing with PCs without fibers (see Figure 6). Different types of fibers were used (N=Nylon, PP=Polypropylene, and P=Polyester) at varying percentages (0.3, 0.4 and 0.5 vol. %) and with similar dimensions (40-60 μm of diameter and 10-20 mm long).

The lowest values of compressive strength were observed for PC with Marble + Calcium Bentonite, independently of the Nylon-fiber percentage. The Nylon fibers have a rigid shape, which differs from the polypropylene or polyester fibers having a more elastic shape. In our studies the highest values have been found in formulations combining two mineral aggregates ($CaCO_3$ and $SiO_2$) and one fiber. Thus the compressive strength depends on the material type, that is to say either rigid or elastic.

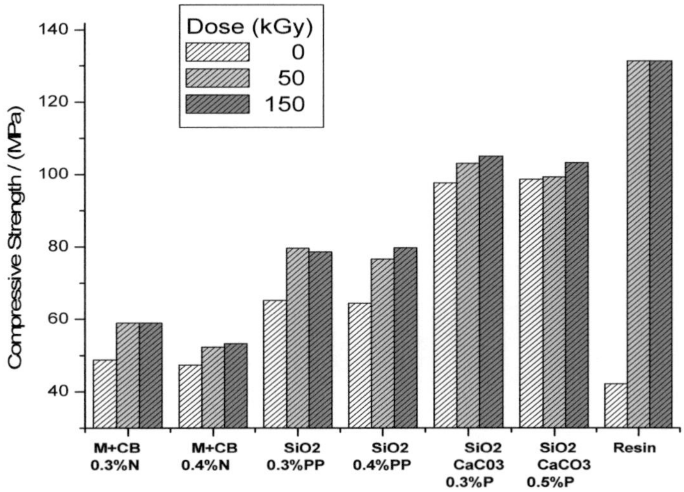


Fig. 7. Compressive strength of fiber-reinforced polymer concrete compounded with different mineral aggregates.

Another important mechanical feature of the PCs is related to the compressive strain at the yield point, as seen in Figure 8. The highest compressive strain values are for PC with Marble. Except PC with $SiO_2$ all specimens have higher values respect to the standard values reported in the literature (0.01 mm/mm) [Martínez-Barrera et al., 2009b]. On the other hand, lower values are observed for PC containing $SiO_2$. Different phenomena are observed, for example for certain PCs specimens the values increase up to a certain dose and afterward decrease (PC with $CaCO_3$, PC with M+CB and for 100% resin). One conclusion is that when using one mineral aggregate the compressive strain is influenced more by the resin than by the mineral aggregates.

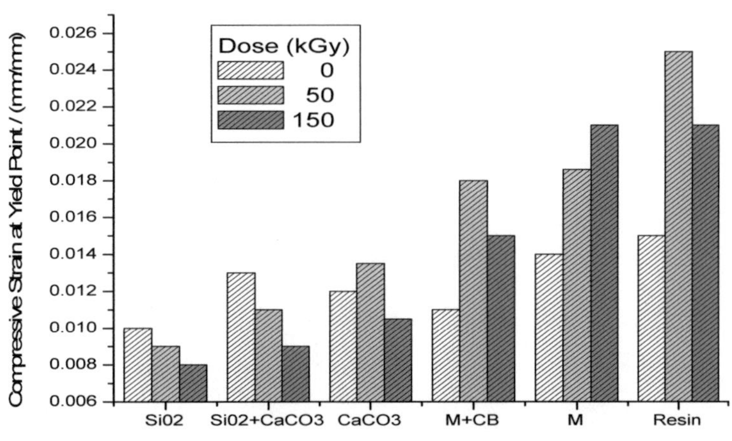


Fig. 8. Compressive strain at yield point of polymer concrete compounded with different mineral aggregates.

In the case of fiber-reinforced PCs, the compressive strain values increase notably for specimens with two mineral aggregates rather than just one (Figure 9). Something notable is that when comparing these compressive strain values to the standard value reported in the literature for PC (0.01 mm/mm) [Martínez-Barrera et al., 2009b]: a) for PC with $SiO_2$ there is 60 % improvement; b) for PC with M+CB up to 180 %, and c) for PC with $CaCO_3$ and $SiO_2$ up to 390 %. So, it is worth point out that the combination of two minerals, one fiber, and specific gamma radiation dose allows higher values of compressive strain.

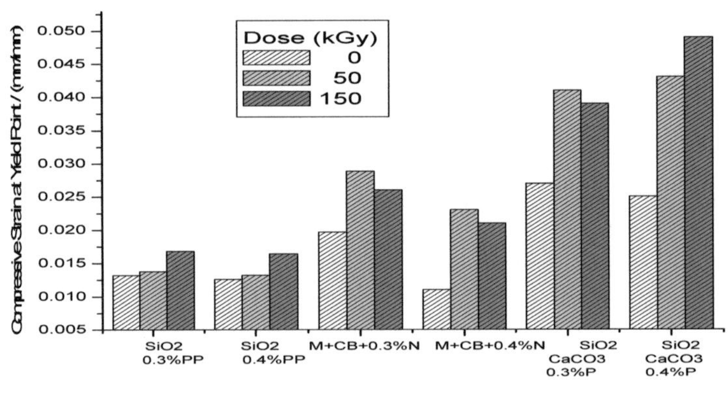


Fig. 9. Compressive strain at yield point of fiber-reinforced polymer concrete compounded with different mineral aggregates.

A third mechanical feature studied was the Young's modulus. Excepting only PC with M+CB, the values are higher than the standard value for polyester-based PCs, namely 6.7 GPa (see Figure 10) [Tavares et al., 2002]. Moreover, the improvement above that standard is notable: a) 139 % for PC with $SiO_2$, b) 122 % for PC with $CaCO_3$, and c) 108 % for PC with $CaCO_3$+$SiO_2$. Generally the higher the gamma irradiation the higher the Young's modulus and the harder the PC becomes.

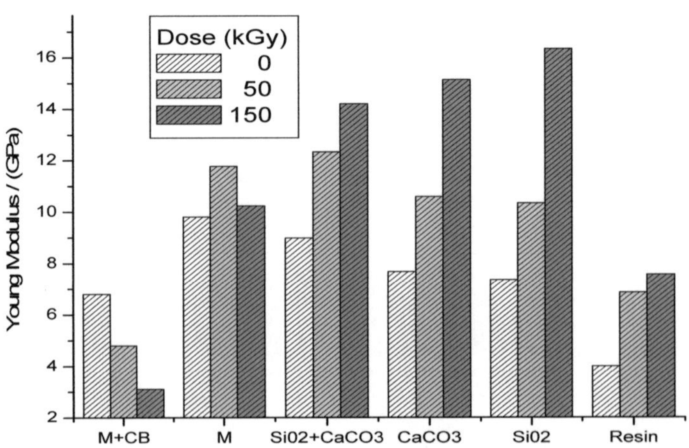


Fig. 10. Young's Modulus of polymer concrete elaborated with different mineral aggregates

In the case of fiber-reinforced PCs, the behavior of the Young's modulus is significantly varied (Figure 11). The PC with $SiO_2$ has higher values with respect to the standard reported for polyester-based PCs (6.7 GPa) [Tavares et al., 2002]. Nevertheless, it is possible to obtain low values for PCt with $CaCO_3$+$SiO_2$, namely 2.8 GPa, which represents a diminution of 58 % with respect to the standard. It is therefore also possible to get a more ductile PC, which may be desirable for certain applications. For the PC with combined $SiO_2$ and $CaCO_3$ the irradiation has little effect, likely due to competing interactions and effects in these materials.

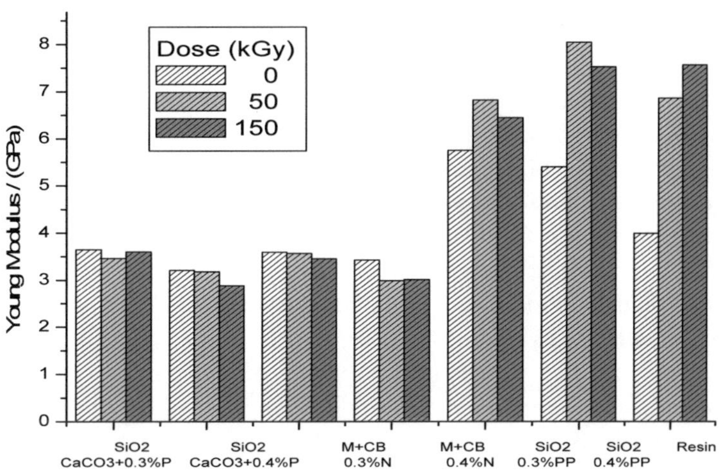


Fig. 11. Young's Modulus of fiber-reinforced polymer concrete compounded with different mineral aggregates

Improvements of E described here have wider implications and may be indicative of improvements or modifications to other properties not directly tested. It is therefore evident that the use of gamma irradiation can be another strategic tool to modify the mechanical properties of polymer concretes.

## 5. Acknowledgements

To Autonomous University of the State of Mexico by grant # UAEM 3053/2011SF. Mr. Miguel Martínez López and Ms. Elisa Martínez Cruz graduated students at the Materials Science Program (Autonomous University of the State of Mexico) have participated in the experiments.

## 6. References


Ajji Z. (2005). Preparation of polyester/gypsum/composite using gamma radiation, and its radiation stability. *Radiat. Phys. Chem.* Vol. 73, pp. 183–187.

Ajji Z., Alhassanieh O. (2011). Preparation of poly(methyl acrylate)/phosphate/ composites and its possible use as storage medium for radioactive isotopes. *J. Radioanal. Nucl. Chem.* Vol. 287, pp. 69–75.

Balabanovich A.I., Levchik S.V., Levchik G.F., Schnabel W., Wilkie C.A. (1999). Thermal decomposition and combustion of γ-irradiated polyamide 6 containing phosphorus oxynitride or phospham, *Polym. Degrad. Stab.* 64, pp. 191-195.

Barkhudaryan V.G. (2000a). Effect of γ-radiation on the molecular characteristics of low-density polyethylene, *Polymer* 41, pp. 575-578.

Barkhudaryan V.G. (2000b), Alterations of molecular characteristics of polyethylene under the influence of γ-radiation, *Polymer* 41, pp. 2511-2514.

Bittner B., Mäder K., Kroll C., Borchert H.H., Kissel T. (1999). Tetracycline-HCl-loaded poly(dl-lactide-co-glycolide) microspheres prepared by a spray drying technique: influence of γ-irradiation on radical formation and polymer degradation, *Journal of Controlled Release,* Vol. 59, pp. 23-32

Bobadilla-Sánchez E.A., G. Martínez-Barrera, W. Brostow, T. Datashvili. (2009). Effects of polyester fibers and gamma irradiation on mechanical properties of polymer concrete containing CaCO3 and silica sand, *eXPRESS Polymer Letters* Vol. 3, pp. 615-620.

Dole M. (1950). Chemistry and physics of radiation dosimetry, *Report of Symposium IX, Army Chemical Center,* Maryland.

Carlsson D.J., Dobbin C.J.B., Jensen J.P.T., Wiles D.M. (1985). Polymer stabilization and degradation, *ACS Symposium Series* Vol. 280, pp. 25.

Carlsson D.J., Chmela S., Mechanisms of polymer degradation and stabilisation, In: Polymers and high-energy irradiation: degradation and stabililisation, (Scott G., Ed.), Elsevier: London, chap. 4. 1990.

Chapiro A. (2002). Polymer irradiation: past–present and future. *XIIth International Meeting on Radiation Processing* Avignon, March 2001. *Rad. Phys. Chem.* 63 207-209.

Charlesby A. (1952). Cross-Linking of Polythene by Pile Radiation. *Proc. R. Soc. Lond. A25,* Vol. 215, 1121 (November, 1952) pp. 187-214.

Charlesby A. (1969). Atomic Radiation and Polymers, Pergamon Press, Oxford.

Czayka M., Fisch M., Uribe R.M., Vargas-Aburto C. (2007). Radiation-thickening of iso-polyester resin, *Radiat. Phys. Chem.* 76, pp.1058–1068

Delahaye N., Marais S., Saiter J.M., Metayer M., (1998). Characterization of unsaturated polyester resin cured with styrene, *J. Appl. Polym. Sci.* Vol. 67, pp. 695-703

Delley C.W., Woodward A.E., Saver J.A. (1957). Effect of Irradiation on Dynamic Mechanical Properties of 6–6 Nylon, *J. Appl. Phys.* 28 pp. 1124-1130.

Dies J., de las Cuevas C., Tarrasa F., Miralles L., Pueyo J.J., Santiago J.L., (1999). Thermoluminescence Response of Heavily Irradiated Calcic Bentonite, *Radiat. Protection Dosimetry* Vol. 85, pp. 481-486.

Gupta M.C., Deshmukh V.G (1983). Radiation effects on poly(lactic acid), *Polymer* 24 pp. 827-830.

Hoseman R., Loboda-Cackovic J., Cackovic H. (1972). Affine deformation of linear polyethylene during stretching and affine transformation to the original shape in the liquid state, *J. Mater Sci.* Vol. 7, pp. 963.

Idrissi B.O.B., Chabert B., Guillet J. (1985). *Macromol. Chem.* Vol. 186, pp. 881.

Ismail M.R., Ali M.A., El-Milligy A.A., Afifi M.S. (1998). Physico-chemical studies of gamma irradiated polyester - impregnated cement mortar composite, *J. Radioanal. Nucl.Chem.* Vol. 238, pp. 111-117.

Jenkins H., Keller A. (1975). Radiation-induced changes in physical properties of bulk polyethylene. I. Effect of crystallization conditions. *J. Macromol. Sci. B* Vol.11, No. 3, pp. 301-323.

Jurkin T., Pucic I. (2006). Post-irradiation crosslinking of partially cured unsaturated polyester resin, *Radiat. Phys. Chem.* 75, pp. 1060–1068.

Kagiya T., Nishimoto S., Watanabe Y., Kato M. (1985). Importance of the amorphous fraction of polypropylene in the resistance to radiation-induced oxidative degradation. *Polym. Degrad. Stab.* Vol. 1, No. 3, pp. 261-275.

Kholyou F., Katbab A.A. (1993). Radiation degradation of polypropylene. *Radiat. Phys. Chem.* Vol 42, No. 1-3, pp. 219-222.

Levitt M., McGahan D.J., Hills P.R. (1973). Comparision of concrete polymer composites produced by high energy radiation, *PCI Journal* Vol. 18, pp. 35 – 41.

Li B., Zhang L. (1997). γ-Radiation damage to nylon 1010 containing neodymium oxide, *Polym. Degrad. Stab.* 55 pp. 17-20.

Loboda-Cackovic J., Cackovic H., Hoseman R. (1974). Structural changes in paracrystallites of drawn polyethylene by irradiation. *Colloid and Polymer Sci.* Vol.252, No.9, pp 738-742

Lotz B., Wittman J.C. (1986). The molecular origin of lamellar branching in the α (monoclinic) form of isotactic polypropylene. *J. Polym. Sci. Part B: Polym. Phys.* Vol. 24, No. 7, pp. 1541-1558.

Malek M.A., Renreng A., Chong Ch.S. (2001), Mechanistic model for bond scission in a polymeric system by radiation, *Radiat. Phys. & Chem.* 60, pp. 603-607.

Martakis N., Niaonakis M., Pissimissis D. (1994). Gamma-sterilization effects and influence of the molecular weight distribution on the postirradiation resistance of polypropylene for medical devices. *J. Appl. Polym. Sci.* Vol. 51, No. 2, pp. 313-328

Martínez-Barrera G., H. López., Castaño V.M., Rodríguez, R. (2004). Studies on the rubber phase stability in gamma irradiated polystyrene-SBR blends by using FT-IR and Raman spectroscopy. *Rad. Phys. & Chem.* Vol. 69, No. 2, pp. 155-162.

Martínez-Barrera G., Menchaca-Campos C., Hernández-López S., Vigueras-Santiago E., Brostow W. (2006). Concrete reinforced with irradiated nylon fibers, *J. Mater. Res.* Vol. 21, pp. 484-491.

Martínez-Barrera G., M.E. Espinosa-Pesqueira, W. Brostow. (2007). Concrete + polyester + CaCO3: Mechanics and morphology after gamma irradiation, *e-Polymers* No. 083, pp. 1- 12.

Martínez-Barrera G., U. Texcalpa-Villarruel, E. Vigueras-Santiago, S. Hernández-López, W. Brostow. (2008a). Compressive strength of Gamma-irradiated polymer concrete, *Polym. Compos.* Vol. 29, pp. 1210-1217.

Martínez-Barrera G., L.F. Giraldo, B.L. López, W. Brostow. (2008b). Effects of gamma radiation on fiber reinforced polymer concrete, *Polym. Compos.* Vol. 29, pp. 1244-1251.

Martínez-Barrera G., A.L. Martínez-Hernández, C. Velasco-Santos, W. Brostow. (2009). Polymer concretes improved by fiber reinforcement and gamma irradiation, *e-Polymers* No. 103, pp. 1-14.

Martínez-Barrera G. and W. Brostow, Fiber-reinforced polymer concrete: Property improvement by gamma irradiation, in *Gamma radiation effects on polymeric materials and its applications*, Research Signpost, Kerala India, p. 27 (2009b).

Martínez-Barrera G., W. Brostow. (2010). Effect of marble-particle size and gamma irradiation on mechanical properties of polymer concrete, *e-Polymers* No. 061, pp. 1-14.

Menchaca C., Álvarez-Castillo A., Martinez-Barrera G., López-Valdivia H., Carrasco H., Castaño V.M. (2003). Mechanisms for the modification of nylon 6,12 by gamma irradiation. *IJMPT* 19 pp. 521-529.

Menchaca-Campos C., Martínez-Barrera G., Resendiz M.C., Lara V.H., Brostow W., (2008). Long term irradiation effects on gamma-irradiated nylon 6, 12 fibers. *J. Mater. Res.* 23 pp. 1276-1281.

Menchaca-Campos C., Martínez-Barrera G., Alvarez-Castillo A., Lara V.H., Lopez-Valdivia H., Carrasco-Abrego H., Effect of gamma radiation on PA 6,12 in argon atmosphere, In: Gamma radiation effects on polymeric materials and its applications, (Barrera-Díaz C., Martínez-Barrera G.: Eds.), Research Signpost, Kerala India, 2009, pp 87 - 102.

Menchaca C., Nava J.C., Valdéz S., Sarmiento-Martínez O., Uruchurtu J. (2010). Gamma-Irradiated Nylon Roughness as Function of Dose and Time by the Hurst and Fractal Dimension Analysis. *Journal of Material Science and Engineering*, Vol. 4, No. 9, pp. 50-58

Menchaca C., Martínez-Barrera G., Fainleib A., (2011). Nylon 6,12 Fibers Under Low-Dose Gamma Irradation. *Journal of Polymer Engineering* Vol. 31, pp. 457–461.

Mykiake A. (1960). Infrared spectra and crystal structures of polyamides. *J. Polym. Sci.*, Vol. 43, (may, 1960), pp. 223-232.

Nishimoto S., Kagiya K., Watanabe Y., Kato M., (1986). Material design of radiation resistant polypropylene: Part II—Importance of the smectic structure produced by quenching treatment. *Polym. Degrad. Stab.* Vol. 14, No. 3, pp. 199-208

Nishimoto S., Kitamura K., Watanabe Y., Kagiya T., (1991). The correlation between the morphology and radiation resistance of polypropylene solid materials. *Radiat. Phys. Chem.* Vol. 37, No.1, pp. 71-75

Nishimoto S., Kagiya T., Radiation degradation of polypropylene, In: Handbook of polymer degradation, (Hamid S.H., Amin M.B., Maadhah A.G.: Eds.), Marcel Dekker,New York, chap. 1, 1992.

Nishiura T., Nishijima S., Okada T. (1999). Creep behavior of epoxy resin during irradiation at cryogenic temperature. *Radiat. Phys. Chem.* Vol. 56, No. 5-6, pp. 605-609.

Norton D.R., Keller A., (1985). The spherulitic and lamellar morphology of melt-crystallized isotactic polypropylene. *Polymer,* Vol. 26, No. 5, pp. 704-716.

Olley R.H., Bassett D.C., (1989). On the development of polypropylene spherulites. *Polymer.* Vol. 30, No. 3, pp. 399-409.

Padden Jr. F.J., Keith H.D., (1959). Evidence for a Second Crystal Form of Polypropylene. *J. Appl. Phys.* Vol. 30, pp. 1479-1484.

Padden Jr. F.J., Keith H.D. (1973). Mechanism for lamellar branching in isotactic polypropylene. *J. Appl. Phys.* Vol. 44, No. 3, pp. 1217-1223.

Patel G.N., Keller A. (1975). Crystallinity and the effect of ionizing radiation in polyethylene. I. Crosslinking and the crystal core. *J. Polym. Sci. Polym. Phys. Ed.* Vol. 13. No. 2, pp. 303-321.

Patel M., Morrell P.R., Murphy J.J., Skinner A., Maxwell R.S., (2006). Gamma radiation induced effects on silica and on silica–polymer interfacial interactions in filled polysiloxane rubber. *Polym. Degrada. Stab.* Vol. 91, No. 2, pp. 406-413.

Ramesh C. (1999). Crystalline transitions in Nylon 12, *Macromolecules* Vol. 32, pp. 5704-5706.

Rezaei-Ochbelagh D., Gasemzadeh Mosavinejad H. Molaei M., Azimkhani S., Khodadoost M. (2010). Effect of low-dose gamma-radiation on concrete during solidification, *Internat. J. Phys. Sci.* Vol. 5, pp. 1496-1500.

Rolando R.J. (1993). Radiation Resistant Polypropylene- New Developments. *J. Plastic Film Sheeting,* Vol. 9, No. 4, pp. 326-333.

Tavares C.M.L., Ribeiro M.C.S., Ferreira A.J.M., Guedes R.M. (2002). Creep behaviour of FRP-reinforced polymer concrete. *Comp. Struc.* Vol. 57, No. 1-4, pp. 47-51.

Tawfik M.E., Eskander S.B. (2006). Polymer concrete from marble wastes and recycled Poly(ethylene terephthalate), *J. of Elastomers and Plastics* Vol. 38, pp. 65–79.

Thanki P.N, Ramesh C, Singh R.P (2001). Photo-irradiation induced morphological changes in nylon 66, *Polymer* Vol. 42, No. 2, pp. 535-538.

Timus D.M., Cincu C., Bradley D.A., Craciun G., Mateescu E. (2000). *Appl. Rad. and Isotopes,* Martakis et al., 1994, pp. 937-944.

Ungar G., Keller A. (1980). Effect of radiation on the crystals of polyethylene and paraffins: 1. Formation of the hexagonal lattice and the destruction of crystallinity in polyethylene. *Polymer,* Vol. 21, No. 11, pp.1273-1277.

Valenza A., Piccarolo S., Spadaro G., (1999). Influence of morphology and chemical structure on the inverse response of polypropylene to gamma radiation under vacuum. *Polymer,* Vol. 40, No. 4, pp. 835-841.

Varga J., Polypropylene: structure, blends and composites. London: Chapman and Hall, 1995.

Wilson J.E., (1974). *Radiation Chemistry of Monomers, Polymers and Plastics,* Marcel Dekker, Inc.,

Woods, R.J., Pikaev, A.K. (1994). Applied Radiation Chemistry. Wiley, New York, pp. 281–283.

Yu J., Li B., Zhang L., (1998). Effect of the fold surface of the lamellae on the behavior of the trapped radicals in irradiated Polyamide-1010. *J. Appl. Polym. Sci.,* Vol. 67, No.7, pp. 1335-1339.

Zhang X.C., Cameron R.E. (1999). The morphology of irradiated isotactic polypropylene. *J. Appl. Polym. Sci.* Vol. 74, No. 9, pp. 2234-2242.

7

# Synthesis of Polyaniline HCl Pallets and Films Nanocomposites by Radiation Polymerization

M. A. Ali Omer[1, 2*], E. Saion[2], M. E. M. Gar Elnabi[1] and Kh. Mohd. Dahlan[3]
*[1]Sudan University of Science and Technology, College of Medical Radiologic Science,*
*[2]Department of Physics, Faculty of Science, University Putra Malaysia, Selangor,*
*[3]Nuclear Agency Malaysia (NAM), Bangi, Selangor,*
*[1]Sudan*
*[2,3]Malaysia*

## 1. Introduction

The sources of radiation are so varies, some of them are natural and others are man-made. Also the types of radiation can be categorized according to their wave length or energy or even to the ability of ionizing the media.

Non-ionizing radiation is electromagnetic radiation that does not have sufficient energy to remove the electrons from the outer shell of the atom. Types of non-ionizing radiation are: ultra violet (U/V), Visible light, infrared (IR), microwave (radio and television), and extremely low frequency (ELF, or as they called EMF or ELF-EMF). Non-ionizing radiations produced by a wide variety of sources at homes and in the workplaces, form lasers to power lines, tanning beds to household appliances, cellular phones to home radios (Smith F. A. 2000).

Ionizing Radiation: refer to the types of radiation that has capability to ionize the media directly or indirectly such as X-ray, γ-ray and neutron.

## 2. Natural sources

The natural sources represented in the following:

i. Cosmic radiation: represent the radiation comes from outside our solar system as positively charged ions (protons, irons, nuclei, helium…) which are interact with atmospheric layer (air) around the ground to produce secondary radiation as (X-ray, Muons, Protons, Alpha particles, Pions, Electrons and Neutrons).
ii. External terrestrial sources: these represent the radioactive materials, which are found naturally in the earth crust, rocks, water, air and vegetation. The major radio-nuclides found in the earth crust are (Potassium-40, Uranium-235, and Thorium-210).

## 3. Artificial sources

The main sources of manmade radiation that expose the public are from (Medical Procedures, as in diagnostic X-ray, radiation therapy, nuclear medicine and sterilization).

The common radioactive elements are I-131, Tc-99m, Co-60, Ir-192, St-90 and Cs-137). Other sources exemplar in occupational and consumption products, these implies the radiation in mines, combustible fuel (gas, coal), ophthalmic glasses, televisions, luminous, watch's dial (tritium), X-ray at air-port (detectors), smoke detectors (americium-241), road construction materials, electrons tubes, and fluorescent lamp starters, nuclear fuel cell, nuclear accidents and nuclear weapons in marshal island and war. The yield of artificial sources either as quantum represented in X-ray and gamma radiation (γ) or as particles with high energy as beta particles (β), alpha particles (α), neutrons and electrons. The common artificial sources are accelerators and nuclear reactors (Smith F. A. 2000).

All of the above radiation types were used in researches; today the most common radiation sources applied in researches and in man serves are:

i. Co-60, as artificial source for gamma (γ) radiation.
ii. Linear accelerators for photon and electron beams, with energy range of (0.3-10 MeV and up to 20 MeV).

These energies are insufficient to initiate nuclear reaction; hence the irradiated element does not exhibit any radioactivity, see the table of radiation sources (1).

Table (1) shows the sources of radiation (Smith, 2000)

| Category | Source |
|---|---|
| Nuclear power | $^{235}U$ fission products, $^{90}Sr$, $^{137}Cs$ |
| Occupational exposure | X-ray, Isotopes for (γ) ray |
| Weapons tests | $^{235}U$, $^{239}Pu$, fission products |
| Every day sources | Coal, Tobacco and Air-travel |
| Medical tests & treatment | X-ray, (γ)radiation & electrons |
| Cosmic rays | Protons, electrons, neutrons |
| Food | 40K, 137Cs, 14C and 131I |
| Rocks & building | 235U, 238U, and 232Th |
| Atmosphere | 222Rn and 137Cs |

Table 1.

Ionizing radiation is a broad energetic spectrum of electromagnetic waves or high velocity atomic or subatomic particles. The radiation can be categorized according to their ability to ionize the media. Non-ionizing radiation is electromagnetic radiation that does not have sufficient energy to remove an electron of the atom. The various types of non-ionizing radiation are ultra violet (UV), visible light, infrared (IR), microwaves (radio and television), and extremely low frequency (ELF, or as they called EMF or ELF-EMF). Ionizing radiation is electromagnetic radiations, such as X-rays, γ-rays and charged particles (electrons, β-particles and α-particles) which possess sufficient energy to ionize an atom by removing at least an orbital electron. According to the 1996 European Guideline of the European Atomic Energy Community (EURATOM), electromagnetic radiation with a wavelength of 100 nm or less is considered as ionizing radiation which is corresponds to ionizing potential of 12.4 eV or more (Smith, 2000). The ionization potential is dependent on the electronic structure of the target materials and generally in the order of 4 - 25 eV.

The International Commission of Radiation Units (ICRU) has subdivided the ionizing radiation into direct and indirect ionizing radiation, based on the mechanisms by which they ionize the atom. *Direct ionizing radiations* are fast charged particles, such as alpha particles, electrons, beta particles, protons, heavy ions, and charged mesons, which transfer their energy to the orbital electron directly and ionize the atom by means of Columbic force interactions along their track. *Indirect ionizing radiations* are uncharged quantum, such as electromagnetic radiations (X-rays and γ-rays), neutrons, and uncharged mesons, which undergo interactions with matter by indirectly releasing the secondary charged particles which then take turn to transfer energy directly to orbital electrons and ionize the atom. Some properties of ionizing radiation are shown in Table 2. Table (2) shows the properties of different ionizing radiation.

| Characteristics | Alpha | Proton | Beta or electron | Photon | Neutron |
|---|---|---|---|---|---|
| Symbol | ${}_{2}^{4}\alpha$ or $He^{+2}$ | ${}_{1}^{1}p$ or $H^{+}$ | ${}_{-1}e$ or β | γ- or X-rays | ${}_{0}^{1}n$ |
| Charge | +2 | +1 | -1 | Neutral | Neutral |
| Ionization | Direct | Direct | Direct | Indirect | Indirect |
| Mass (amu) | 4.00277 | 1.007276 | 0.000548 | - | 1.008665 |
| Velocity (m/s) | $6.944 \times 10^{6}$ | $1.38 \times 10^{7}$ | $2.82 \times 10^{8}$ | $2.998 \times 10^{8}$ | $1.38 \times 10^{7}$ |
| Speed of light | 2.3% | 4.6% | 94% | 100% | 4.6% |
| Range in air | 0.56 cm | 1.81 cm | 319 cm | 820 m | 39.25 cm |

1 atomic mass unit (amu) = $1.6 \times 10^{-27}$ kg.
Speed of light c = $3.0 \times 10^{8}$ m/sec.

Table 2. The properties of different ionizing radiation

## 4. Gamma ray (γ-ray) interaction and attenuation coefficients

In general the characteristic of radiation interaction with matter represented in photoelectric (*Predominates for photons in the low energy range between 10 keV and 200 keV*), Compton (*Predominated at energies of 100 keV - 10 MeV.* (McGervey, 1983)), Pair production (*Predominated at energies greater than twice the rest mass of an electron,* i.e. $2m_0c^2$ = 1.022 MeV, *where m refers to mass of electron and c refers to speed of light* (Johns and Cunningham 1983)), Triplet production process (*occurs when the incident photon have an energy of* $4m_0c^2$, i.e. *it implies both the pair production at the nucleus level plus triplet production*) and Raleigh scattering (*predominant for photons at low energy range from 1 keV to 100 keV*) table (3), is that each individual photon is absorbed or scattered from the incident beam in a single event. The photon number removed $\Delta I$ is proportional to the thickness traveled through $\Delta x$ and the initial photon number $I_0$, i.e. $\Delta I = -\mu I_0 \Delta x$, where, $\mu$, is a constant proportionality called the attenuation coefficient. In this case, upon integrating, we have the following equation (1)

$$I = I_0 e^{-\mu x} \tag{1}$$

The attenuation coefficient is related to the probability of interaction per atom, i.e. the atomic cross section $\sigma_a$ is given by equation (2)

$$\mu = \frac{\sigma_a N_A \rho}{A} \tag{2}$$

where $A$ is the mass number and $N_A$ the Avogadro's number (6.022 x $10^{23}$ mol/1).

Table 3 briefly summarized the entire γ-radiation photon interactions with their possible energies required to initiate the reactions (Smith, 2000; Siegbahn, 1965).

| Process | Type of interaction | Other names | Approximate *E* of Maximum importance. | Z dependence |
|---|---|---|---|---|
| Photoelectric | With bonded electrons, all *E* given to electron | | Dominant at low *E* (1-500) KeV, cross section decrease as *E* increase | $Z^3$ |
| Scattering from electrons coherent | With bond atomic electron, with free electrons | Rayleigh electron, resonance scattering, Thomson scattering | <1MeV and greatest at small angles. Independent of energy | $Z^2$, $Z^3$<br>Z<br>Z |
| Incoherent | With bond atomic electron, with free electrons | Compton scattering | <1MeV least at small angle. Dominate in region of 1 MeV, decreases as *E* increase | Z<br>$Z^2$ |
| Pair Production | In Coulomb field of Nucleus | Elastic Pair production | Threshold ~1MeV, *E* > 5MeV. Increase as *E* increase. | Z |
| Pair production Delbruk scattering | In coulomb field of electron & nucleus | Triplet production inelastic pair production. Nuclear potential scattering | Threshold at 2 MeV increases as *E* increases. Real Max > imaginary, below 3 MeV(both increase as *E* increases) | $Z^4$ |

Table 3.

The essence of γ-radiation interaction with molecules and the induction of physical and chemical characteristics that leading to form new compound is ascribed to the amount of energy being transferred, which will create ion, free radicals and excited molecule. Such interaction process is termed ionization and excitation of the molecules, which can cause chemical changes to the irradiated molecule. This is due to the fact that all binding energy for organic compound in the range of 10 – 15 eV. In case of low transferred energy by photon, the molecule undergoes excitation state before returning to the rest state by emitting X-ray photons or break down to release free radicals which in turn undergoes polymerization.

The ejected electron from the irradiated molecule ($A^+$) is subjected to the strong electric field of the formed positive charge. Therefore the recombination is a frequently occur, either during irradiation or after the end of irradiation to create energetic molecule ($A^{**}$). Such highly energetic excited molecule will break down into free radicals and new molecule (Denaro, 1972). The fundamental of this reaction can be shown in the following scheme Figure (1).

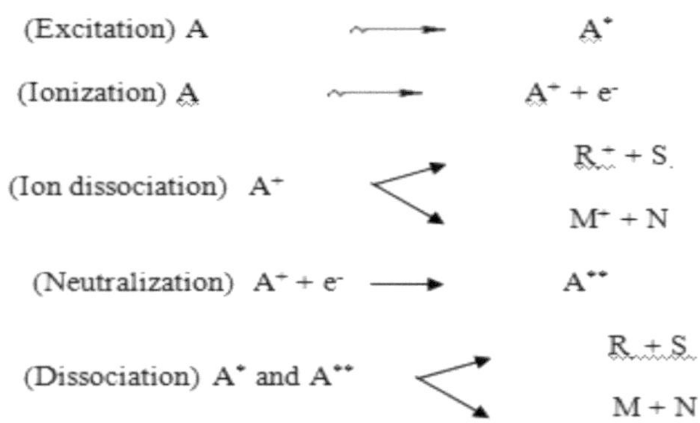


Fig. 1. The expected irradiation results of the organic molecules, where R. and S. are free radicals and M and N are molecular products.

## 5. Radiation polymerization

Radiation polymerization is a process in which the free radicals interact with the unsaturated molecules of a low molecular unit known as monomer to form high molecular mass polymer or even with different monomers to produce crosslink polymer. The formed polymer can be in different forms called homopolymer and copolymer depending on the monomer compositions link together. Radiation-induced polymerization process can be achieved in different media whether it is liquid or solid unlike the chemical polymerization which can only accomplished in aqueous media. It is also temperature independent. Radiation polymerization often continues even after removing away from the radiation source. Such condition is known as post-polymerization (Lokhovitsky and Polikarpov, 1980). Since radiation initiation is temperature independent, polymer can be polymerized in the frozen state around aqueous crystals. The mechanism of the radiation induced polymerization is concerning the kinetics of diffusion-controlled reactions and consists of several stages: addition of hydroxyl radicals and hydrogen atoms to carbon-carbon double bond of monomer with subsequent formation of monomer radicals; addition of hydrated electrons to carbonyl groups and formation of radical anion of a very high rate constant and the decay of radicals with parallel addition of monomer to the growing chain.

## 6. Cross linking

The process of crosslink occurs due to interaction between two free radical monomers which combine to form intermolecular bond leading to three dimensional net of crosslinked highly molecular polymer, more likely dominate in unsaturated compound or monomer. The crosslinked polymer show strong mechanical strength and high thermal resistance.

## 7. Radiation grafting

Radiation grafting is a process in which active radical sites are formed on or near the surface of an exciting polymer, followed by polymerization of monomer on these sites. Grafting is accompanied by homopolymerization of the monomer; the material to which the monomer is grafted is described as the backbone, trunk or support. Radiation grafting is used to modify the polymers texture such as film, fibers, fabrics and molding powders. The process of grafting can be expressed as follow; suppose the polymer A is exposed to γ-rays, thus the active free radical sites $A^*$ created randomly along the polymer backbone chain, this free radical initiate a free radical on the monomer B then undergoes grafting polymerization at that active sites. The extension of the attached monomer B upon the base polymer A is termed as the degree of grafting DOG which refers to the mass of the grafted polymer as a percentage of the mass of the original base polymer. Such process can be expressed in schematic Figure (2).

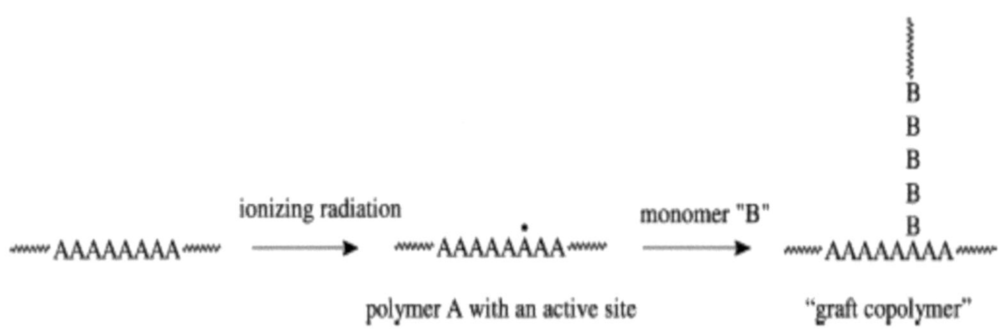


Fig. 2. Schemes for grafting process for polymer A with monomer B using gamma radiation.

Conducting polymers and their composites exhibit excellent optical, electrical, and electrochemical properties and therefore they have potential applications in enhancement the electrode performance of rechargeable batteries and fuel cells, electric energy storage systems in supercapacitors, solar energy conversion, photoelectrochromics, corrosion protection, electromagnetic interference shielding and biosensors (Malinauskas *et al.*, 2005).

In this work attempts are made to produce conducting polyaniline (PANI) formed in pallets and dispersed in PVA matrix (films) then their structure, optical properties and electrical conductivity are investigated. However, for the first time the polymerization of pure PANI is fully achieved by ionizing radiation (Mohammed, 2007). The prime advantage of radiation processing in this work is that no oxidizing agent is used to polymerize the conducting PANI i.e. giving pure product.

## 8. Conducting polyaniline nanoparticles

PANI has high electrical conductivity that can be controlled by oxidation or protonic doping mechanism during synthesis. PANI is known for its excellent thermal and environmental stability but poor processibility due to insolubility in most common solvents and brittleness that limits its commercial applications. In the composites form with another water soluble polymers such as PVA, poly(vinyl pyrrolidone), poly(acrylic acid) and poly(styrene sulfonic acid) (PSSA) which used as stabilizers, the processibility of PANI could be improve and a functionalized protonic acid can be added into the composites to chemically polymerize PANI. The PANI dispersion can then be cast to form composite film containing PANI nanoparticles. To improve the conductivity further, chemically and electrochemically PANI/ polymer composites have been irradiated with x-rays, gamma radiation, and electron beams (Bodugoz *et al.*, 1998; Sevil *et al.*, 2003; Wolszczak *et al.*, 1996 a and b; Angelopous *et al.* 1990). When ionizing radiation interacts with polymer materials active species such as ions and free radicals are produced and thus, improved the PANI conductivity.

Conducting PANI has been synthesized by chemical and electrochemical methods, which the later is considered the common one because of better purity. Chemically and electrochemically synthesized polyaniline are subjected to many shortcomings such as impurities, solvent toxicity, long tedious process, poor compatibility, insoluble, expensive, low production and difficult in their preparation, etc. However, report on synthesis of PANI nanoparticles using only $\gamma$-irradiation has not been reported until the date of 2007. The advantages of radiation processing is that no metallic catalyst, no oxidizing or reducing agent is needed, synthesis in a solid-state condition, fast and inexpensive, and controllable acquisitions. The synthesis of PVA/PANI nanoparticles by $\gamma$-irradiation doping is proposed in this work.

## 9. Methodology

### 9.1 Materials and equipments

The materials used for preparing the samples in this study, namely as polyvinyl alcohol PVA, aniline hydrochloride AniHCl, $\gamma$-radiation as an effective tool for polymerization process and reducing agent, Petri-dishes, micrometer, UV-spectroscopy, Raman spectroscopy and LCR-meter.

### 9.2 Method

The Aniline hydrochloride AniHCl monomer as 2.5, g (28.6 W/V) has been dissolved in distill deionized water of 100 ml under nitrogen atmosphere and bubbling in the solution with continuous stirring using magnetic stirrer for 3 hour. Then the solution has been irradiated with $\gamma$-radiation receiving 10, 20, 30, 40 and 50 kGy. The polymerized AniHCl i.e. polyaniline PANI-HCl has been precipitated filtered and collected in a form of powder. The powder (2.5g) pressed by 10 tons to form pallets.

On the other hand a polyvinyl alcohol (PVA) was supplied by SIGMA (Mw = 72,000 g/mol, 99 - 100% hydrolyzed) has been prepared by dissolving 30.00 g PVA powder in 600 ml distilled deionized water at controlled temperature of 80 $^{o}$C in the water-bath. The solution was magnetically stirred throughout at that temperature for 3 hours and then left to cool at room temperature. After cooling to room temperature, a weight of AniHCl, 2.5, g was added into 100 ml PVA solution, which gives the AniHCl concentrations as 28.6, wt%, by weight in comparison to the PVA. The mixtures were stirred continuously for 10 hours using a magnetic stirrer in nitrogen atmosphere. Then the PVA/AniHCl blend solution has been irradiated by $\gamma$-radiation receiving 10, 20, 30, 40 and 50 kGys and after irradiation the solution has been divided into Petri-dishes, each contains 20 ml and left to dry at ambient temperature and dark room for 3 days to evaporate the water. The casting film was pealed off and cut into several pieces which were eventually packed in a sealed black plastic bag.

Fig. 3. The UV-visible spectrophotometer model Camspec M530.
Faculty of Science, Department of Physics-UPM

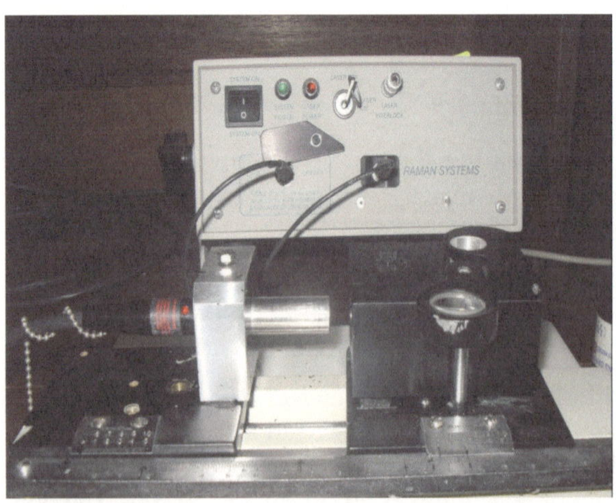


Fig. 4. Raman system and its accessories for sample set up and characterization,
Faculty of Science, department of Physics - UPM

The thickness of the films was determined by a digital micrometer model Mitutoyo no: 293-521-30-Japan, The average thickness of the films was 2 mm, and then the products (Films of PANI-HCl and pallets) have been characterized using the following instrument:

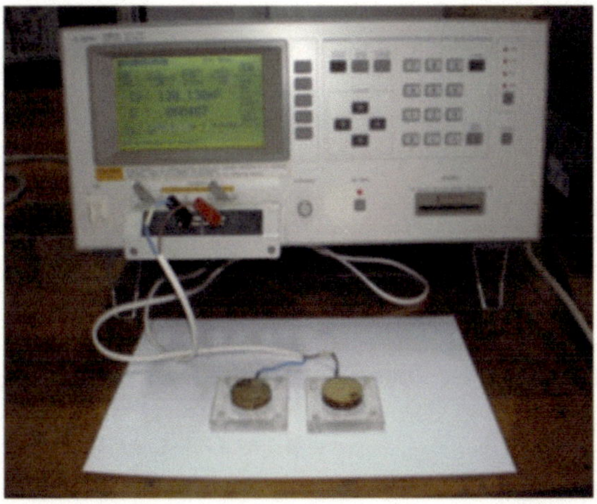

Fig. 5. The LCR-meter model HP 4284A with the sample set up for conductivity measurement. Faculty of Science, department of Physics - UPM

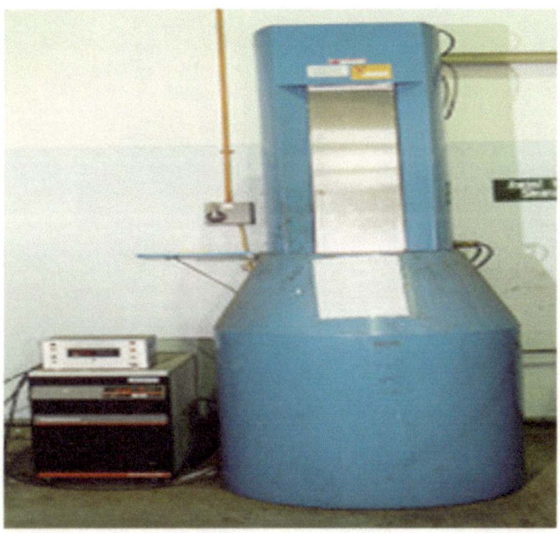

Fig. 6. γ-irradiation system model (J. L. Sherperd) at the Malaysian Nuclear Agency, Bangi - Malaysia UPM

## 10. Results and discussion

Figure 7 shows the prepared PVA solution (a), AniHCl\PVA solution (b). And AniHCl\PVA solution irradiated with 50 kGy γ-radiation doses (c). It shows that the PVA is a soluble in water appears as clear glycerin like material and after the dissolving of AniHCl it shows the oily color and after an irradiation with 50 kGy the color turned to dark green solution, which is the color of polyaniline PANI. While Fig. 7-d shows the pur pallets of PANI-HCl and the formed films of PVA\ PANI-HCl (e). These obtained materials have been subjected for further characterization.

Upon irradiation, the PVA/AniHCl blend films with doses up to 50 kGy, γ-rays interacts with the PVA binder liberating electrons by photoelectric effect and Compton scattering and followed by ions of $H^+$ and $OH^-$ from the bond scission. However, the contribution of these ions to the final product is not very significant. On the other hand, the interaction of γ-rays with the AniHCl is dominant due to the fact that HCl is easily dissociated to $H^+$ and $Cl^-$ ions by radiation. The protonation of aniline monomer by $Cl^-$ produced conducting PANI nanoparticles which can be visualized by the change of color of the un-irradiated PVA/AniHCl blend film from colorless to dark green at 50 kGy, as illustrated by the photograph pictures in Figure 5.14. As mention earlier, the formation of C=N double bonds of imines group produced green colour of PANI and the intensity increases with increasing of dose (*see Raman spectrum*). Before irradiation, all PVA/AniHCl blend films were colourless even exposed in air, suggesting the UV-visible radiation has no influence in the formation of conducting PANI. Only after irradiation with the dose between 20 kGy and 50 kGy, the green colour became intense.

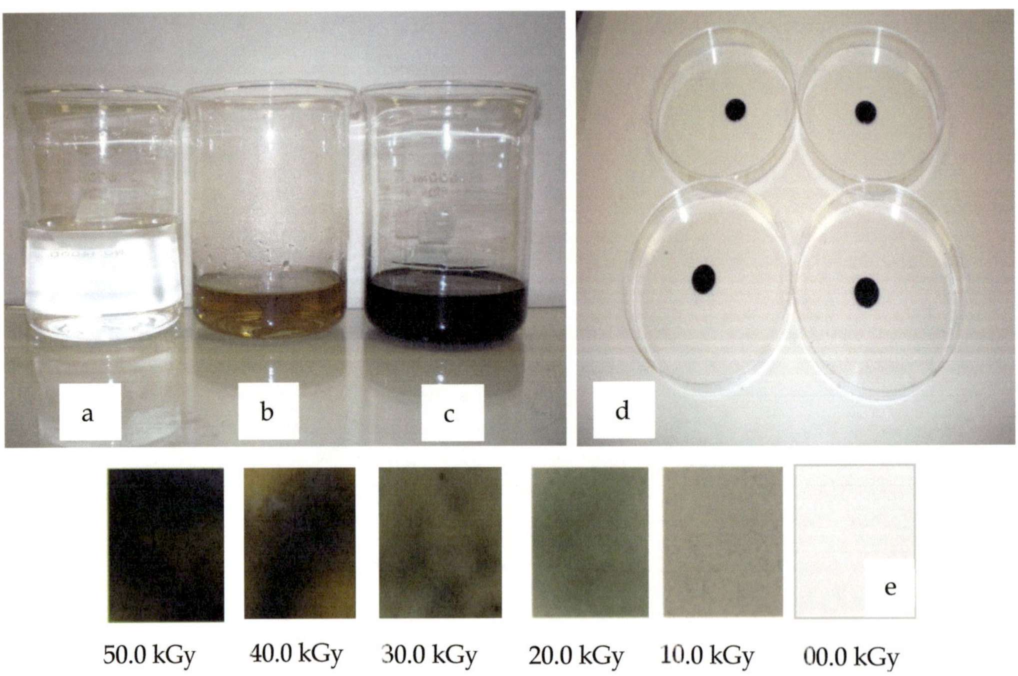


Fig. 7. Shows the prepared PVA solution (a), AniHCl\PVA solution (b), AniHCl\PVA solution irradiated with 50 kGy γ-radiation dose (c), the PANI-HCl pallets (d) and the formed films of PANI-HCl at different radiation doses.

Figure 8 shows the UV-visible absorption spectra of an irradiated AniHCl\PVA composite at different radiation doses at 0, 10, 20, 30, 40, and 50 kGy and for a concentration of 28.6

wt% AniHCl formed as films. The optical absorption spectra of the irradiated films were measured by using UV-Visible double beam spectrophotometer with air as a reference. The optical absorption is a useful tool to study electronic transitions in molecules, which can provide information on band structure and band gap energy. The basic principle is that photons from UV-visible light source with energies greater than the band gap energy will be absorbed by the materials under study. The absorption is associated with the electronic transitions from highly occupied molecular orbital (HOMO) π-band to lowly unoccupied molecular orbital (LUMO) π*-band of electronic states (Arshak and Korostynska, 2002). The electronic transitions between the valence band (VB) and the conduction band (CV) start at the absorption edge, which corresponds to the minimum energy of band gap $E_g$ between the lowest minimum of the CB and the highest maximum of the VB.

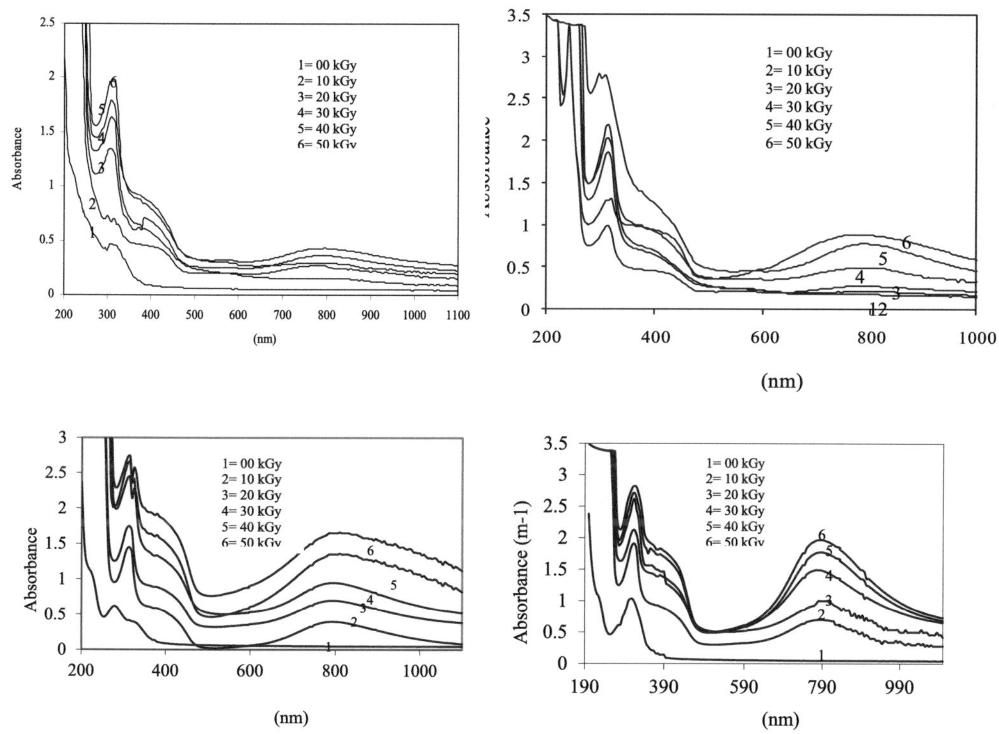


Fig. 8. Shows the UV-visible absorption spectra of PANI nanoparticles dispersed in PVA matrix for AniHCl monomer concentrations of (a) 9, (b) 16.7, (c) 23, and (d) 28.6 wt%.

The spectra of irradiated films reveal two prominent absorption peaks at 315 and 790 nm assigned to the electronic transitions of chlorine $Cl^-$ and C=N bond respectively. The absorbance corresponds to the excitation of outer electrons through π-π* electronic transitions at the bands of 315 nm (3.95 eV) and 790 nm (1.57 eV). The absorbance increases with the increase of dose and AniHCl concentration and both peaks become sharper with dose increase, indicating the amount of $Cl^-$ and polarons formed (represented by C=N) have increased with dose increase. Both peaks shifted slightly to higher wavelengths with the increase of dose but were not very significant.

The absorbance at 790 nm is due to the creation of C=N double bond of imines group representing the polarons in conducting PANI that gives the green colour. This result is in agreement with previous study carried out by Rao, *et al.* (2000), in which the absorption band for the chemically prepared conducting PANI salt peaking in the range of 420 - 830 nm depending on the degree of oxidation. Earlier Malmonge and Mattoso (1997) found that the absorption band of chemically synthesized PANI was 630 nm and when exposed to X-rays, the peak became sharper and shifted to 850 nm leading to an increase of the conductivity. Recent study by Cho *et al.* (2004) showed that the absorption bands were peaking at 740 - 800 nm for PANI chemically prepared by hydrochloric acid doping and dispersed in PVA matrix.

The unirradiated PVA/AniHCl film showed a broad peak at 315 nm because of the presence of $Cl^-$ in AniHCl monomer and no other peak is visible in UV region. The peak increases in intensity at higher concentration of AniHCl monomer. As the dose increases the absorbance at 315 nm increases due to increased formation of chlorine $Cl^-$ ions from the dissociation of HCl. Solid phase of HCl was present as the residual of radiation doping of imines group which can be seen from SEM micrographs in Figure 5.4. De Albuqerque, *et al.* (2004) measured UV-Visible spectra of emeraldine salt solution and found two absorption peaks at 320 nm and 634 nm. The presence of the absorption peak at 315 nm has been reported by Azian (2006) for irradiated PVA/AniHCl composites below 20 kGy and was confirmed by the UV-Visible spectroscopy measurements on HCl solution.

## 11. Quantitative analysis formation of PANI composites

Figure 9 shows the absorbance at 790 nm band for conducting PANI composites that increases exponentially with dose and can be fitted to the theoretical relationship of the form:

$$y = y_0 \exp(D / D_0) \tag{3}$$

where $y$ is the absorbance at dose $D$, $y_0$ is the absorbance at zero doses and $D_0$ is the dose sensitivity parameter.

The exponential increase of absorbance of PANI nanoparticles turns out to be of similar trend with the exponential increase of C=N formation determined from the Raman scattering measurement. This indicates the same phenomenon measured by two different methods produces almost similar result. Thus, the quantitative analysis of polarons could be extracted by either the Raman scattering or the optical absorbance method. The values of $D_0$ from the absorbance of PANI composites at different AniHCl concentrations were determined from the inverse of the gradient ln $y$ vs. dose, as shown in Figure 9 and plotted for different AniHCl concentrations. The result shows a decrease in the $D_0$ value with the increase of AniHCl concentration. Thus, the PVA/PANI composites became more radiosensitive at higher AniHCl concentration as shown in Figure 10. The linear relationship between $D_0$ and AniHCl concentration $C$ is $D_0 = -0.29C + 23.7$.

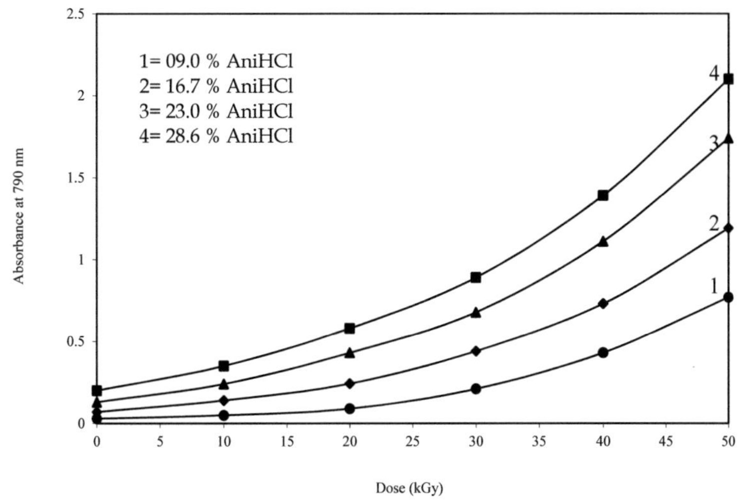


Fig. 9. Shows the exponentially increment of absorbance at 790 nm due to the formation of PANI

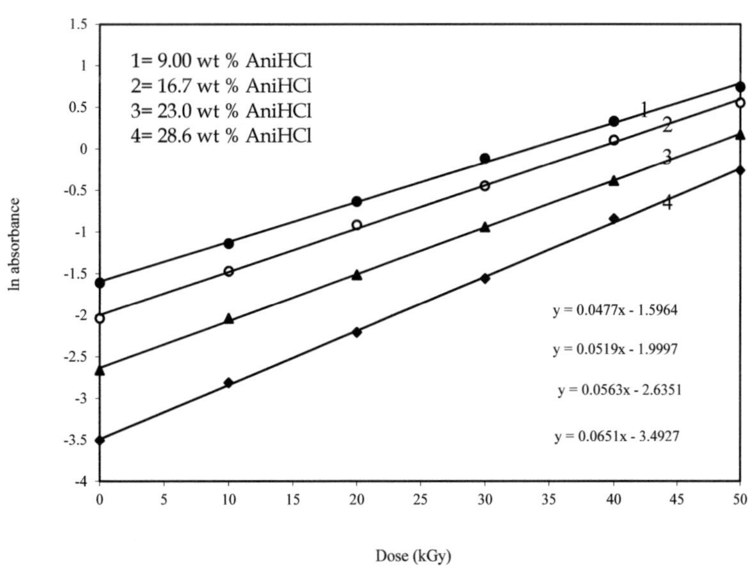


Fig. 10. Shows the ln (log$_e$) absorbance (ln $y$) vs. dose at different AniHCl concentrations and the gradient is used to determine the dose sensitivity $D_0$.

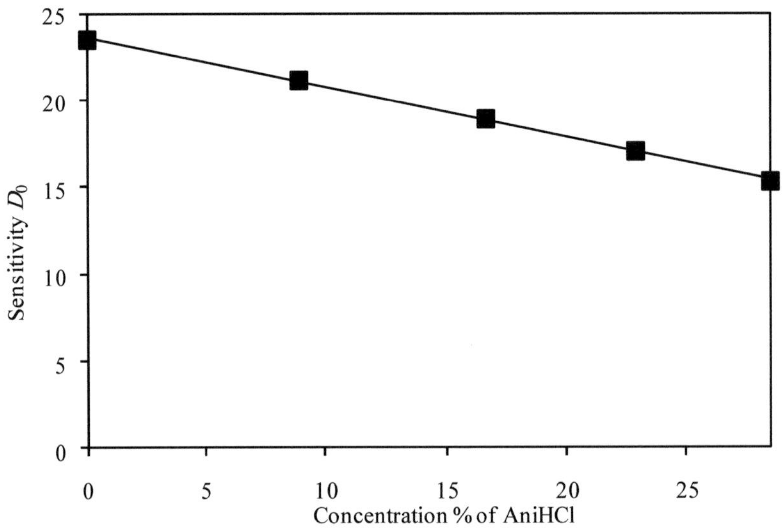


Fig. 11. Shows the deduction of Dose sensitivity $D_0$ of PANI nanoparticles versus AniHC concentration

## 12. Quantitative analysis of HCl formation

Figure 12 shows the absorbance at 315 nm band due to the formation of HCl versus radiation dose. The absorbance increases exponentially following the radiation dose increment and leading to saturation at doses higher than 50 kGy, indicating chlorine ions $Cl^-$ were being consumed for the formation of conducting PANI composites. The relation between the absorbance of $Cl^-$ and dose could be fitted to the relation of the form:

$$y = A_0(1 - \exp(-D/D_0)) \tag{4}$$

where $y$ is the absorbance at the applied dose $D$ for each concentration, $A_0$ is the difference between the absorbance at 50 kGy and 0 Gy for each concentration. The values of $D_0$ for the formation of crystalline HCl at different AniHCl concentrations can be determined from the inverse of the gradient $\ln\left(1-\frac{y}{A_0}\right)$ versus dose, as shown in Figure 13. The values of $D_0$ at different AniHCl concentrations are shown in Figure 14 which can be written in the form $D_0 = 15.75\ C + 14.456$.

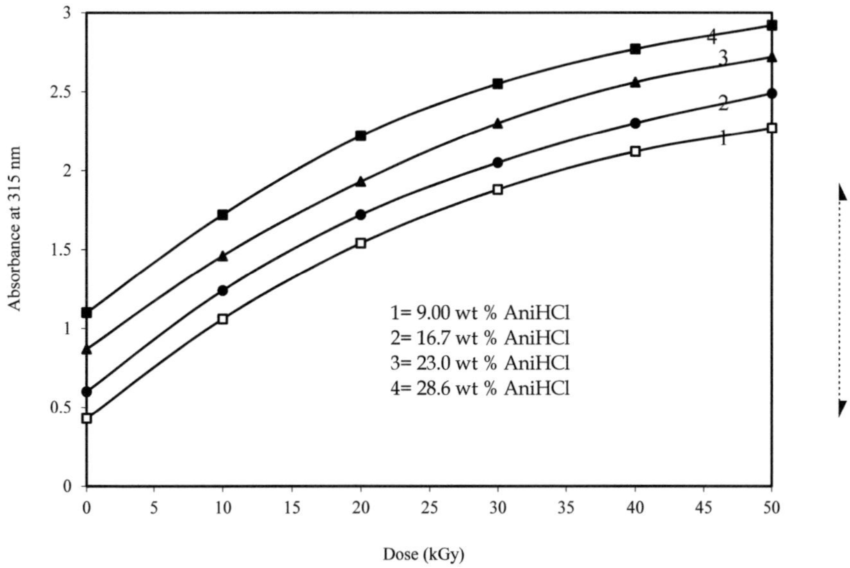


Fig. 12. Shows the absorbance at 315 nm for the consumption of Cl- in composite PVA/PANI nanoparticles vs. radiation dose.

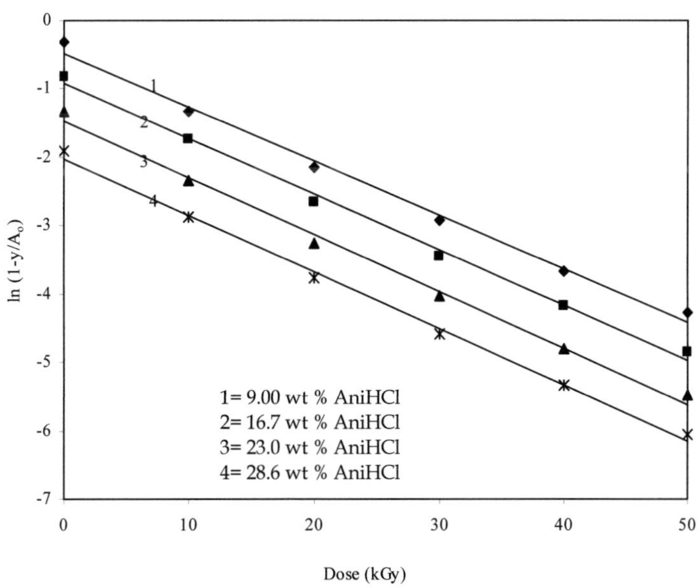


Fig. 13. Shows the $\ln\left(1-\frac{y}{A_0}\right)$ versus dose for consumption of Cl- at different AniHCl concentrations to deduce the dose sensitivity $D_0$.

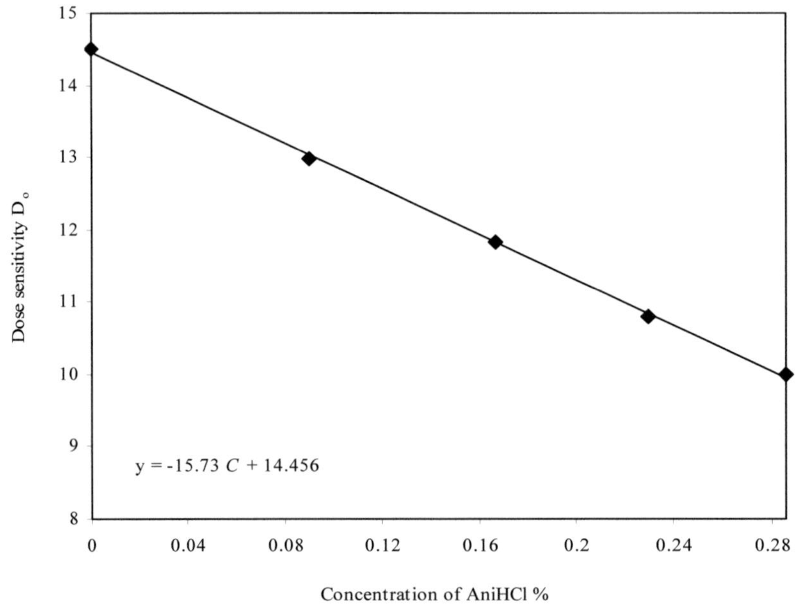


Fig. 14. Shows the dose sensitivity $D_0$ of composite of PVA/PANI nanoparticles versus monomer concentration for consumption of $Cl^-$

## 13. Band gap of PANI nanoparticles

At high absorption level, $\alpha > 10^4$ $cm^{-1}$, the absorption coefficient $\alpha(\nu)h\nu$ is related to the band gap $E_g$ according to the Mott and Davis (1979) using the following relation:

$$\alpha(\nu)h\nu = B(h\nu - E_g)^m \tag{5}$$

where, $h\nu$ is the energy of the incidence photon, $h$ is the Planck constant, $E_g$ is the optical band gap energy, $B$ is a constant known as the disorder parameter which is dependent on composition and independent to photon energy. Parameter $m$ is the power coefficient with the value that is determined by the type of possible electronic transitions, i.e.1/2, 3/2, 2 or 1/3 for direct allowed, direct forbidden, indirect allowed and indirect forbidden respectively. The band gap denotes the energy between the valence bands (VB) and the conduction band (CB). The direct allowed band gap at different doses were evaluated from the plot of $(a(v)hv)^2$ vs. $hv$. By extrapolation a straight line of $a(v)hv)^2$ versus $hv$ curves for $(a(v)hv)^2 = 0$, the band gap can be determined as shown in Figure 15. The results showed that band gap $E_g$ value decreases with the increase of the radiation dose shown in Figure 16. The decrease in the band gap energy with increasing dose is attributed to more conducting PANI nanoparticles formed and as more polarons in the irradiated composite reduce the band gap between VB and CB for the $\pi - \pi^*$ electronic transition. We found that when the doses were increased from 10 to 50 kGy the band gap decreases from 1.36 to 1.18 eV for 9 wt %, from 1.28 to 1.09 eV for 16.7 wt %, from 1.21 to 1.04 eV for 23 wt % and from 1.12 to 1.00 eV for 28.6 wt %.

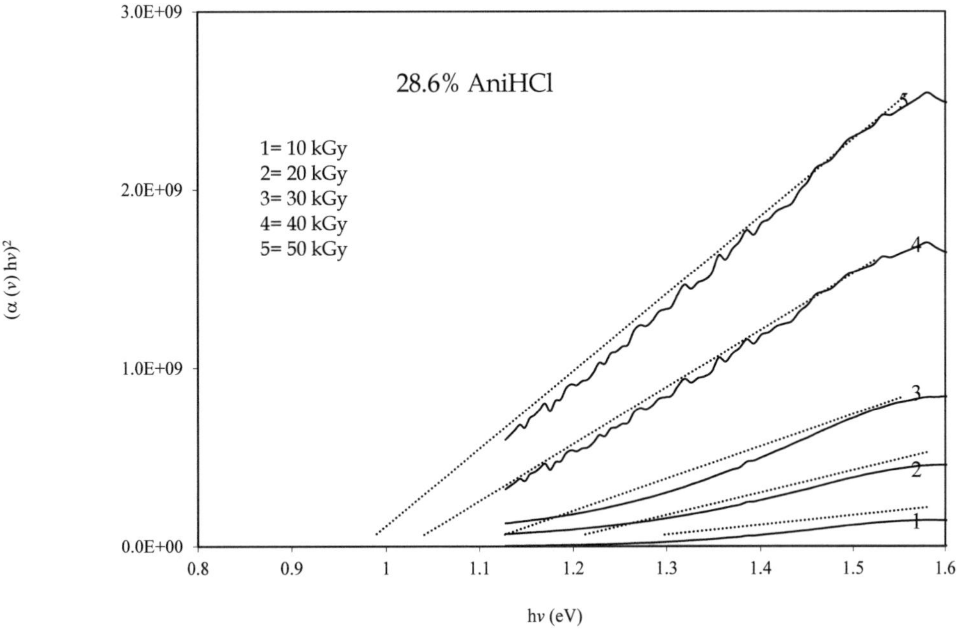


Fig. 15. Shows Variation of direct allowed energy gap for AniHCl monomer concentrations of 28.6 wt% at different doses (example for plot of $(a(v)hv)^2$ vs. $hv$. By extrapolation a straight line of $a(v)hv)^2$ versus $hv$ curves for $(a(v)hv)^2 = 0$)

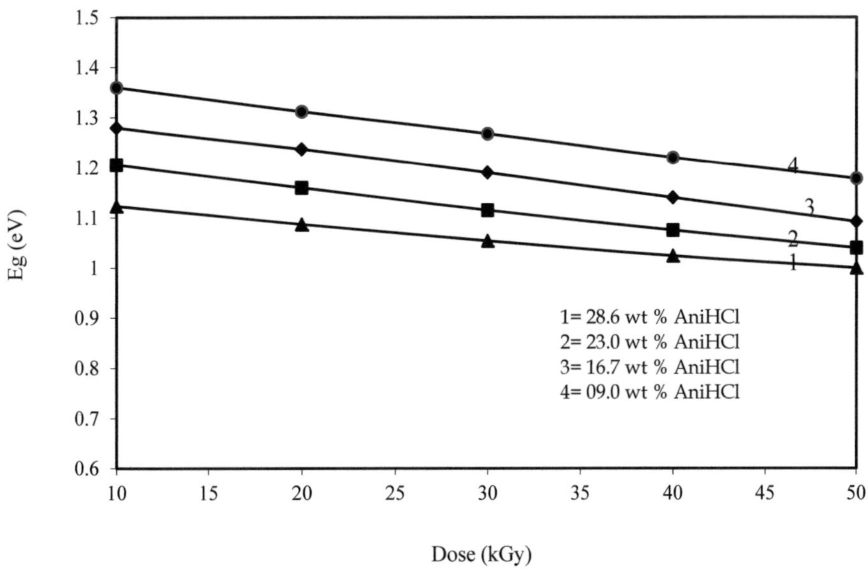


Fig. 16. Shows the band gap energy $E_g$ vsersus dose for PANI composites at different monomer concentrations

## 14. Electrical conductivity of composite of PVA/PANI nanoparticles

Polymers are commonly insulators as they have no significant mobile charges to serve the electrical conductivity. One of the requirements for polymers to exhibit good conductivity is the existence of π-electrons, which overlaps along the conjugated chain to form π-conjugated band. The conductivity of conjugated polymers or pure polymers can be increased after suitable oxidization or reduction process (Kanazawa *et al.*, 1979; Blythe, 1979) by doping or blending with charge donors of several organic groups (El-Sayed *et al.*, 2003) like hydroxyl, amine, carboxylate, sulfonate, and quaternary ammonium (Blanco *et al.*, 2001) or by radiation induced doping (Park *et al.*, 2002). In this work, the PVA was first blended with the organic monomer, AniHCl and then followed by $\gamma$ irradiation to oxidize the monomer into the conducting PANI.

The conductivity of polymer composites, generally consist of free or weakly bound electronic and ionic charges and trapped ionic charges in the polymer matrix. The free charges are free to move in electrical field, independent of frequency and contribute to the direct current (dc) conductivity. While charge carriers that are trapped in the polymer matrix require alternating electric field at certain frequency to liberate the ions from one site to another site in succession by hopping mechanism and contribute to the alternating current (ac) conductivity. Realizing this, the electrical conductivity of un-irradiated and irradiated PVA will be measured and discussed first. This allows us to determine the conductivity values and identify the type of charge carriers in the un-irradiated and irradiated PVA before blending the PVA with AniHCl monomer at various concentrations and undergo $\gamma$irradiation.

## 15. Conductivity of PVA/AniHCl composite at various concentrations

Figure 17 shows the conductivity measured at different frequencies from 20 Hz to 1 MHz for the PVA/AniHCl composites with different concentrations of AniHCl. At low AniHCl concentrations both the dc and ac conductivity are clearly seen. The dc conductivity is frequency-independent served by weakly bound electrons, $H^+$, and $Cl^-$ and those of phonon assisted tunneling process that gain charge mobility at room temperature. The $H^+$ and $Cl^-$ ions were derived from dissociation of HCl which is weakly attached to the phenyl group of aniline monomer. The ac conductivity at high frequencies is due to trapped $H^+$ and $Cl^-$ ions in PVA matrix that required alternating electric field at given frequency and contributes to the conductivity by hopping between the localized sites.

The conductivity increases with the increase of AniHCl concentration until 28.6 wt% before the conductivity drops to the lower values at higher concentrations of 33.0 and 38.0 wt%. The conductivity of higher concentrations is mainly the dc conductivity contributed from weekly bound $H^+$ and $Cl^-$ ions.

Figure 18 shows the dc conductivity component at various AniHCl monomer concentrations. The conductivity increases significantly from $6.61\times10^{-8}$ S/m at 0 wt% to $1.04\times10^{-4}$ S/m at 28.6 wt% and there is subsequently dropped in conductivity at 33.0 wt% and 38.0 wt%. A decrease in conductivity may be due to the increase of crystallinity in the polymer matrix as more crystalline chlorine are present within the composite films. It may also be due to high viscosity and caused resistance or impedance to oppose ion mobility in

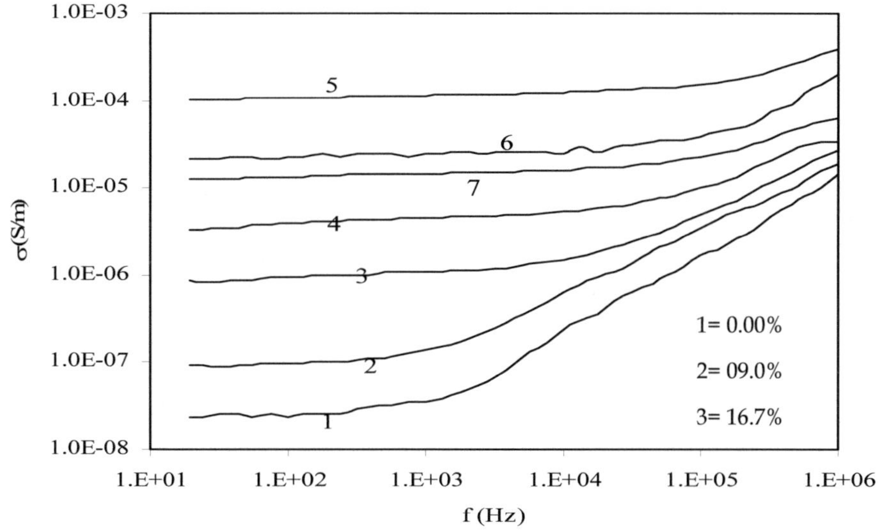


Fig. 17. Conductivity of the PVA/AniHCl composites versus frequency at different concentrations of AniHCl.

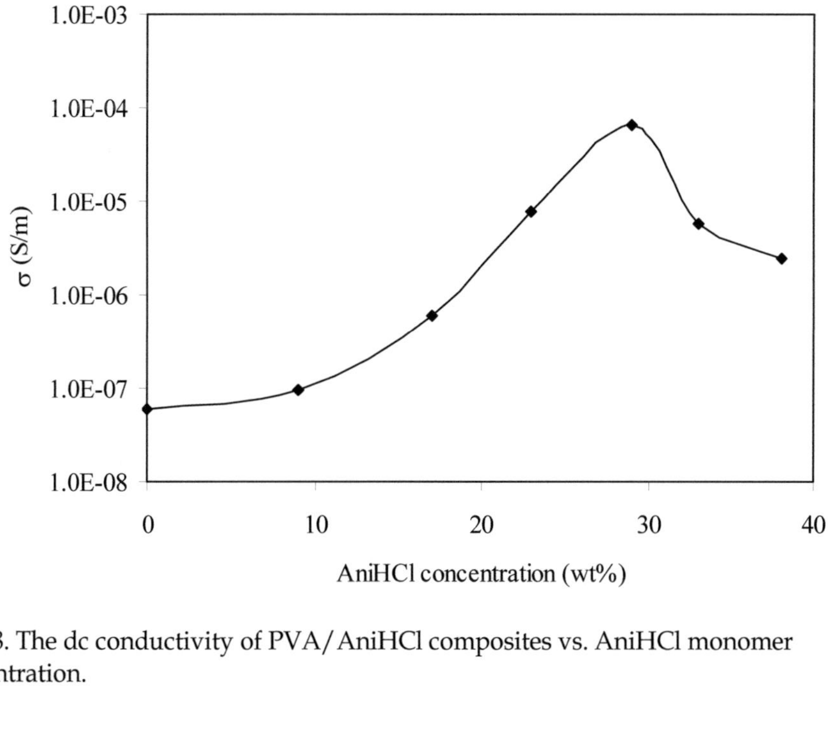


Fig. 18. The dc conductivity of PVA/AniHCl composites vs. AniHCl monomer concentration.

the composites (Guo *et al.*, 2004; Bidstrup, 1995). It has been shown that the ionic mobility is inversely proportional to viscosity as in the theoretical relation given by Richard (2002). For this reason, the conductivity of PVA/AniHCl composites decreased in values at monomer concentrations of 33.0 and 38.0 wt%. Subsequently, the analysis of these samples was discarded from the further discussion.

## 16. Conductivity of PANI composites at various doses

Figure 19 shows the conductivity of PANI composites dispersed in PVA matrix polymerized at doses up to 50 kGy for various AniHCl concentrations from 9.0 to 28.6 wt%. The results show that the conductivity increases with the increase of dose and monomer concentration. As the dose and AniHCl concentration increased more polarons were formed and thus, increase the conductivity of conducting PANI composites. Moreover, as the dose increased the band gap of conducting PANI decreases to about 1.0 eV for 28.6 wt% AniHCl concentration and radiation dose at 50 kGy. This is closed to the silicon semiconductor band gap of about 0.8 eV. The conductivity comprises of the dc and ac components given equation (6).

$$\sigma(\omega) = \sigma_{dc}(0) + \sigma_{ac}(\omega) \tag{6}$$

At low doses below 10 kGy, the composites behave like insulators, where the dc and ac components are due to weakly bound and trapped $H^+$ and $Cl^-$ ions in PVA/AniHCl matrix respectively. The ac conductivity at higher doses follows the universal power law of the form $\sigma_{ac}(\omega) = A\omega^s$ (Johnscher, 1976). Since the ac component is limited to the lower concentrations of AniHCl and at lower doses as shown in Figure 18, we suspected that the conductivity is not related to polarons in this situation. The ac component occurs at higher frequency region and becomes less important at higher doses. This indicates that at higher doses the conductivity is dominated entirely by the dc conductivity due to polarons. Therefore, detail analysis of the ac conductivity will not be discussed further. The species of polarons are considered the main criteria of conducting polyemeraldine salt that results in a remarkable shift of the dc conductivity to higher values with increasing dose and monomer concentration up to 28.6 wt%. Detail analysis of the dc conductivity is given in the following subsection.

## 17. The dc conductivity of PANI composites determined from direct extrapolation method

The dc conductivity $\sigma_{dc}(0)$ of conducting PANI composites was deduced from direct extrapolation of dc portion Figures 19 and from calculation using the resistance $Z_0$ obtained from the Cole-Cole plots. Figure 20 shows the dc conductivity $\sigma_{dc}(0)$ of PANI composites deduced by the direct extrapolation. We found that the dc component for 9 wt% AniHCl monomer increases from $6.31 \times 10^{-7}$ S/m at 0 kGy to $1.12 \times 10^{-3}$ S/m at 50 kGy, while for 16.7 wt % monomer, the dc conductivity increases from $3.63 \times 10^{-6}$ S/m at 0 kGy up to $5.75 \times 10^{-3}$ S/m at 50 kGy. As for 23 wt % monomer the conductivity increases from $4.02 \times 10^{-6}$ S/m at 0 kGy to $2.40 \times 10^{-2}$ S/m at 50 kGy. The highest conductivity measured

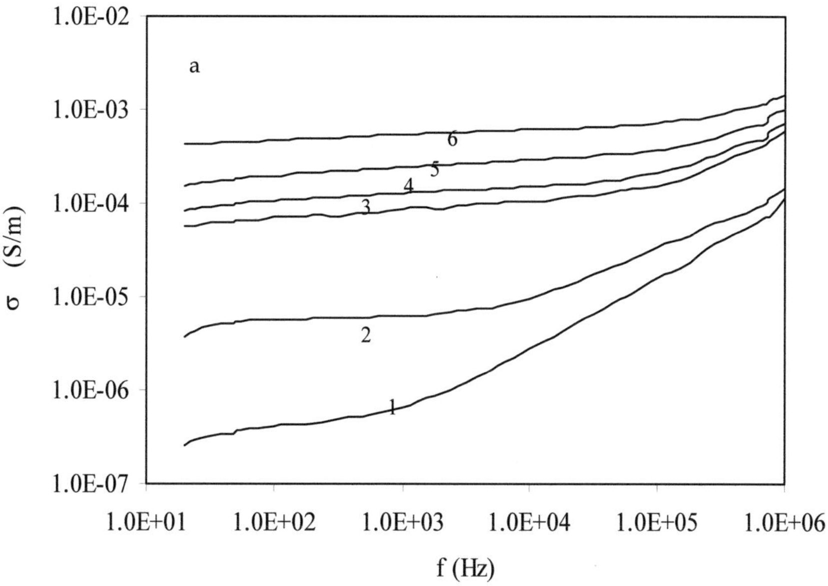
a
1.0E-02
1.0E-03
1.0E-04
1.0E-05
1.0E-06
1.0E-07
σ (S/m)
1.0E+01
1.0E+02
1.0E+03
1.0E+04
1.0E+05
1.0E+06
f (Hz)
1
2
3
4
5
6

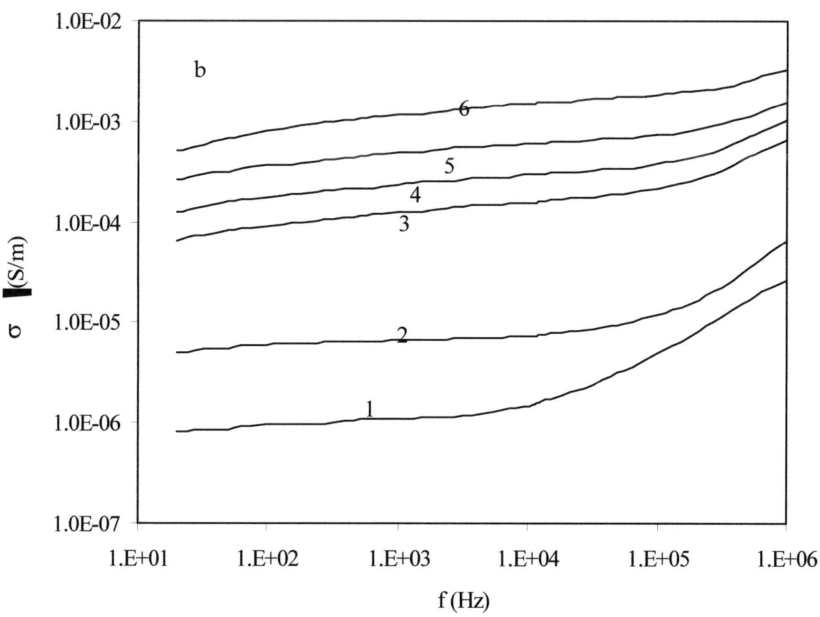
b
1.0E-02
1.0E-03
1.0E-04
1.0E-05
1.0E-06
1.0E-07
σ (S/m)
1.E+01
1.E+02
1.E+03
1.E+04
1.E+05
1.E+06
f (Hz)
1
2
3
4
5
6

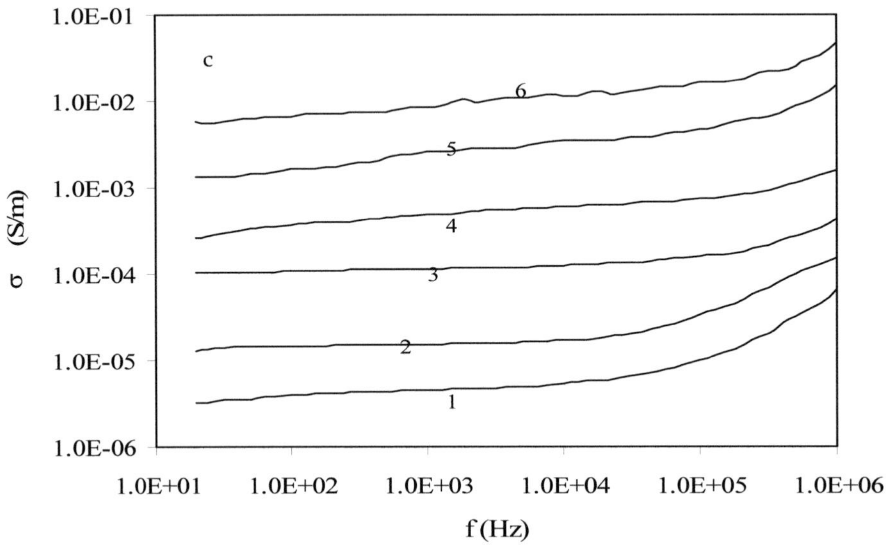


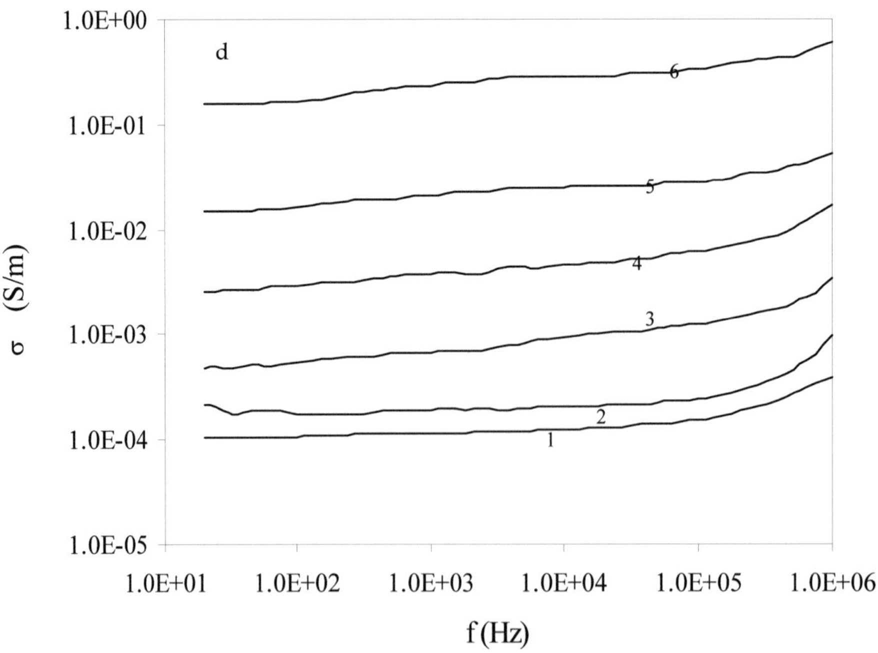


Fig. 19. Shows the conductivity vs. frequency of PVA/PANI nanocomposites irradiated up to 50 kGy for various monomer concentrations (a) 9.0, (b) 16.7, (c) 23.0, and (d) 28.6 wt%.

was for 28.6 wt % monomer at which the dc conductivity increases from $1.04 \times 10^{-4}$ S/m at 0 kGy up to $1.17 \times 10^{-1}$ S/m at 50 kGy. The obtained values were compared with previously published data for chemical and electrochemical doping methods. MacDiarmid *et al.* (1987) have successfully prepared conducting PANI by HCl doping and obtained a conductivity of 1.0 S/cm or $1.0 \times 10^{+2}$ S/m. Recently Blinova *et al.*, (2006) have successfully measured the conductivity of 15.5 S/cm or $1.55 \times 10^{+3}$ S/m for PANI prepared by chemical doping with 1 M phosphoric acid. The PVA/PANI-HCl composites of polyaniline were prepared and the maximum conductivity achieved was $2.0 \times 10^{-3}$ S/m at 60 wt% PANI (Cho *et al.*, 2004). The difference in conductivity between PANI-Radiation doping and PANI-(chemical/electrochemical doping) is that radiation interaction occurs randomly i.e. not all AniHCl got polymerized and the effect of binder impedance.

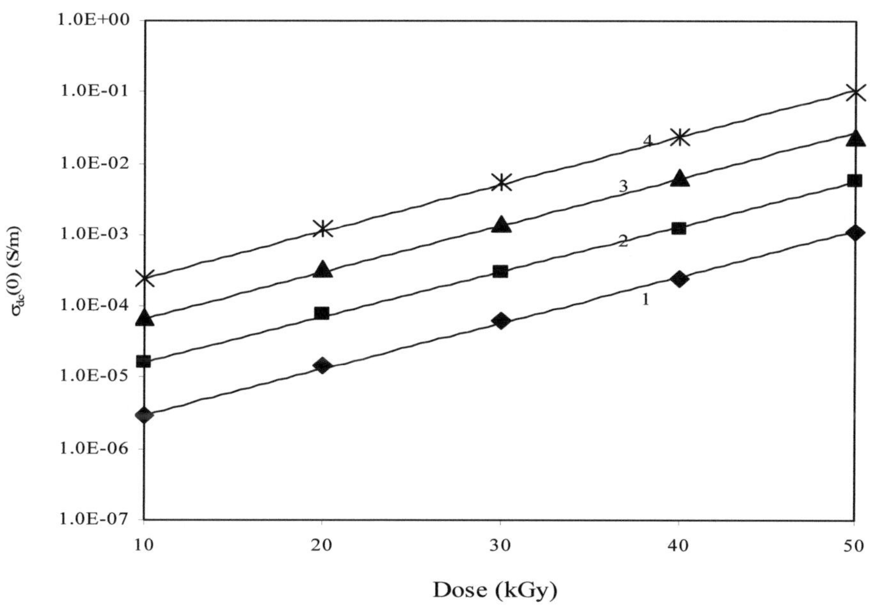


Fig. 20. Shows the dc conductivity $\sigma_{dc}(0)$ by extracted from extrapolation method for PANI composites in PVA matrix at different doses and monomer concentrations.

The dc conductivity of conducting PANI composites seems to begin at dose of 10 kGy. Referring to the absorption spectra Figure (2), the absorbance at 790 nm band for conduction PANI showed up at 10 kGy for all monomer concentrations, confirming that the formation of PANI begins at 10 kGy as measured by conductivity measurement. This minimum dose might be the threshold of radiation dose to start polymerizing the conducting PANI for all AniHCl concentrations. The general relationship between the dc conductivity and the dose is in the form: $\sigma_{dc}(0) = \sigma_0 \exp(D / D_0)$, $D_0$ is the dose sensitivity that can be deduce from the gradient linear slope of ln $\sigma_{dc}(0)$ versus dose.

## 18. The dc conductivity of PANI determined from the Cole-Cole plots

The dc conductivity, $\sigma_{dc}(0)$ can be calculated from the resistance $Z_0$ obtained from the Cole-Cole plots. Figure 21 shows the Cole-Cole plot curves for various AniHCl monomer concentrations that display similar semicircle characteristics, a typical impedance spectra of synthetic-metal or metallic-polymer film composites (Vorotyntsev *et al.*, 1999; Tarola, *et al.*, 1999). At low frequency region for certain dose and monomer concentration, there is a straight line spike due to interstitial effect of the electrodes. It has been reported by Mariappan and Govindaraj. (2002) that the depressed semicircle at the low frequency region is related to characteristics of parallel combination of the bulk resistance and capacitance phase element of the samples. While Chen *et al.* (2003) ascribed the presence of straight line at low frequency region due to the capacitive characteristics of conducting polymer film.

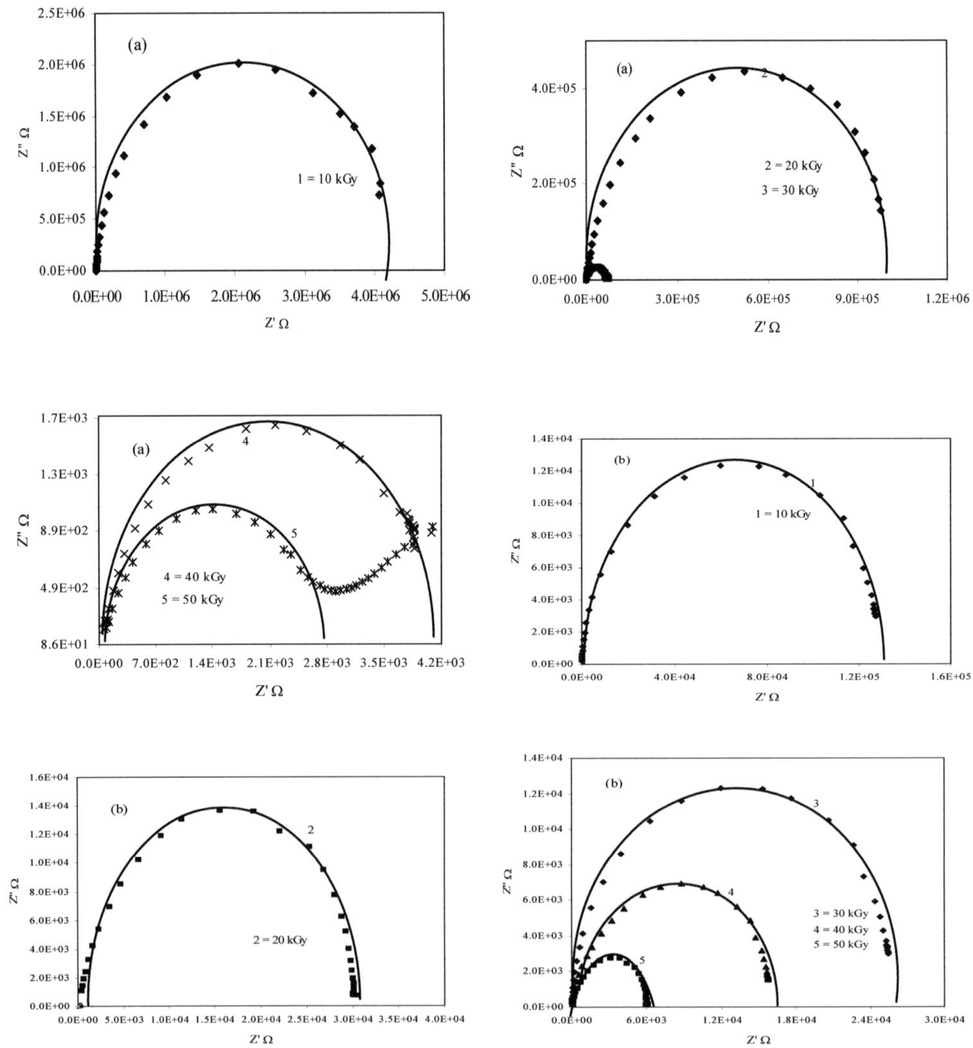

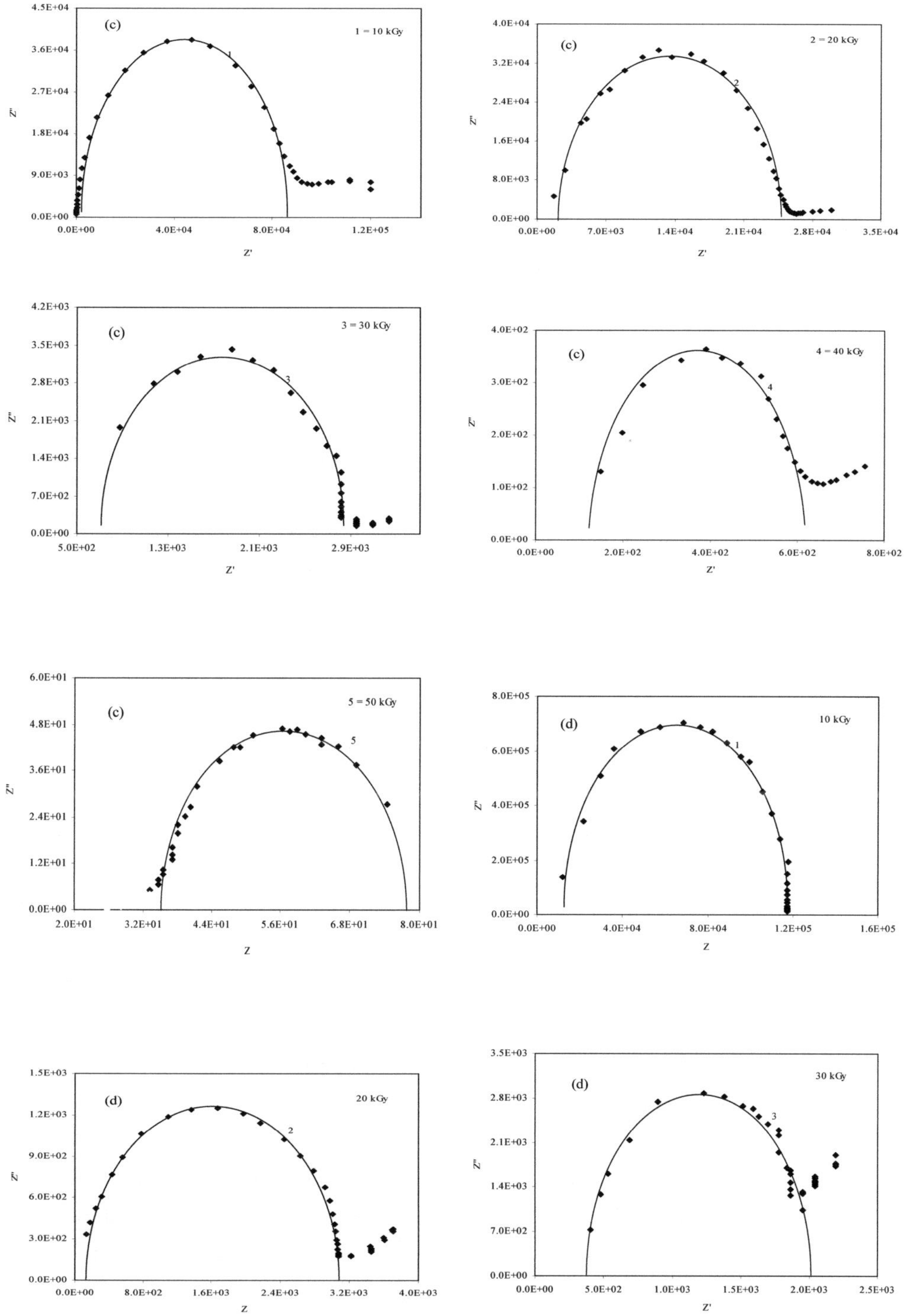
(c)
1 = 10 kGy
Z''
Z'
0.0E+00
9.0E+03
1.8E+04
2.7E+04
3.6E+04
4.5E+04
4.0E+04
8.0E+04
1.2E+05
(c)
2 = 20 kGy
8.0E+03
1.6E+04
2.4E+04
3.2E+04
4.0E+04
7.0E+03
1.4E+04
2.1E+04
2.8E+04
3.5E+04
(c)
3 = 30 kGy
7.0E+02
1.4E+03
2.1E+03
2.8E+03
3.5E+03
4.2E+03
5.0E+02
1.3E+03
2.9E+03
(c)
4 = 40 kGy
1.0E+02
2.0E+02
3.0E+02
4.0E+02
6.0E+02
8.0E+02
(c)
5 = 50 kGy
1.2E+01
2.4E+01
3.6E+01
4.8E+01
6.0E+01
2.0E+01
3.2E+01
4.4E+01
5.6E+01
6.8E+01
8.0E+01
Z
(d)
10 kGy
2.0E+05
4.0E+05
6.0E+05
8.0E+05
1.2E+05
1.6E+05
(d)
20 kGy
3.0E+02
6.0E+02
9.0E+02
1.2E+03
1.5E+03
8.0E+02
1.6E+03
2.4E+03
3.2E+03
4.0E+03
(d)
30 kGy
5.0E+02
1.0E+03
1.5E+03
2.0E+03
2.5E+03

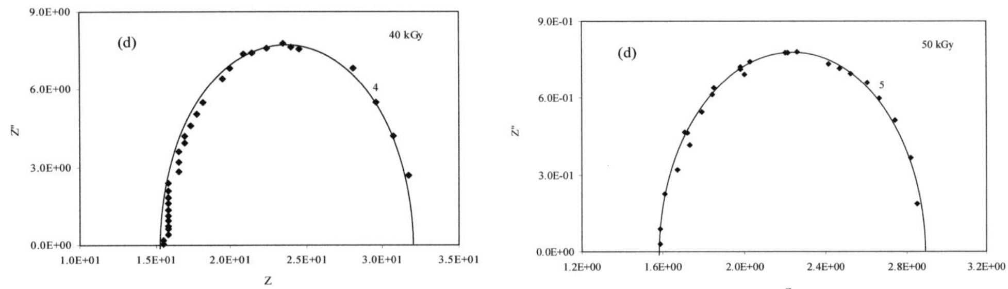


Fig. 21. Shows the Cole-Cole plots for PANI nanoparticles in PVA matrix at (a) 9 wt %, (b) 16.7 wt %, (c) 23 wt % and (d) 28.6 wt % of AniHCl monomer.

In such spectra the semicircles radius decreases with dose increment, indicating that the resistance $Z_0$ of the polymer composites decreases with dose, hence the dc conductivity $\sigma_{dc}(0)$ increases with dose (Kobayashi *et al.*, 2003). The increase of dc conductivity of the PANI composites is due to polaron species caused by γradiation beginning at 10 kGy. The inclined straight line appear at the end of the semicircles was due to electrode polarization or space effect (Hodge *et al.*, 1976 and Mariappan and Govindaraj., 2002), while Lewandowski *et al.* (2000) ascribed it to non secured verticality of electrode spikes as well as to capacitance interface between the electrode and the dielectric.

We found that the dc conductivity obtained from the Cole-Cole plots are quite typical with those deduced from the direct extrapolation method. The dc conductivity is $5.75\times10^{-6}$ S/m at 10 kGy and $1.32\times10^{-3}$ S/m at 50 kGy for 9.0 wt %. It is $1.0\times10^{-5}$ S/m at 10 kGy and $2.95\times10^{-3}$ S/m at 50 kGy for 16.7 wt %, while for 23.0 wt % it is $2.40\times10^{-5}$ S/m at 10 kGy and $1.26\times10^{-2}$ S/m at 50 kGy. For the concentration of 28.6 wt% it is $7.76\times10^{-5}$ S/m at 10 kGy and $1.17\times10^{-1}$ S/m at 50 kGy. The results are slightly different from the values determined by the extrapolating method. Previously Dutta, *et al.* (2001) measured ac and dc conductivity of chemically doped PVA/PANI blends and obtained the highest dc conductivity of $4.8\times10^{-2}$ S/m.

Figure 22 shows the dc conductivity $\sigma_{dc}(0)$ versus radiation dose of conducting PANI composites for different monomer concentrations. The relation between the radiation dose $D$ and the dc conductivity $\sigma_{dc}(0)$ can be fitted to the empirical exponential relation of the form $\sigma_{dc} = \sigma_0 \exp(D / D_0)$ where, $\sigma_0$ is the conductivity at zero doses, $D$ is the absorbed dose and $D_0$ is the dose sensitivity of the composites to radiation effect. In order to determine the dose sensitivity $D_o$ of the composites for irradiation, we followed 'Arrhenius type' plot of $\ln\sigma_{dc}$ versus dose, as the gradient of the linear regression plot gives $1 / D_0$, where $D_0$ is the dose sensitivity of the composites.

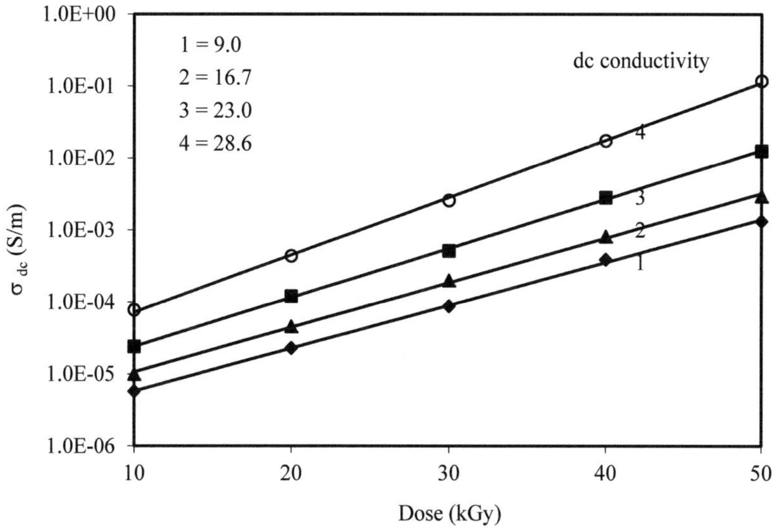


Fig. 22. Shows the dependence of dc conductivity ($\sigma_{dc}$) on the applied radiation dose theoretical method. The conductivity obeys the relation of the following form $\sigma_{dc} = \sigma_0 \exp(D / D_0)$

Figure 23 shows the variation of $\ln\sigma_{dc}(0)$ as a function of dose for different monomer concentration of PANI nanoparticles. The linear regressions of the "Arrhenius type plot" $\ln\sigma_{dc}(0)$ versus dose give the slope of $1/D_0$ from which the dose sensitivity value can be determined, as shown in Table 4. The study reveals that as the monomer concentration increases the dose sensitivity decreases.

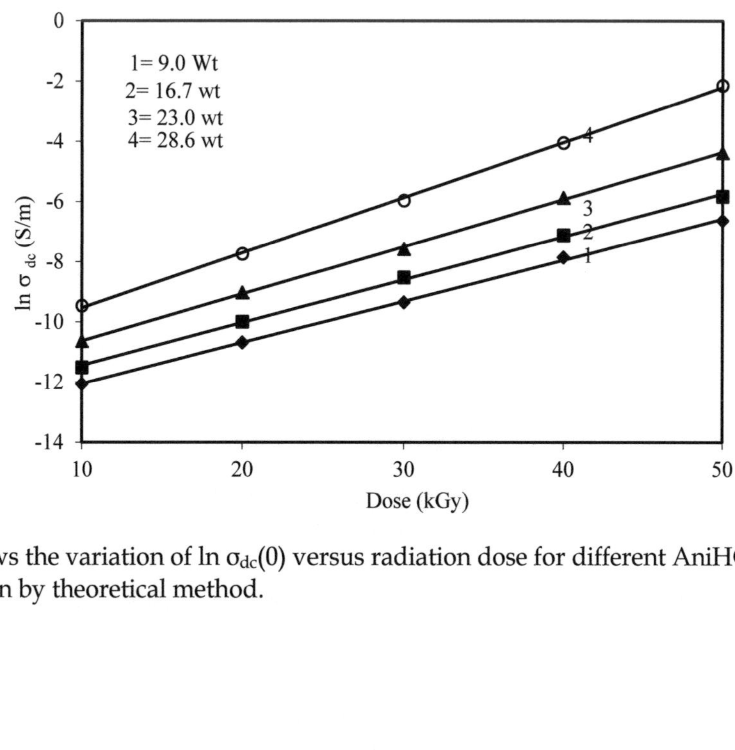


Fig. 23. Shows the variation of ln $\sigma_{dc}(0)$ versus radiation dose for different AniHCl concentration by theoretical method.

| AniHCl concentration (Wt %) | $D_0$ (the Cole-Cole method) (kGy) | $D_0$(extrapolating method) (kGy) |
|---|---|---|
| 9.0 | 7.3 | 7.3 |
| 16.7 | 7.0 | 6.8 |
| 23.0 | 6.7 | 6.1 |
| 28.6 | 5.6 | 5.7 |

Table 4. Shows the relation between monomer concentration and dose sensitivity $D_0$

Figure 24 shows the dose sensitivity versus AniHCl concentration which reveals a decrease in dose sensitivity with the increase of monomer concentration i.e. as the dose increases the composites become more radiosensitive to produce conducting PANI nanoparticles. The correlation between the dose sensitivity and the concentration of monomer is given by the formula: $D_0$ = -5.1$C$ + 7.8, where $C$ refers to the AniHCl concentration. The increasing of radiosensitivity by increasing the AniHCl concentration is attributed to higher density of the monomer to be irradiated, thus, producing more polarons in conducting PANI.

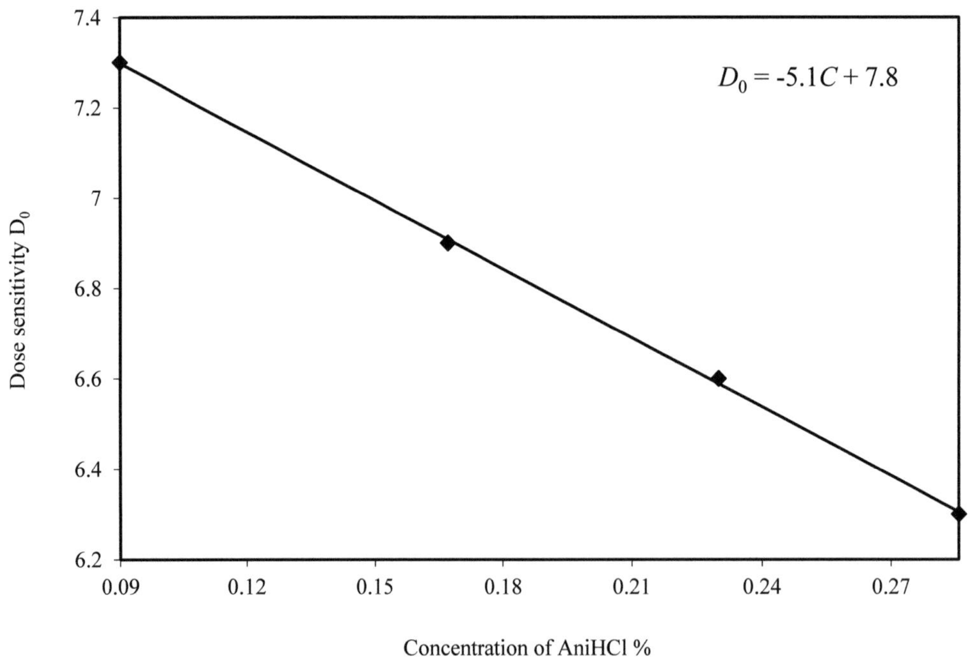


Fig. 24. Shows the variation of dose sensitivity $D_0$ versus the concentration of AniHCl within the PVA film by theoretical method.

## 18. Raman scattering analysis of PANI nanoparticles

The Raman scattering analysis was performed on the PVA/PANI nanocomposites before and after γ-irradiation up to 50 kGy and for all monomer concentrations. The significant of Raman spectroscopy study is that it can be used to investigate particular covalent bonds of some molecular species where the amount is expected to change after γ-irradiation. In this method, the vibrational transitions of particular molecular bonds could provide information on the chemical structure of the materials, which might be modified by ionizing radiation. Thus, Raman scattering (inelastic scattering) method is vital for the identification of substances by targeting at particular bonds which can become a chemical finger printing and provide quantitative information of the samples of interest (Barnes, 1998).

Figure 25 shows Raman spectra of 28.6%-AniHCl composites of PVA/PANI composites at different doses and reveal the prominent peak originated at Raman shift 1637 $cm^{-1}$ assigned to C=N bond stretching of imines group which gives the PANI color and represents the polaron species. In addition to the formed polarons of imines group, the Raman spectra also show Raman shift at 2100 $cm^{-1}$ and 2527 $cm^{-1}$ assigned for the π-bonds between double bond carbon C=C stretching within the aromatic ring and C=O stretching of aldehyde derivative from PVA bond scission respectively. Also shown is the weak intensity of Raman shift 3023 $cm^{-1}$ assigned to C-H bending.

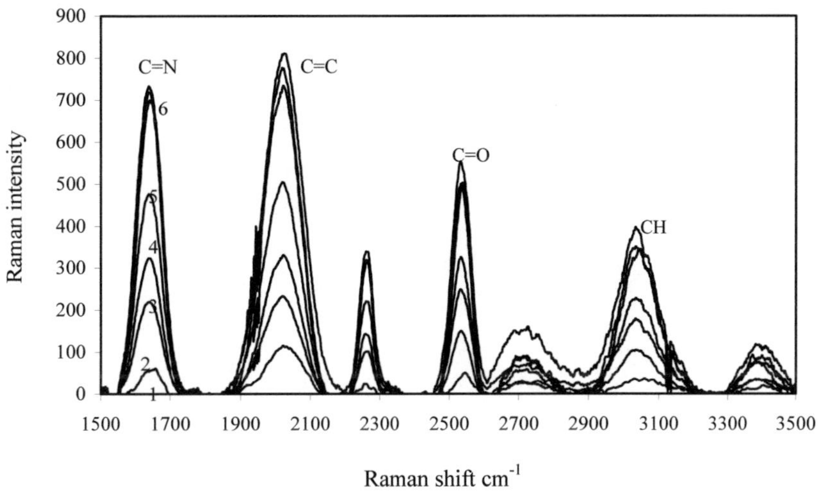


Fig. 25. Shows the Raman spectra showing Raman shifts of covalent bond species in the 28.6%-AniHCl nanocomposites of PVA/PANI nanoparticles induced by radiation doping at different doses

## 19. The SEM morphology of PANI nanoparticles

The morphology and particle size of the conducting PANI nanoparticles were studied by means of a scanning electron microscope (SEM). Figure 26 shows the SEM image of PANI nanoparticles polymerized 'in situ' by radiation doping with the dose of 50 kGy for AniHCl concentration of 28.6 wt%. The micrograph was taken at the electron operating voltage of 15 kV and 10,000 times magnification. It reveals the formation of conducting PANI nanostructures distributed almost uniformly and the diameter of spherical PANI nanoparticles was estimated to be in the range of 50 - 100 nm. The micrograph also reveals some fibrous clusters made up from aggregates of many PANI nanoparticles. The PANI cluster size is about 100 - 200 nm in diameter and 300 - 400 nm in length. There have been reported that the diameters of the PANI nanoparticles polymerized chemically with hydrochloric acid were about 100 to 150 nm for PVA/PANI nanocomposites (Cho *et al.*, 2004) and 40 nm for PVP/PANI nanocomposites (Dispenza *et al.*, 2006). This suggests that the type of binder determined the diameter of spherical nanoparticles.

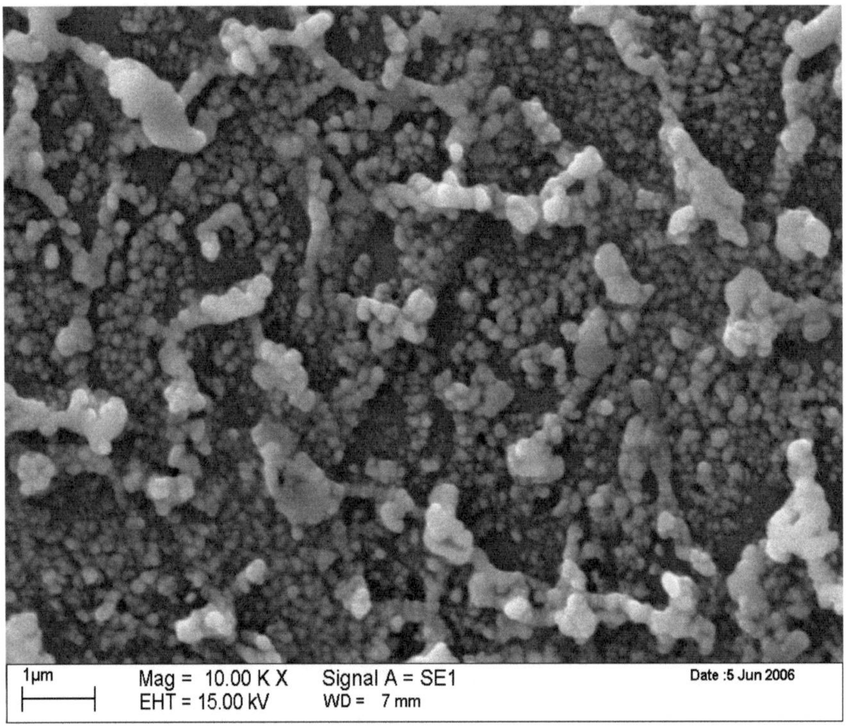


Fig. 26. Shows SEM image of PANI nanoparticles polymerized by 50-kGy Co-60 γ-rays for 28.6 wt% monomer.

The formed pallets of pure PANI-HCl (Fig. 1) were characterized with Voltmeter and LCR-meter. It is conductivity was obviously higher than that of PVA\PANI-HCl, which is ascribed to the presence of PVA within the composites, the conductivity was 1 S/m and it is UV-spectrum was peaked at 790 nm which is same as in PANI\PVA composites.

## 20. Conclusion

The ionizing γ-radiation could be used successfully to obtain the polymerization of monomers such as aniline hydrochloride AniHCl and the induced properties for the new product could be controlled by adjusting the amount of monomers and the applied radiation dose with consideration to technical aspects.

## 21. References

Albuquerque, J.E. de, L.H.C. Mattoso, R.M. Faria, J.G. Masters and A.G. MacDiarmid. 2004. Study of the interconversion of polyaniline oxidation states by optical absorption spectroscopy. *Synthetic Metals* 146, (1): 1-10.

Angelopoulos, J. M. Shaw, W. S. Huang and R. D. Kaplan. In-Situ Radiation Induced Doping. 1990. Molecular Crystal Liquid Crystal 189 No. 1: 221-225.

Arshak, A., S. Zleetni and S.K. Arshak. 2002. $\gamma$-irradiation sensor using optical and electrical properties of manganese phthalocyanine (MnPc). Thick Film Sensor 2: 174-184.

Azian, Osman. 2006. Ionizing Radiation Effects on Poly(Vinyl Alcohol)/(Aniline Hydrochloride blend Films. M.Sc. dissertation, Physics Department, UPM, Malaysia.

Barnes, A., A. Despotakis, P.V. Wright, T.C.P. Wong, B. Chambers, A.P. Anderson. 1998. Electrochim. Acta 43: 1629-1635.

Bidstrup, S., Simpson J. 1995. Light Scattering Study of Vitrification during the Polymerization of Model Epoxy Resins. Journal of Polymer Physics Ed. 33-43.

Blanco, J. F, Q. T. Nguyen, P. Schaetzel. 2001. Novel hydrophilic membrane materials: sulfonated polyethersulfone Cardo. Journal of Membrane Science186: 267-279.

Blinova, Natalia V., Jaroslav Stejskal, Miroslava Trchová and Jan Prokeš. 2006. Polyaniline prepared in solutions of phosphoric acid: Powders, thin films, and colloidal dispersions, Polymer 47 (1): 42-48.

Blythe, A.R. 1979. In Electrical Properties of Polymers. Cambridge university Press P. 90.

Bodugoz, H. Sevil, U. A. Güven O. 1998. Radiation induced conductance in blends of poly(aniline-base) with poly(vinyl chlrodie)-co(vinyl acetate). Macromolecule Symposium 169: 289-295.

Chen, W.C., T.C. Wen, H.S. Teng. 2003. Polyaniline-deposited porous carbon electrode for supercapacitor. Electrochim. Acta 48: 641-649.

Cho, Min, Seong, S.Y. Park, J.Y. Hwang, H.J. Choi. 2004. Synthesis and electrical properties of polymer composites with polyaniline nanoparticles. Material Science Engineering C 24: 15-18.

Denaro, A. R. and Jayson, G. G. 1972. Fundamental of radiation chemistry. Butterworths, London.

Dispenza, C., M. Leone, C.Lo. Presti, F. Librizzi, G. Spadaro, V. Vetri. 2006. Optical properties of biocompatible polyaniline nano-composites. Journal of Non-Crystalline Solids 352: 3835–3840.

Dutta, P., Biswas S., Ghosh M., De S. K., Chatterjee S. 2001. The dc and ac conductivity of the polyaniline-polyvinyl alcohol blends. Synthetic Metals 122: 455-461.

El-Sayed, S., M., Arnaoutry M. B., Fayek S. A. 2003. Effect of grafting, gamma irradiation and light exposure on optical and morphological properties of grafted low density polyethylene films. Polymer Testing 22: 17-23.

Guo, Z., J. Warner, P. Christy, D. E. Knanbuehl, G. Boiteux, G. Seytre. 2004. Ion mobility time of flight measurements: isolating the mobility of charge carriers during an epoxy-amine reaction. Polymer 45: 8825-8835.

Hodge, IM, M. D. Ingram, A. R. West. 1976. Impedance and modulus spectroscopy of polycrystalline solid electrolytes. Journal of Electroanalytical Chemistry, 74: 125-143.

Johnsher, A., K., M. S. Frost, 1976. Weekly frequency dependant of electrical conductivity in chalogenide glass. Thin solid films, 37, 103-107.

Johns, H., E., J.R. Cunningham. 1983. The Physics of Radiology, fourth ed., Thomas, Sprinfield, IL.

Kanazawa K. K., A. F. Diaz, R. H. Geiss, W. D. Gill, J.F. Kwak, J.A. Logan, J.F. Rabolt and G.B. Street. 1980. J. Chem. Soc. Chem. Commun. 854.
Kobayashi, N., Chinone H., Miyazaki A. 2003. Polymer electrolyte for novel electrochromic display. Electrochimica Acta 48: 2323-2327.
Lewandowski, A., Skorupska K., Malinska J. 2000. Novel (polyvinyl alcohol) -KOH-H2O alkaline polymer electrolyte. Solid state ionic 133: 265-271.
Lokhovitsky, V.I. and V.V. Polikarpov. 1980. Technology of radiation emulsion polymerization. Atomizdat, Moscow-Russia.
MacDiarmid, A.G., J. Chiang, A.F. Richther. 1987. Polyaniline: a new concept in conducting polymers. Synthetic Metals 18: 285-290.
Malmonge, J A, Mattoso L H C. 1997.Doping of polyaniline and derivatives induced by X-ray radiation. Synthetic Metals 84: 779-780.
Malinauskas A. Malinauskiene J. and Ramanavicius A. (2005). Conducting polymer-based nanostructurized materials: electrochemical aspects. Nanotechnology, 16: R51-R62.
Mariappan, C. R., Govindaraj G. 2002. Ac conductivity, dielectric studies and conductivity scaling of NASICON materials. Materials Science and Engineering B94: 82-88.
McGervey, J., 1983. Introduction to Modern Physics, 2nd. edition, Academic Press, New York.
Mohammed A. Ali Omer, Elias S., Khairulzaman Hj M. Dahlan. 2007. Radiation Synthesis and Characterization of Conducting Polyaniline and Polyaniline/silver nanoparticles. Ph. D. thesis, Physics Department, UPM, Malaysia.
Mott, N., F. and Davis E.A. 1979. Electronic Process in Non-crystalline Materials, 2nd. Edition. Clarendon Press, Oxford, UK.
Park, H. B., S. Y. Nam, J. W. Rhim, J. M. Lee, S. E. Kim, R. Kim, Y. M. Lee. 2002. Copolymerization of Styrene onto Polyethersulfone Films Induced By Gamma Ray Irradiation. Journal of Applied Polymer Science 86: 2611-2614.
Rao, P. S., Anand J., Palaniappan S. and Sathyanarayana D. N. 2000. European Polymer Journal 36: 915-920.
Richard, A., Pethrick, David Hayward Pro. 2002. Study of ageing of adhesive bonds with various surface treatments. Polymer Science 27: 1983 - 2017.
Sevil, U. A., O. Güven, A. Kovács, I. Slezsák. 2003. Gamma and electron dose response of the electrical conductivity of polyaniline based polymer composites. Radiation Physics and Chemistry 67: 575–580.
Siegbahn, K., 1965. Alpha, Beta and Gamma Ray Spectroscopy, North-Holland, Amsterdam, Netherlands.
Smith, F., A. 2000. A Primer in Applied Radiation Physics. World Scientific Publishing Co. Pte. Ltd.
Tarola, A., D. Dini, E. Salatelli, F. Andreani, F. Decker. 1999. Electrochemical impedance spectroscopy of polyalkylterthiophenes. Electrochim. Acta 44: 4189-4193.
Vorotyntsev, M.A., J.P.Badiali, G. Inzelt. Electrochemical impedance spectroscopy of thin films with two mobile charge carriers: effects of the interfacial charging. 1999. J. Electroanal.Chem. 472: 7-19.

# Part 4

# Radiation Biology

8

# Radiation Induced Radioresistance – Role of DNA Repair and Mitochondria

Madhu Bala
*Radiation Biology Department, Institute of Nuclear Medicine and Allied Sciences*
*Brig. S K Mazumdar Marg, Delhi,*
*India*

## 1. Introduction

The bio-positive effects of exposure to small doses of environmental stressors such as radiation, chemicals and mutagens have been reported since long. A number of studies and reviews (Bala & Mathew, 2000; Luckey, 2008; Pandey et al., 2006; Sasaki et al., 2002) document that exposure to small doses of ionizing radiation enhanced the tolerance towards the detrimental effects of lethal doses of ionizing radiation given subsequently. Such phenomenon was observed in prokaryotes as well as in eukaryotes. Some of the laboratory studies with human lymphocytes are summarized in Table 1. The information on beneficial effects of low dose irradiation also poured in from the epidemiological studies (reviewed by Bala & Mathew, 2000; Dasu & Denekamp, 2000; Luckey, 2008). The populations exposed to high background radiation showed long term beneficial effects such as increased life span, enhanced immune system, decreased cancer mortality and cancer risk (Calabrese et al., 2001; Cohen, 1999; Nambi & Soman, 1987, UNSCEAR, 2000). Among the A-bomb survivors from Hiroshima and Nagasaki, those, who received doses lower than 200 mSv, showed no increase in cancer deaths. Further, the population which received doses below 100 mSv, showed decrease in the mortality caused by leukemia in comparison to the age-matched control cohorts (UNSCEAR, 1994).

Often in epidemiological studies the exposure to low levels of radiation was for longer duration, while in laboratory studies the exposure to low level radiation was for a shorter duration (sometimes even a pulse exposure). Nonetheless, beneficial effects were observed in short as well as in prolonged exposures. This strongly suggested that the low dose radiobiological studies could have bearing in diverse and important applications such as radiation protection, risk assessment and radiotherapy. It was, therefore, considered important to initiate investigations for understanding more about the mechanisms of radioprotective effects caused by pre-exposure to low doses of radiation. It was reported that the resistance to lethal doses of ionizing radiation could be induced not only by low doses of radiation but also by variety of agents other than radiation, *viz.* heat, pH, nutrients, UV rays, though, the genes and the molecular pathways affected in these cases differed with the inducing agent. (Bala & Goel 2007; Boreham & Mitchel, 1991; 1994; Boreham et al.,2000). Further, it was believed that the induced resistance to lethal doses of radiation by short term pre-exposure to low doses of radiation was transient in nature and the radiation doses required to induce beneficial effects varied qualitatively as well as quantitatively from organism to organism. To understand the genetic basis of the phenomenon of low dose induced radioresistance in a comprehensive manner, we chose two different model systems i.e. the microbe *Saccharomyces cerevisiae* and cultured human peripheral blood lymphocytes. The unicellular eukaryote *Saccharomyces cerevisiae* was used to carry out the basic studies and perform the genome wide search for identifying the affected genes. The cultured human peripheral blood lymphocytes were employed to study the role of select genes. In order to minimize the load of mutations, it was considered important to select the smallest possible doses of ionizing radiation for pre-exposure. The term radiation-induced radioresistance (RIR) was introduced (Bala & Goel, 2007) to explain the phenomenon of radioresistance to lethal doses of ionizing radiation, which was (i) specifically induced by a single pre-exposure to sub-lethal doses (causing not more than 10% death of population, $\leq LD_{10}$ ) of ionizing radiation, and (ii) was transient in nature. The conventional term 'radio-adaptive response' was avoided because the term 'adaptation' in one of the several senses referred to the evolutionary transformations, where new stable behavioral patterns evolved due to prolonged exposure to the environmental stress. This review chapter presents some of our important findings.

| References | Important findings |
|---|---|
| Olivieri *et al.*, 1984 | First *in vitro* experiment with human lymphocytes, reduction in chromosomal aberration was greater at higher pre-irradiation dose. |
| Sanderson & Morley, 1986 | Pre-exposure of human lymphocytes to [$^3$H] dThd reduced the number of mutations at *hprt* locus by 1.5 or 3.0 Gy of X-rays. |
| Shadley & Wolff, 1987 | Only stimulated and not $G_0$ lymphocytes, pre-irradiation with low dose of X-rays showed survival benefit against high doses of X-rays. |
| Shadley & Wiencke, 1989 | The beneficial effects of low dose depended upon total dose, dose rate of pre-irradiated dose but not on dose rate of challenge dose in human lymphocytes. |
| Boothman *et al.*, 1996 | Elevated level of PCNA, cyclin D1, cyclin A in human cell line pre-irradiated with gamma-rays, may play a role in cell cycle regulation and DNA repair. |
| Carette *et al.*, 2002 | Implication of PBP74 in low dose irradiated human tumour cell lines HT29 and MCF-7 with gamma-rays within 30 min after irradiation. |
| Seo *et al.*, 2006 | Role of p27Cip/Kip in the induction of radio-adaptive response in gamma-irradiated RIF cells. |

Table 1. Summary of some important studies performed to understand low dose response in human lymphocytes.

## 2. Materials and methods

The studies were performed sequentially with two different types of cells. Initial studies were executed with the unicellular eukaryotic microbe, *Saccharomyces cerevisiae,* where the focus was to identify the effects of inducing radiation doses and dose rates; the beneficial effects induced in terms of survival, mutagenesis and recombinogenesis; for conducting the genome wide search to identify the affected genes; and reconfirming the role of cell cycle, DNA repair and mitochondrial genes in inducing beneficial effects. The subsequent studies were conducted with cultured human lymphocytes to investigate some of the key events that were observed in *S. cerevisiae.*

### 2.1 Studies with *Saccharomyces cerevisiae*

The diploid strain D7 of *Saccharomyces cerevisiae* with genotype a/α: *trp5-12/trp5-27, ilv1-92/ilv1- 92* (Zimmermann *et al.*, 1975) was used because it allowed quick detection of mutants, recombinants and survivors. The strain D7 of *S. cerevisiae* had functional defects in *TRP* gene (heteroallelic) and *ILV* gene (homoallelic) making it auxotrophic for tryptophan and isoleucine. Presence of two different inactive alleles within tryptophan locus (*trp5-12/trp5-27*) caused nutritional requirement for tryptophan which could be recovered by gene conversion to form fully active wild type gene, thereby alleviating the need for tryptophan requirement. The resultant colonies after gene conversion could be scored on synthetic medium lacking tryptophan. The presence of two defective copies of alleles at isoleucine locus *(ilv1-92/ilv1-92)* caused the nutritional requirement of isoleucine which could be corrected by reverse mutation. The resultant colonies could then be scored on the synthetic medium lacking isoleucine. The usefulness of this organism to understand radiation responses and their modification has been demonstrated (Bala & Jain, 1994, Bala & Jain, 1996; Bala & Goel 2004; Bala & Goel 2007).

Cultures of *S. cerevisiae* were grown on yeast extract peptone dextrose medium (YPD; 1% yeast extract powder, 2% peptone, 2% dextrose, HiMedia, India) at 30$\pm$1 $^0$C. The cells were harvested, washed and suspended in phosphate buffer (PB, 67 mM, pH 6.0; 4x$10^7$cells/ml). The cell suspension was cooled to 4 $^0$C, and irradiated with $^{60}$Co- gamma- radiation using Gamma Cell-220 (Atomic energy, Canada; dose rate 0.0078 Gy/s) or Gamma Cell-5000 (BRIT, India; dose rate 1.26 Gy/s). After low dose irradiation, cell suspension was maintained at 30$\pm$1 $^0$C till the subsequent exposure to lethal doses of radiation ($LD_{50}$, 400 Gy). The survivors, gene convertants and revertants were estimated using defined synthetic complete (DSC), tryptophan omission (TRP-) and isoleucine omission (ILV-) medium, respectively as described (Bala & Goel, 2007). While survivors were expressed as a fraction of unirradiated controls, the gene convertants and revertants were expressed as fraction of CFUs on respective omission medium to the CFUs on DSC medium after the corresponding treatment. The RIR was calculated in terms of percent changes in survivors, convertants and revertants as below:

$$\%\ \text{Change}_{(s,c,r)} = \frac{s/c/r\ \text{pre}-\text{irradiated} - (s/c/r\ \text{non pre}-\text{irradiated})\ X\ 100}{s/c/r\ \text{from non}-\text{pre}-\text{irradiated cells}}$$

Where, s=survivors; c=convertants/$10^6$ survivors; r=revertants/$10^6$ survivors (after 400 Gy)

RNeasy Mini Kit and OneStep rt-PCR Kit (Qiagen, Germany) was used to isolate total RNA and carry out rt-PCR respectively (Bala & Goel, 2007). Table 2 enlists the gene specific primers (synthesized from IDT, Coralville). Reverse transcription was at 50 $^o$C for 30 min, followed by incubation at 95 $^o$C for 15 min. The amplification was for 25 cycles. The denaturation was at 94 $^o$C for 45 s; annealing at 60 $^o$C for 45 s; extension at 72 $^o$C for 60 s and the final extension was at 72 $^o$C for 10 min. The PCR products were separated on 1% agarose gel, stained with ethidium bromide and quantified using Lab Works software, version 4.0

(UVP Inc., U.K.). Real-time one-step rt-PCR kit with SYBR green as flourophore (Qiagen, Germany) was used as per manufacturer's protocol to perform quantitative rt-PCR using iCycler (Bio-RAD, US, software version 2.1). The fold changes were determined by calculating the fold change in threshold cycle (Δ Ct').

$$\text{Fold change} = 2^{-(\text{Ct values of control- Ct value of irradiated sample})}$$

Where Ct: threshold cycle

| Gene | Forward primer | Reverse primer |
|---|---|---|
| *XRS2* | AGCAACAATACTGAGAAGG | TGAAATTGGAAATACTCGGA |
| *MRE11* | GTCACTCTACCAAGTACTGA | CCATATCACCATATCCAGGAA |
| *RAD50* | GGCTTTCATCTCTCAGGA | ATTCCTGGGTGAGGGGAA |
| *SSC1* | GTCCCACAAATCGAAGTCAC | GGCATTGTTGCCGTTGTTG |
| *OXI3* | GAAGTATCAGGAGGTGGTGAC | TCCCACCACGTAGTAAGTATCG |
| *OGG1* | CAGGATGAAAGTGAGCTATGT | CAGATCTATTTTTGCTTCTTTG |

Table 2. Primers for genetic studies with *Saccharomyces cerevisiae*

For microarray studies, the labeled cDNA was synthesized from total RNA by using CyScribeTM First-Strand cDNA Labeling Kit (Amersham Biosciences). Either Cy3-dUTP or Cy5-dUTP was incorporated into the cDNA of samples under comparison. The cDNAs were dried in a vacuum trap. Pre-printed DNA microarrays with complete set of 6400 Open Reading Frames (ORFs) of *S. cerevisiae* genome (Microarray Centre of the University Health Network, Toronto, Canada), were used in this study. The slides were first hybridized in pre-hybridization solution (6x SSC, 0.5% SDS, 1% bovine serum albumin) for 1 h and then hybridized overnight with labeled probe at 42 °C in a water bath. Before using as a hybridization probe, the labeled cDNA was re-suspended in 40 μl of hybridization solution (50% Formamide, 6x SSC, 5x Denhardt's, 0.5% SDS, 20 μg of poly(A) and salmon sperm, Invitrogen). For each test, cDNAs from the un-irradiated control and from the stress dose irradiated samples were together hybridized on to one chip. Further, for each test, two different hybridizations were performed by swapping the fluorochromes to cross check the transcriptional changes, if any, due to experimental procedures. At least two DNA microarrays were analyzed for each test condition. The chips were scanned at a resolution of 10 μm and data was analyzed using GenePix Pro 4.0 analysis software (Axon Instruments, Union City, CA).

To study the DNA damage in individual chromosomes by pulsed field gel electrophoresis, the samples were prepared as described earlier (Bala & Jain 1996, Bala & Mathew, 2002). In brief, the cell suspension was washed with PB, centrifuged; pellet was treated with lyticase enzyme and then immobilized in low melting agarose plugs using the mould provided by BioRad USA. The plugs were first treated with LET buffer [0.5 *M* EDTA pH 8.0, 0.01 *M* Tris(hydroxymethyl)-aminomethane pH 7.0, 7.5% β–Mercaptoethanol] for 20 h at 37 °C. The LET buffer was removed, plugs were washed two times with NDS buffer [0.01 *M* Tris(hydroxymethyl)-aminomethane pH 7.0,7.5% EDTA pH 8.0, 1% n-luaryl sarcosine] . The plugs were then treated with NDS buffer containing 2mg/ml Proteinase K for 20 h at 48 °C. Sufficient washings were given in EDTA (0.5 *M*, pH 8.0) thereafter. The plugs were stored at 4 °C before electrophoresis. The pulsed-field gel electrophoreses (PFGE) was for 20 h (60 sec for first 13 h and 90 sec pulse for next 7h) at 200 V, using CHEF DRII (BioRad, USA), to resolve genomic DNA into a number of chromosomal bands.

### 2.2 Studies with cultured human Peripheral Blood Mononuclear Cells (PBMC)

Heparinized vacutainers (Griener, Astria) were used to draw 3-5 ml venous blood from healthy, non-smokers, non-alcoholic male donors (age 25-30 years). The blood was layered on the ficoll-histopaque column (Sigma Aldrich Chemicals, USA) and centrifuged at low speed at 26$\pm$2°C, the interface between plasma and histopaque comprising PBMCs was collected and washed three times with serum free RPMI-1640 (HiMedia, India). The washed cells were suspended @$1\times10^6$ cells/ml in complete RPMI-1640 containing 10% fetal bovine serum, 100 units/ml penicillin sodium salt, 100 µg/ml streptomycin sulphate, 2 mg/ml sodium bicarbonate. Phytohemagglutinin (PHA, Difco, Hamburg, Germany) was added to stimulate the cells proliferation. The cultures were setup in 96 well flat-bottomed micro titer plates (Tarson, India) at 37°C, 5% $CO_2$. Each well had 150 µl volume containing $1.5\times10^5$ cells. The 22-24 hour old cultured PBMCs were irradiated first with low dose of $^{60}$Co-γ-radiation (0.07 Gy, using Gamma Cell GC 220, Canada dose rate 0.0078 Gy/s) and then after suitable time interval with lethal dose of $^{60}$Co-γ-radiation (5.0 Gy, using Gamma Cell-5000, BRIT, India; dose rate 1.26 Gy/s).

The cell proliferation was quantified using Hoechst 33342. The cells were washed at least three times with saline in microtiter plate, freshly prepared Hoechst 33342 solution in serum free RPMI (10 µg/ml) medium was added and the suspension was incubated at 37°C for 30 min. (Blaheta et. al., 1991). Fluorescence was measured at $\lambda_{ex}$ 355 nm and $\lambda_{em}$ 460 nm in fluorescence spectrophotometer (Varion, Australia).

To score the micronuclei, cytochalasin B (Sigma Aldrich Chemicals, USA) was added at 44 hour after initiation of human PBMC culture and the cells were harvested at 72 hour. $1\times10^6$ cells were washed, the cell pellet was suspended in 200 µl carnoy solution (methanol: acetic acid, 3:1) and incubated at 4°C for 2 hour. This cell suspension was laid on the chilled slides, dried overnight at 26 ± 2°C and stained with hoechst 33342 (10 µg/ml) at 26 ± 2°C for 30 min. in dark. Micronuclei were counted at $\lambda_{ex}$ 355 nm and $\lambda_{em}$ 460nm as per criteria described (Fenech, 1993). At least 1000 cells per sample were scored at 1000× magnification under oil immersion.

### 2.3 Western blotting

Protein extraction and Western blotting was as per procedures standardized in our laboratory (Bala & Goel, 2007). Briefly, $4\times10^7$ cells/ ml of *S. cerevisiae* were lysed and treated with 160 ml of 50% trichloroacetic acid (TCA), washed with 1.5 ml of chilled acetone, re-suspended in 100 ml of extraction buffer (4% SDS; 0.16 M Tris-Cl, pH 6.8; 20% Glycerol; 0.38 M b-mercaptoethanol) and heated for 4 min at 95 °C. For extracting proteins from PBMCs, standard protocol was used. Briefly $5\times10^6$ cells were suspended in PB containing protease inhibitors for 1.5 hours at 4 °C. The cells were ruptured by sonication and soluble proteins were collected after centrifugation in cold. Total soluble proteins were quantified by using Bradford's reagent and resolved by one dimensional SDS-polyacrylamide gel electrophoresis (SDS-PAGE) using Mini-PROTEAN II (BIO-RAD, US). Gels were stained with Coomassie brilliant blue R-250. Electro-blotting was on nitrocellulose membrane

(Millipore) and treatment with primary and secondary antibody was as described earlier (Bala & Goel, 2007).

### 2.4 Statistical analysis

Each experiment based on CFUs assay, had three replicates and was repeated at least three times. The data was presented as the average of three experiments $\pm$ S.D. For estimating differential gene expression, DNA damage and protein expression, the data was analyzed using paired t-test. For cell survival, mutagenesis and recombinogenesis the data was analyzed using two-way ANOVA. $P \leq 0.05$ was considered significant.

## 3. Results and discussion

### 3.1 Studies with *Saccharomyces cerevisiae*

#### 3.1.1 RIR inducing doses, survival and mutagenesis

Systematic study with cultures grown to different phases (mid-log phase, late log phase and stationary phase) showed that the stationary phase cultures did not show any RIR. Mid-log phase cultures showed 25% increase, while late log phase cultures showed only 12% increase in survivors in comparison to the non-pre-irradiated cultures. This comparison was made at pre-irradiation dose 20 Gy ($LD_{10}$), challenge dose 400 Gy ($LD_{50}$) and the time interval between stress dose irradiation and challenge dose irradiation 4.5 h (Sharma & Bala, 2002). This was in agreement with earlier reports (Cai & Liu, 1990), where mitogen stimulated human lymphocyte cultures showed far better radio-adaptive response than the resting cells. The RIR, since was maximum with mid-log phase cells, further studies were planned with the mid-log phase cells. Pre-treatment with three different doses of $^{60}$Co-gamma-ray viz. 4, 10 and 20 Gy ($\leq LD_{10}$) showed that RIR increased with increase in the pre-irradiation dose. In comparison to non-pre-irradiated controls, the 4 Gy pre-irradiated samples showed maximum 13% increase in survivors, 10 Gy pre-irradiated samples showed maximum 27% increase in survivors and 20 Gy pre-irradiated samples showed maximum 32% increase in survivors after lethal irradiation (400 Gy). The time of maximal increase in survivors was delayed at higher stress doses and was approximately 10 h after irradiation at 20 Gy, 6 h after irradiation at 4 or 10 Gy (Figure 1). However, there was no linear correlation between the pre-irradiation $^{60}$Co-gamma-ray dose and increase in survival due to RIR. These studies suggested that priming of cells with small radiation doses may induce some signaling events which may lead to RIR. The pre-irradiation (stress) dose (20 Gy), thereafter, was delivered at two different dose rates i.e. 0.0078 Gy/s and 1.26 Gy/s to the mid-log phase cells. It was observed that in comparison to non-pre-irradiated cultures, the cultures pre-irradiated (20Gy) at lower dose rate (0.0078 Gy/s) and lethally irradiated with 400 Gy, showed a maximum of 32% increase in survivors while cultures pre-irradiated at higher dose rate (1.26 Gy/s) showed a maximum of 25% increase in survivors after lethal irradiation. Further, in comparison to non-pre-irradiated cultures, the pre-irradiated cultures showed decrease in gene convertants and revertants when irradiated with lethal dose (400 Gy, Dwivedi et al., 2001). The dose rate also impacted the mutations and gene conversion quantum and time kinetics. Pre-irradiation dose (20 Gy), delivered at lower dose rate decreased the gene conversions and mutations for a longer time period in comparison to the same dose delivered at higher dose rate (Figure 1).

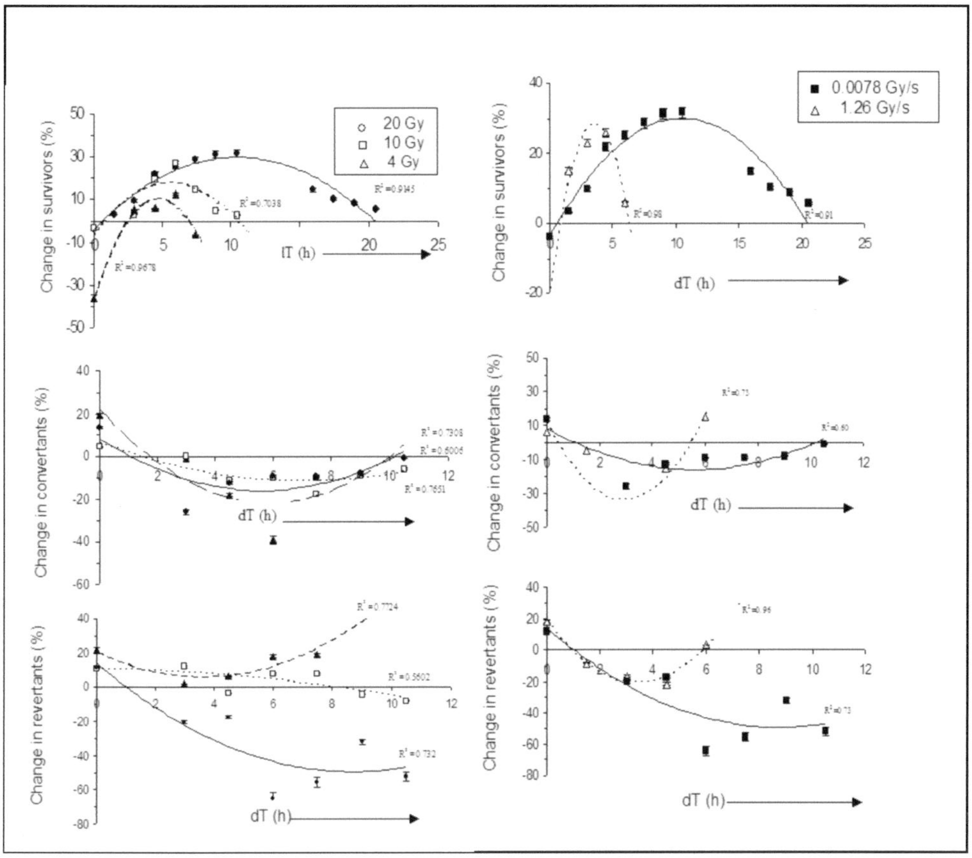


Fig. 1. Effect of different pre-irradiation doses (4,10, 20 Gy) and of dose rates of stress dose (20 Gy) on RIR in *Saccharomyces cerevisiae.* dT (h): duration in hours between pre-irradiation and lethal irradiation (400 Gy). The values are average $\pm$ S.D. of three experiments. (From Dwivedi et al., 2008).

There are reports to show that there is difference in the nature and quantum of DNA damage by stress doses delivered at different dose rates (Chaubey et al., 2006). The higher RIR (survival) by stress doses delivered at lower dose rate as compared to the same stress dose delivered at higher dose rate suggested that nature of damage generated by stress dose, is an important determinant of induction of protective mechanisms.

### 3.1.2 Alteration in gene expression after irradiation with low dose of $^{60}$Co-gamma-radiation - whole genome analysis

As many as 110 open reading frames (ORFs) displayed more than 2 fold increase in transcription at 4.5 h after the low dose irradiation (20 Gy) and some of the annotated once are listed in Table 3.

The functional groups of the up-regulated genes were DNA damage, repair, synthesis, energy generation, metabolism and stress response. Besides this, many transcripts with unknown function (not listed in Table 3), were also up-regulated. Some genes such as IRE1, HSP12 were down-regulated 4.5 h after irradiation (20 Gy) (Table 4). Sahara et al., 2002 reported that Hsp12p might play a role in protein binding in yeast. The Ire1p and Hac1p participate in the "unfolded protein response" (UPR) pathway. It is predicted that in our study the UPR pathway was down regulated. Other genes that were down regulated were DDR48, MSN2. Further studies are planned to understand the role of these genes and the pathways in which they participate to induced RIR.

| ORF | Gene | Function |
|---|---|---|
| **Stress response** | | |
| | *HSP78* | Heat shock protein |
| *YOLO53C* | *DDR2* | DNA damage stress response |
| *YDL229W* | *SSB1* | cytosolic HSP70 |
| *YER103W* | *SSA4* | cytosolic HSP70 |
| **Signal Transduction** | | |
| *YKL161C* | *MLP1* | Serine-threonine kinase |
| *YKL168C* | *KKQ8* | Serine threonine protein kinase |
| *YOL016C* | *CMK2* | Ca/calmodulin dependent kinase |
| *YHR030C* | *SLT2* | MAP kinase pathway |
| **Energy and Metabolism** | | |
| *YDL022W* | *GPD1* | Glycerol-3P-dehydrogenase |
| *YJL155C* | *FBP26* | Fructose/mannose metabolism |
| *YPL088W* | *ALD4* | Putative aryl dehydrogenase |
| *YOR374W* | *HOR2* | Mitochondrial Aldehyde |
| *YER062C* | *PRB1* | dehydrogenase |
| *YEL060C* | *GTT1* | DL-glycerol-P-phosphase |
| *YIR038C* | *CIT1* | Protein degradation |
| *YNR001C* | *PDC1* | Glutathione transferase |
| *YLR044C* | *ADH1* | TCA cycle; citrate synthase |
| *YOL086C* | *RHR2* | Glycolysis, pyruate decarboxylase |
| *YIL053W* | *FAS2* | Alcohol dehydrogenase |
| *YPL231W* | *COX4* | Glycerol metabolism |
| *YGL187C* | *STF2* | Fatty acid biosynthesis |
| *YGR008C* | *TPS2* | Cytochrome C Oxidase |
| *YDR074W* | *PHO5* | ATPase stabilizing factor |
| *YBR093C* | *NCA3* | Glucose, fructose metabolism |
| *YJL116C* | | Phosphate metabolism |
| *YLR327C* | | Mitochondrion biogenesis |
| | | Organelle biogenesis |
| **Transcription and Translation** | | |
| *YKR062W* | *TFA2* | RNA pol II transcription factor |
| *YNL301C* | *RP28B* | Constituent of ribosome |
| *YNL178W* | *RPS3* | Constituent of ribosome |
| *YFR031CA* | *RPL5B* | Component of ribosome |
| *YDL083C* | *RPS16B* | Constituent of ribosome |
| *YGL031C* | *RPL30A* | Component of ribosome |
| *YOR204W* | *DED1* | mRNA processing helicase |
| *YDR088C* | *SLU7* | mRNA processing |
| *YLR325C* | *RPL38* | ribosomal protein L38 |
| *YDR280W* | *RRP45* | 3'->5' exoribonuclease |
| **DNA damage, repair, synthesis** | | |
| *YER164W* | *CHD1* | chromodomain-helicase-DNA-binding (CHD) family |
| *YML061C* | *PIF1* | DNA helicase |
| *YER095W* | *RAD51* | Recombinase |
| *YER018C* | *SPC25* | spindle pole body component |
| *YDR201W* | *SPC19* | spindle pole body component |
| *YNL222W* | *SSU72* | nuclear protein |
| *YNL206C* | unknown | similar to SSRP proteins (DNA structure-specific) |
| *YMR137C* | *SNM1 /PSO2* | required for inter-strand crosslink repair |
| *YNL250W* | *RAD50* | DNA binding protein |
| *YOL090W* | *MSH2* | MutS homolog; mismatch repair |
| *YCR014C* | *POL4* | DNA polymerase IV |
| *YER088C* | *DOT6* | Nuclear protein with Myb DNA-binding domain |
| *YBL019W* | *APN2* | exonuclease III homolog (AP endonuclease) |
| *YCR088W* | *ABP1* | actin binding protein |
| *YDR545W* | *YRF1* | Y' helicase (subtelomerically-encoded) |
| *YJL065C* | unknown | similar to DNA polymerase epsilon subunit c |
| *YNL088W* | *TOP2* | DNA topoisomerase II |
| *YJR021C* | *REC107* | ds break formation complex subunit |
| *YKL113C* | *RAD27* | ssDNA endonuclease |
| *YGL163C* | *RAD54* | DNA-dependent ATPase |
| *YDR225W* | *HTA1* | Transcription; Chromatin binding |
| *YGLO37C* | *PNC1* | Cell aging |
| **Cell cycle regulatory** | | |
| *YER059W* | *PCL6* | cyclin (Pho85p) |
| *YLL065W* | *GIN11* | growth inhibitor |
| *YDR285W* | *ZIP1* | synaptonemal complex protein |
| *YAL063C* | *FLO9* | *cell wall protein* |
| *YBR211C* | *AME1* | microtubule associated |
| *YNL289W* | *PCL1* | G1/S cyclin |
| *YPR120C* | *CLB5* | G1/S cyclin |
| *YDR217C* | *RAD9* | Cell cycle, DNA damage |
| *YFL029C* | *CAK1* | Cdk-activating kinase |

Table 3. Up regulated (> 2.0 folds) transcripts, 4.5 h after the irradiation (20 Gy). Categorization of ORFs into functional groups is based on SGD Library.

| Gene | Function |
|---|---|
| *HSP26* | Heat shock protein |
| *GPD1* | Glycerol-3-phosphate dehydrogenase ($NAD^{+}$),cytoplasmic |
| *HSP12* | Heat Shock protein |
| *HSP30* | Heat Shock protein |
| *SSA1* | Cytosolic HSP70 |
| *YGP1* | Secreted glycoproteins |
| *ECM32* | DNA dependent ATPase/DNA helicase B |
| *ATH1* | Acid trehalase, vacuolar |
| *ARN1* | Ferrichrome-type siderophore transporter |
| *DDR48* | Heat Shock Protein |
| *MSN2* | Stress-responsive regulatory protein |

Table 4. Some important genes down regulated (≥1.5 folds) after the 20 Gy irradiation

### 3.1.3 Confirmation of stress dose induced time dependent changes in selected transcripts as well as associated genes

#### 3.1.3.1 The MRX complex

Significant over expression of genes from the DNA damage, response, repair complex, prompted us to perform real time quantitative PCR for the *MRX* complex (*MRE11, RAD50* and *XRS2)* of which *RAD50* is an essential gene. The β-actin gene, though considered as a house keeping gene, showed differences in the stress dose irradiated cultures in comparison to the non-pre irradiated cultures, suggesting that β-actin gene could not be used as a house keeping gene. The experimental data was, therefore, compared with reference to the un-irradiated control at the corresponding time. The results obtained from the real time quantitative rt-PCR (Figure 2) confirmed the significant increase in *RAD50* transcripts at 4.5 h in stress dose irradiated cultures and supported the information obtained by microarray (Table 3).

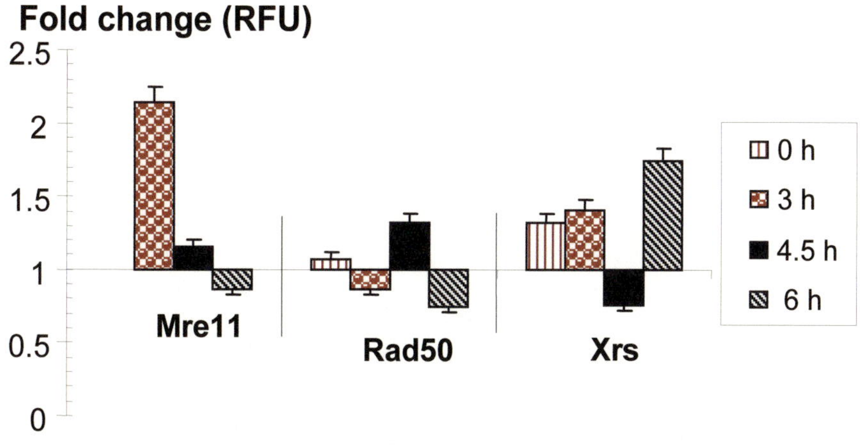


Fig. 2. Effect of stress dose irradiation on relative time dependent changes in gene expression of MRE11, RAD50 and XRS2 as studied by Real Time- reverse transcription PCR; the value of untreated control was assigned as one. RFU: relative fluorescence units (from Dwivedi et al., 2008)

The *RAD50/MRE11* complex possesses single-strand endonuclease activity and ATP-dependent double-strand-specific exonuclease activity. Rad50 provides ATP-dependent control of mre11 by unwinding and/or repositioning DNA ends into the MRE11 active site. The rt-PCR studies showed that in non-irradiated controls, there was significant increase in *MRE11* transcript level from 0 h to 4.5 h. In comparison to the non-irradiated controls, the stress dose irradiated samples showed significantly higher level of *MRE11* transcripts at 3 h time interval after irradiation. Further, in comparison to non-irradiated controls, the stress dose irradiated samples showed significantly higher levels of *XRS2* up to 3 h and reduced levels at 4.5 h after irradiation. The Mre11 complex influences diverse functions in the DNA damage response. The complex comprises the globular DNA-binding domain and the Rad50 hook domain, which are linked by a long and extended Rad50 coiled-coil domain. Recently it is reported that functions of *MRE11* complex are integrated by the coiled coils of Rad50 (Hohl et al., 2011). *MRE11* is reportedly involved in DNA double-strand break repair and possesses single-strand endonuclease activity and double-strand-specific 3'-5' exonuclease activity. Its role in meiotic DSB processing is also reported (Smolka et al., 2007).

### 3.1.3.2 The heat shock proteins

The western blotting studies with members of *HSP70* family showed that in untreated controls, level of *Kar2 p* did not increase significantly from 0 h to 4.5 h. In comparison to non-irradiated controls, the stress dose irradiated samples showed significantly higher *Kar2 p* level at 3 and 4.5 h. The *Ssa1p* transcript level did not change in the untreated control from 0 h till 4.5 h. In stress dose irradiated samples, the *Ssa1p* level increased up to 3 h but decreased significantly at 4.5 h, in comparison to the non-irradiated control (Bala & Dwivedi 2005). By microarray technique also the *SSA1* level were found to be lower in the stress dose irradiated samples as compared to the untreated control at 4.5 h (Table 4). In stress dose irradiated cultures, the *Ssa2p* level was significantly higher than the non-irradiated control at 0h and 3 h (Figure 3).

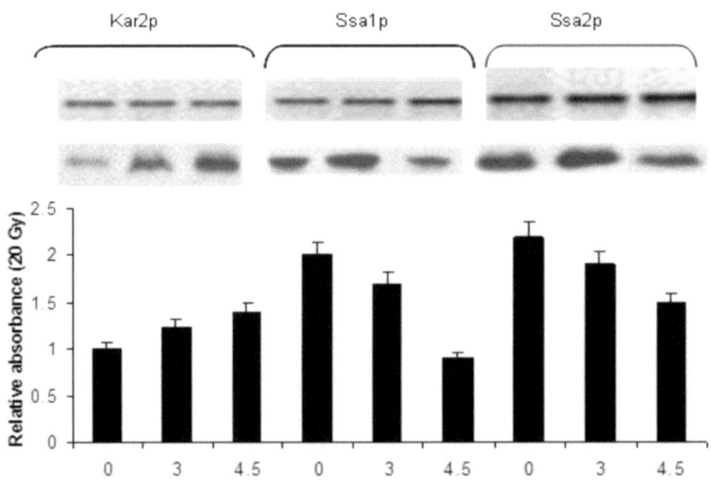


Fig. 3. The effect of low dose irradiation (20 Gy) on expression of Kar2p, Ssa1p and Ssa2p. The top strip of membrane blot shows the protein expression in unirradiated controls.

### 3.1.3.3 The mitochondrial genes

*Saccharomyces cerevisiae* is an excellent eukaryotic model system to study DNA repair mechanisms because DNA repair pathways are highly conserved between human and yeast. Furthermore, yeast and human mitochondria resemble each other in structure and function. Mitochondria are the major sites of energy (ATP) production in the cell. Mitochondria also perform many other cellular functions, such as respiration and heme, lipid, amino acid and nucleotide biosynthesis. Mitochondria also maintain the intracellular homeostasis of inorganic ions and initiate programmed cell death. Mitochondria are the major source of endogenous reactive oxygen species (ROS) in cells as they contain the electron transport chain that reduces oxygen to water by addition of electrons during oxidative phosphorylation. The rt-PCR studies with the mitochondrial genes (*SSC1* gene coding for mtHsp70, *OXI3* gene coding for COX1 respiratory component of complex-IV and *OGG1*gene) showed that the expression of *OXI3* was more than unirradiated controls up to 6 h and that of *SSC1* only at 2 and 10 hours after irradiation (20 Gy, Figure 4a,b). The expression of OGG1 was increased up to 2 hour only, after irradiation [(20 Gy), data not shown]. The mitochondrial genome of eukaryotic cells is extremely susceptible to damage due to constant exposure to significant amounts of reactive oxygen species (ROS) produced endogenously by mitochondria as a by-product of oxidative phosphorylation (Gupta et al., 2005). It is known that inactivation of *OGG1* in yeast leads to spontaneous mutations in the mitochondrial genome. Our analysis revealed that irradiation with low doses of gamma radiation enhanced the expression of expression of OGG1 at 2 h only.

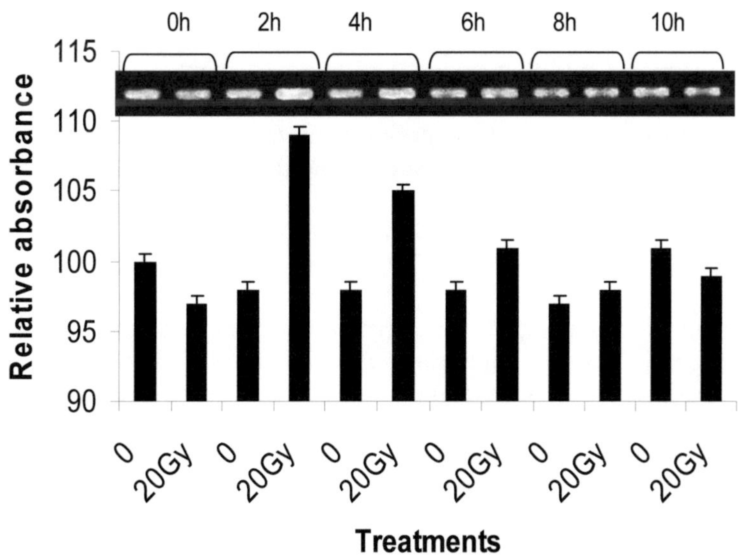


Fig. 4a. Expression of *OXI3* gene after low dose pre-irradiation at different time intervals in *Saccharomyces cerevisiae* (From Arya et al., 2006).

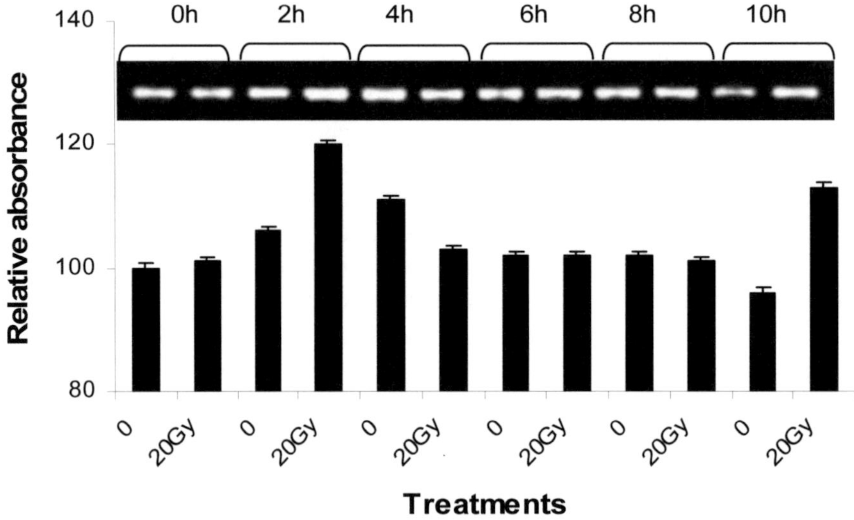


Fig. 4b. Expression of *SSC1* (mt-HSP70) gene after low dose pre-irradiation at different time intervals in *Saccharomyces cerevisiae* (From Arya et al., 2006)

Maintenance of mitochondrial DNA (mtDNA) is essential for ensuring respiratory competence. *MGM101* was identified as a gene essential for mtDNA maintenance in *S. cerevisiae*. The MGM101p binds the DNA. The MGM101 function exclusively in the repair of DNA contained in the mitochondrial organelle, and is predicted to participate in base excision and/or nucleotide excision repair pathways. *Saccharomyces cerevisiae* contain 3 different Hsp70s i.e. *SSC1, SSQ1,* and *SSC3*. Amongst these, *SSC1* is the most abundant constitutively expressed multifunctional Hsp70 and is essential for the viability of yeast cells. It plays a critical role in protein translocation across the mitochondrial inner membrane and folding of almost all pre-proteins targeted to the mitochondrial matrix compartment, thus maintaining protein homeostasis in mitochondria. For proper translocation function, SSC1p moves to the translocation Tim23-channel as a core component of "import motor complex" via the peripheral membrane protein, Tim44. Besides translocation function, SSC1p plays a crucial role in folding of proteins that are imported into the mitochondrial matrix The mitochondrial genome of eukaryotic cells is extremely susceptible to damage due to constant exposure to significant amounts of ROSs produced endogenously by mitochondria as a by-product of oxidative phosphorylation. It is known that inactivation of OGG1 in yeast leads to spontaneous mutations in the mitochondrial genome. Our analysis revealed that irradiation with low doses of gamma radiation enhanced the expression of OGG1 after 2 h. The inactivation of human OGG1 is known to induce both the spontaneous and induced mutations in the mitochondrial genome.

### 3.1.4 Sequencing of mitochondrial genes

The amplified product of two of the mitochondrial genes *COX1* and *SSC1* were sequenced using commercial services. No significant change was observed in DNA sequence of both these genes after low dose (20 Gy) irradiation (Figure 5a and 5b). This suggested that a RIR inducing doses were not inducing any mutations in the gene products.

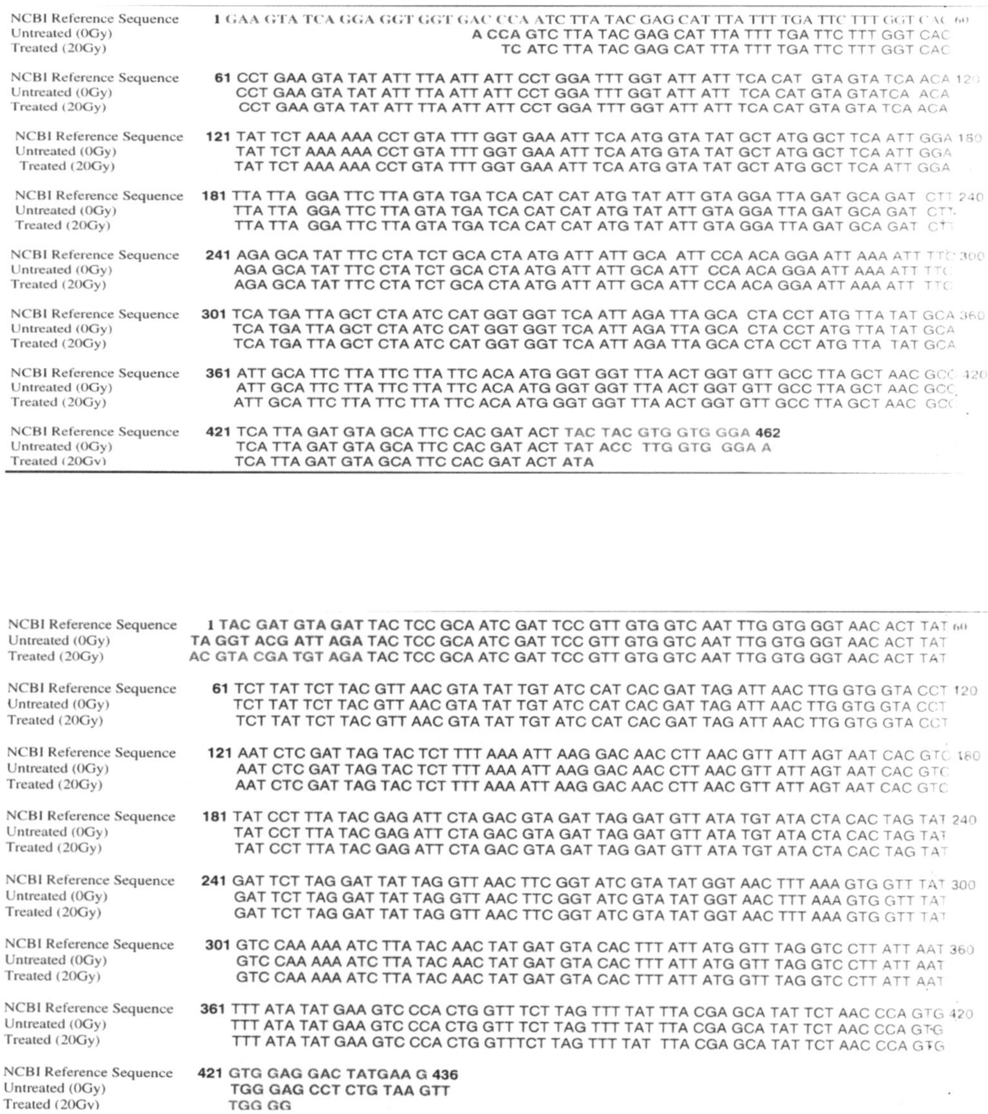


Fig. 5a. Sequencing of COX1 gene PCR product by forward primer (top chart) and reverse primer (bottom chart). Irradiation with low dose (20 Gy) did not induce changes in DNA sequence.

```
NCBI Reference Sequence     1 GTT GTT GTT GTT GTT AGA GTC GTT CTT GTA TAA TTG TTC AAA CAA TTT CAT CGA GGA AGT 60
Untreated (0Gy)               AT GTT GTT GTT GTT CGA GTC GTT CTT GTA TAT TGT TCG AC      TTT TCT CGA GGA AGT
Treated (20Gy)                AT GTT GTT GTT GTT AGA GTC GTT CTT GTA TAT TGT TCA A   CAA TTT CAT CGA GGA AGT

NCBI Reference Sequence    61 TTG CAA TTC TTC GGT CTT GGT CTT TAA CTC CTC AGC GTT AAC CTC TTC GCC ACC TTG TAC 120
Untreated (0Gy)               TTG CAC TTC TTC CGT CTT GGT CTT TT  CTC CTC ATC GTT AAC CTC TTC GCC ACC TTG TAC
Treated (20Gy)                TTG CAA TTC TTC GGT CTT GGT CTT TAA CTC CTC AGC GTT AAC CTC TTC GCC ACC TTG TAC

NCBI Reference Sequence   121 TCT AGC AAC CAA CTC CTT CAA GGA AGT GAT TTG ATC CCT AAC CTT TTG GGC TTC AGC CTT 180
Untreated (0Gy)               TCT AGC AAC CAA CTC CTT CAA GGA AGT GAT TTG ATC CCT AAC CTT TTG GGC TTC AGC CTT
Treated (20Gy)                TCT AGC AAC CAA CTC CTT CAA GGA AGT GAT TTG ATC CCT AAC CTT TTG GGC TTC AGC CTT

NCBI Reference Sequence   181 GTC AAC CTT ACC TTC AAA TTC TTT CAA GGA GTT TTC AGT ATC GTT GGC CAA TTG GTC AGC 240
Untreated (0Gy)               GTC AAC CTT ACC TTC AAA TTC TTT CAA GGA GTT TTC AGT ATC GTT GGC CAA TTG GTC AGC
Treated (20Gy)                GTC AAC CTT ACC TTC AAA TTC TTT CAA GGA GTT TTC AGT ATC GTT GGC CAA TTG GTC AGC

NCBI Reference Sequence   241 CTT GTT GGC AGT TTC GAT GGC TTG TTT TCT AGC TTC ATC TTG AGA CTT GAA TTT TTC AGC 300
Untreated (0Gy)               CTT GTT GGC AGT TTC GAT GGC TTG TTT TCT AGC TTC ATC TTG AGA CTT GAA TTT TTC AGC
Treated (20Gy)                CTT GTT GGC AGT TTC GAT GGC TTG TTT TCT AGC TTC ATC TTG AGA CTT GAA TTT TTC AGC

NCBI Reference Sequence   301 GTC GTT AAC CAT TTG TTC AAT TTC GTT TTC GGA CAA ACC AGA AGA ACC GGC AAC AGT AAT 360
Untreated (0Gy)               GTC GTT AAC CAT TTG TTC AAT TTC GTT TTC GGA CAA ACC AGA AGA ACC GGC AAC AGT AAT
Treated (20Gy)                GTC GTT AAC CAT TTG TTC AAT TTC GTT TTC GGA CAA ACC AGA AGA ACC GGC AAC AGT AAT

NCBI Reference Sequence   361 AGA AGA ATC TTT GTT TGT AGC TTT GTC TCT AGC AGA AAC GTT AAT AAT ACC ATC GGC ATC 420
Untreated (0Gy)               AGA AGA ATC TTT GTT TGT AGC TTT GTC TCT AGC AGA AAC GTT AAT AAT ACC ATC GGC ATC
Treated (20Gy)                AGA AGA ATC TTT GTT TGT AGC TTT GTC TCT AGC AGA AAC GTT AAT AAT ACC ATC GGC ATC

NCBI Reference Sequence   421 GAT GTC AAA AGT GAC TTC   GATT   TGTG   GGAC 450
Untreated (0Gy)               GAT GTC AAA AGT GAC TTCAGATT TTGAC   GGACATAACGACAAT
Treated (20Gy)                GAT GTC AAA AGA GAC ATCAATA  TTGAGAGCACACAACGACAACGGCAACAATGCC
```

```
NCBI Reference Sequence     1 GTC CCA CAA ATC GAA GTC ACT TTT GAC ATC GAT GCC GAT GGT ATT ATT AAC GTT TCT GCT 60
Untreated (0Gy)                                                GTT TTG ATC GAT GCC GAT GGT ATT ATT AAC GTT TCT GCT
Treated (20Gy)                                                 ATT TTG ATC GAT GCC GAT GGT ATT ATT AAC GTT TCT GCT

NCBI Reference Sequence    61 AGA GAC AAA GCT ACA AAC AAA GAT TCT TCT ATT ACT GTT GCC GGT TCT TCT GGT TTG TCC 120
Untreated (0Gy)               AGA GAC AAA GCT ACA AAC AAA GAT TCT TCT ATT ACT GTT GCC GGT TCT TCT GGT TTG TCC
Treated (20Gy)                AGA GAC AAA GCT ACA AAC AAA GAT TCT TCT ATT ACT GTT GCC GGT TCT TCT GGT TTG TCC

NCBI Reference Sequence   121 GAA AAC GAA ATT GAA CAA ATG GTT AAC GAC GCT GAA AAA TTC AAG TCT CAA GAT GAA GCT 180
Untreated (0Gy)               GAA AAC GAA ATT GAA CAA ATG GTT AAC GAC GCT GAA AAA TTC AAG TCT CAA GAT GAA GCT
Treated (20Gy)                GAA AAC GAA ATT GAA CAA ATG GTT AAC GAC GCT GAA AAA TTC AAG TCT CAA GAT GAA GCT

NCBI Reference Sequence   181 AGA AAA CAA GCC ATC GAA ACT GCC AAC AAG GCT GAC CAA TTG GCC AAC GAT ACT GAA AAC 240
Untreated (0Gy)               AGA AAA CAA GCC ATC GAA ACT GCC AAC AAG GCT GAC CAA TTG GCC AAC GAT ACT GAA AAC
Treated (20Gy)                AGA AAA CAA GCC ATC GAA ACT GCC AAC AAG GCT GAC CAA TTG GCC AAC GAT ACT GAA AAC

NCBI Reference Sequence   241 TCC TTG AAA GAA TTT GAA GGT AAG GTT GAC AAG GCT GAA GCC CAA AAG GTT AGG GAT CAA 300
Untreated (0Gy)               TCC TTG AAA GAA TTT GAA GGT AAG GTT GAC AAG GCT GAA GCC CAA AAG GTT AGG GAT CAA
Treated (20Gy)                TCC TTG AAA GAA TTT GAA GGT AAG GTT GAC AAG GCT GAA GCC CAA AAG GTT AGG GAT CAA

NCBI Reference Sequence   301 ATC ACT TCC TTG AAG GAG TTG GTT GCT AGA GTA CAA GGT GGC GAA GAG GTT AAC GCT GAG 360
Untreated (0Gy)               ATC ACT TCC TTG AAG GAG TTG GTT GCT AGA GTA CAA GGT GGC GAA GAG GTT AAC GCT GAG
Treated (20Gy)                ATC ACT TCC TTG AAG GAG TTG GTT GCT AGA GTA CAA GGT GGC GAA GAG GTT AAC GCT GAG

NCBI Reference Sequence   361 GAG TTA AAG ACC AAG ACC GAA GAA TTG CAA ACT TCC TCG ATG AAA TTG TTT GAA CAA TTA 420
Untreated (0Gy)               GAG TTA AAG ACC AAG ACC GAA GAA TTG CAA ACT TCC TCG ATG AAA TTG TTT GAA CAA TTA
Treated (20Gy)                GAG TTA AAG ACC AAG ACC GAA GAA TTG CAA ACT TCC TCG ATG AAA TTG TTT GAA CAA TTA

NCBI Reference Sequence   421 TAC AAG AAC GAC TCT AAC AAC AAC AAC AAC 436
Untreated (0Gy)               TAC AAG AAC GAC TCT AAC AAC AAC AAC AAC AAC AAC GGC CAC CAA TGG CCG AA
Treated (20Gy)                TAC AAG AAC GAC TCT AAC AAC AAC AAC AAC AAC AAC GGN CAC CAA TGC CAC AA
```

Fig. 5b. Sequencing of SSC1 (mt HSP70) gene PCR product by forward primer (top chart) and reverse primer (bottom chrt). Irradiation with low dose (20 Gy) did not induce changes in DNA sequence.

### 3.1.5 Chromosomal DNA damage and repair in *S. cerevisiae* cells showing RIR

Study of chromosomal DNA damage was considered important because of its role in low dose induced responses (Bala & Dwivedi 2005; Collis et al., 2004). *S. cerevisiae* has small genome divided into sixteen chromosomes of sizes ranging from 240 to 2200 kb, which can be easily resolved into discrete bands by pulsed-field gel electrophoresis (PFGE). In our study plan, PFGE could resolve the genomic DNA into several bands (Figure 6 a,b,

Lane 1). No significant change in the fluorescence intensity or mobility of bands could be recorded after low dose irradiation (20 Gy) (Lane 2). However, in comparison to untreated controls (lane1), in samples irradiated with 200 Gy, there was observable decrease in the fluorescence intensity of high molecular weight bands and increase in the smear intensity along the lanes (Figure 6, lane 3). This suggested that 20 Gy was too small a radiation dose to cause sufficiently large number of double strand breaks to be detected by PFGE. Although, presence of other types of DNA damage viz. base damage, DNA cross-links or single strand breaks as predicted by ionizing radiation at this dose could not be ruled out. The increase in the smear along the lanes in 200 Gy irradiated samples was due to settling down of broken DNA fragments along the lanes. This was similar to our earlier observations with X-irradiated (Bala & Jain, 1996) and $^{60}Co$-γ–irradiated yeast cells (Bala & Mathew, 2002) and indicated that radiation dose 200 Gy could cause DNA double strand breaks immediately after irradiation. The samples pre-irradiated with 20 Gy, incubated in PBG for 2 h and then irradiated with 200 Gy showed greater DNA bands intensities in higher molecular weight region (Lane 4) as compared to non-pre-irradiated but 200 Gy irradiated samples (Lane 3). This suggested that cells which were pre-irradiated with 20 Gy and maintained in PBG for 2 h prior to lethal dose (200 Gy) irradiation, suffered considerably lower DNA damage as compared to the lethally-irradiated

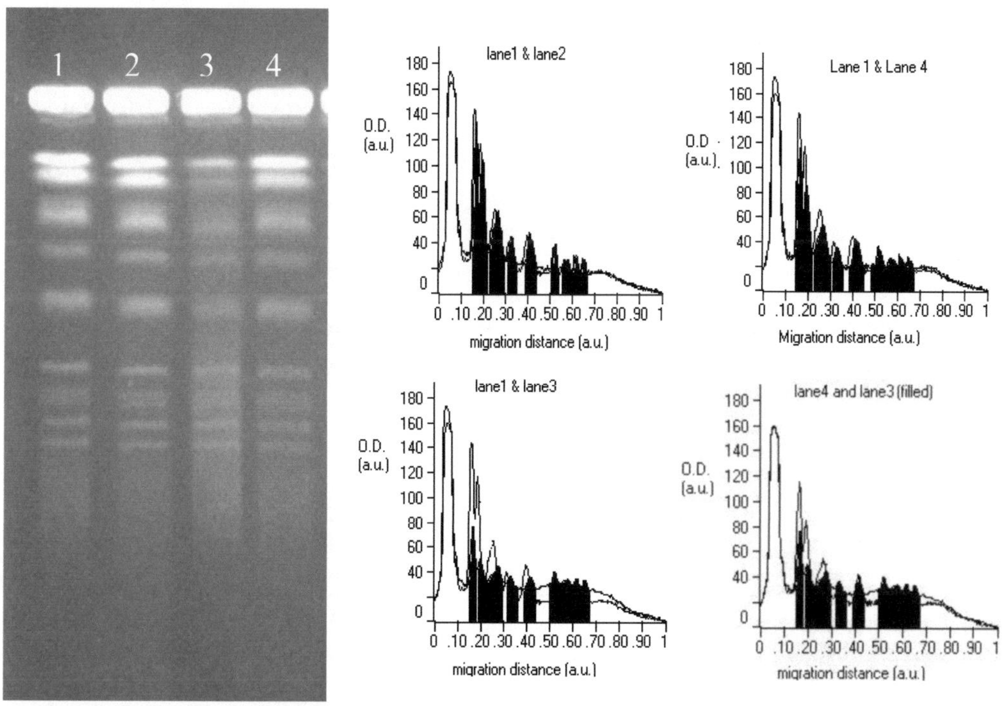


Fig. 6. (Gel picture and corresponding densitometry): The pre-irradiation with low dose (20 Gy) reduces chromosomal DNA damage induced by lethal doses of $^{60}Co$-γ-radiation (200 Gy) as studied by pulsed-field gel electrophoresis. Lane1: untreated control; Lane 2: $^{60}Co$-γ-ray (20 Gy); Lane 3: $^{60}Co$-γ-ray (200 Gy); Lane 4: $^{60}Co$-γ-ray (20 Gy) + incubation in PBG for 2 h + $^{60}Co$-γ-ray (200 Gy).

cells (lane 3). During analysis of pulsed-field gels throughout this study, the intensity changes in the individual bands in the lower molecular weight region were not given much importance because their intensities were influenced by the intensities of DNA fragments settling down as smears in the lower molecular weight regions and this has been shown to create errors in data analysis in our earlier studies (Bala & Jain 1996, Bala & Mathew 2002). Induction of gene transcription or protein expression has been reported after low dose irradiation (Franco et al., 2005). Our studies showed that low dose radiation enhanced DNA repair ability and produced protective proteins to minimize the indirect damaging effects of subsequent high dose radiation.

### 3.2 Studies with human PBMCs

The phenomenon of radioadaptive response has been reported in human lymphocytes in various studies (Table 1). The advantages of the lymphocytes as a model to understand the low dose ionizing radiation response is due to their radiosensitivity. More over, the lymphocytes are found in circulating peripheral blood and therefore, can be easily obtained from peripheral venous blood. These cells involved in cell mediated immunity as well as humoral immunity and cell proliferation and their radiosensitivity is similar to that of proliferating cells of the hematopoetic tissue.

#### 3.2.1 RIR in human PBMCs - cell proliferation, micronuclei formation

Hoechst binds with DNA and an increase in fluorescence is observed with the increase in cell number. This assay was used to determine the effects of low doses of $^{60}$Co-gamma-radiation on proliferation of PBMCs. The cells pre-irradiated with low dose of $^{60}$Co-gamma-rays (0.07 Gy) and 4-5 hours later irradiated with high dose $^{60}$Co-gamma-rays (5 Gy, LD 50) showed significantly higher cell proliferation (RIR), in comparison to non-pre-irradiated but lethally irradiated (5 Gy) cells. The RIR, however, was much less before 4h and after 5h of time interval between low dose and high dose exposure (Bala et al., 2002). The maximum increase was observed if $^{60}$Co-gamma-rays (5 Gy, LD 50) were given 5 h after the low dose (0.7 Gy) irradiation. In comparison to the non-pre-irradiated controls, the pre-irradiated cells showed decrease in micronuclei frequency and the decrease varied between the donors from 23.4% - 31.8% (Table 5, Bala et al., 2002).

| Donors | Micronuclei per 1000 binucleated cells at doses (Gy) | | | | % decrease |
|---|---|---|---|---|---|
| | 0 | 0.07 | 5.0 | 0.07+5.0 | |
| I | 13±0.33 | 15±1.02 | 245±11.08 | 167±8.92* | 31.8±2.29 |
| II | 18±0.89 | 16±1.26 | 166±9.67 | 89±6.65* | 23.4±2.14 |
| X | 12±0.65 | 14±0.88 | 298±15.38 | 206±8.77* | 30.8±2.47 |

Table 5. Micronucei (MN) in human PBMCs stained with hoechst 33342 and total 1000 cells were scored per sample at 400 × magnification in fluorescence microscope. % decrease in MN = ($MN_{low\ dose+high\ dose}$ -$MN_{low\ dose}$) X 100/$MN_{high\ dose;}$ * indicates significant ($p<0.05$) change with respect to 5.0 Gy irradiated controls (from Bala et al., 2002).

#### 3.2.2 Effect of RIR on MRN complex proteins in human PBMCs

The *MRE11/RAD50/NBN* (*MRN*) complex in humans comprises genes, which are homologous/analogous the to *MRX* complex in *S. cerevisiae*. The *MRN* complex is involved in double-strand break (DSB) repair, DNA recombination and cell cycle checkpoint control (Carson et al., 2003). The complex participates in single-strand endonuclease activity and double-strand-specific 3'-5' exonuclease activity. The protein expression of Mre11p in low dose irradiated cells was enhanced (about 1.5 times) as compared to non pre-irradiated cell after 5.0 hours of irradiation. This was similar to the enhanced the expression of *MRE11* in *S. cerevisiae* (Table 3, Figure 2). *NBS1* or *p95* is another component of the *MRN* complex, which has a role in the recruitment of the *MRN* complex to double strand break sites for DNA repair. *NBS1* plays a critical role in the cellular response to DNA damage and the maintenance of chromosome integrity. *NBS1* modulate the DNA damage signal sensing by recruiting PI3/PI4-kinase family members ATM, ATR, and probably DNA-PKcs to the DNA damage sites and activating their functions (Frappart 2005; Stiff et al., 2005). It can also recruit *MRE11* and *RAD50* to the proximity of DSBs by an interaction with the histone H2AX. *NBS1* also functions in telomere length maintenance by generating the 3' overhang which serves as a primer for telomerase dependent telomere elongation. The Nbs1p levels in low dose irradiated cells were significantly reduced (nearly 2 times) as compared to non-pre-irradiated cell after 5.0 hours of irradiation. It is not clear why the protein levels were reduced. *NBS1,* since, is inducible gene, time dependent studies are now planned to understand the role of *NBS1* in RIR. After 5 hour of low dose exposure Rad50p level was similar as in unirradiated cells (Figure 7). *RAD50* is required to bind DNA ends and hold them in close proximity This could facilitate searches for short or long regions of sequence homology in the recombining DNA templates, and may also stimulate the activity of DNA ligases and/or restrict the nuclease activity of *MRE11A* to prevent nucleolytic degradation past a given point (Jager et al., 2001, Waltes et al., 2009). In our study, the levels of RAD50p did not alter 5 hour after low dose irradiation in comparison to the untreated controls.

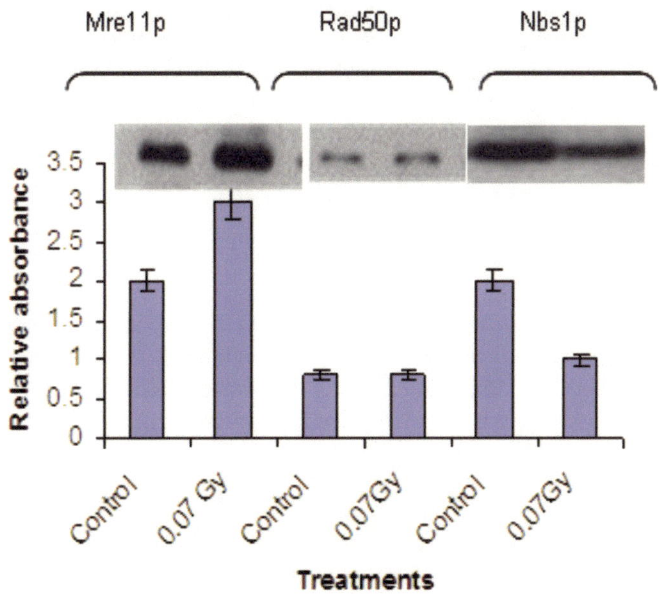


Fig. 7. Change in expression of Mre11p, Nbs1p and Rad50p in human PBMCs 5 hour after irradiation with low dose (0.07 Gy) of $^{60}$Co-$\gamma$-rays. Results were mean ±SD.

## 4. Summary and conclusion

Exposure to low dose radiation could be of significance in clinical evaluation of risk assessment, radiotherapy and radiation protection. Although a number of mechanisms such as enhanced DNA repair, alterations in stress proteins, immuno-modulation, and antioxidant defense system have been proposed to contribute to beneficial effects of low dose exposure, the understanding about the mechanisms inadequate. This is primarily because the reports are scattered, and among the available reports there is variability of dose response, the model systems as well as the experimental design followed in different laboratories. Our studies with a uniform experimental design on two different model systems viz. *Saccharomyces cerevisiae* and human PBMCs, has clearly demonstrated that irradiation with lower doses of ionizing radiation has a beneficial effect on the organism. Moreover, the effect of lower dose of radiation can not be predicted simply by extrapolating the effect of higher doses. The term 'Radiation induced radioresistance' or 'RIR' was suggested in our studies to refer a phenomenon where a single small dose radiation exposure could lead to better tolerance to the subsequently given lethal doses of radiation, and the effect was transient. Although, the RIR caused by low dose irradiation appears to be a complex interplay of many genes, this study shows that the genes of *MRX/MRN* complex, *HSP* family and also mitochondrial gene have a confirmed role in phenomenon leading to RIR. *KAR2* is an integral component of unfolded protein response (UPR) pathway. Up regulation of *KAR2* negatively regulated UPR pathway (Kimata et al., 2003), and, therefore may have caused accumulation of unfolded cytosolic proteins. If this is true then, after low dose irradiation, up regulation of *KAR2* helped the accumulation of proteins that might have unfolded due to radiation stress. The accumulation of unfolded proteins may have been responsible for increased levels of *SSA1* / *SSA2* / *SSA4* similar to that observed in *S. cerevisiae* cells after heat shock (Stone & Craig 1990). Also, it was observed the mitochondrial genes for maintaining the functional integrity of mitochondria; as well as to counter the reactive oxygen species that may have been produced because of oxidative stress produced after irradiation, were up-regulated. Further, the absence of mutation in representative sequences, decrease in revertants as well as tryptophan prototrophs, decrease in the micronuclei frequency together with enhanced levels of error-free DNA repair, strongly suggested that priming with low doses imparted transient radio-resistance to the cells culminating in the survival benefits via error-free mechanisms.

## 5. Acknowledgements

The grant provided by DRDO vide project No RD-P1-2000/INM-289 and by DST for ILTP project A 9.5) for these studies is gratefully acknowledged. The efforts of research fellows (D. Singh, M. Dwivedi, A. Arya and S. Singh); trainees (Ashutosh, Divya, Sugandh); and technical assistants (C. Prakash and A. Sharma) in performing some of experimental studies, are also acknowledged.

## 6. References

Arya, A.K., Garg, A.P. & Bala M. (2006). Upregulation of Cox1 and Shy1 genes following exposure to low dose ionizing radiation in *accharomyces cerevisiae. Indian Journal of Radiation Research.* Vol. 3, No. 4 (Nov). pp 273. ISSN 0973-0168.

Bala, M., & Goel. H.C. (2007). Modification of low dose radioresistance by 2-deoxy-D-glucose in *Saccharomyces cerevisiae:* Mechanistic aspects. *J. Radiat. Res,* Vol. 48, No. 4, pp 335–346.

Bala, M. & Dwivedi, M. (2005). Enhanced transcript of *Hsp70* family and *MRX* complex accompanies low dose ionizing radiation induced radio-resistance. Procedings of 34th Annual Meeting of the European Society of Radiation Biology, European Radiation Research, at University of Leicester, UK. 5-8, Sep, 2005.

Bala, M. & Goel, H.C. (2004). Radioprotective effect of Podophyllotoxin in *Saccharomyces cerevisae. J. Environ. Pathol. Toxicol. Onco,* Vol. 23, pp.139-144.

M., Ashutosh & Singh S. (2002). Effects of Low Dose 60Co-□-irradiation on DNA Repair in Human PMN Cells. Procedings of International conference on radiation damage and its modification, at LASTEC, Delhi 12-15, Nov, 2002.

Bala, M. & Mathew, L. (2002). An *in vitro* approach to study chromosomal DNA damage. *Molecular Biology Reports,*Vol. 28, (December). pp. 199-207.

Bala, M., Sharma, A.K. & Goel, H.C. (2001). Effects of 2-deoxy-D-Glucose on DNA repair and mutagenesis in UV-irradiated yeast. *J. Radiat. Res,* 42, No. 3, pp. 285-294. ISSN

Bala, M. & Mathew, L. (2000). Radiation hormesis and its potential to manage radiation injuries. *Journal of Scientific and Industrial Research,* Vol. 59, (December). pp. 988-994.

Bala, M. & Jain, V. (1996). 2-DG induced modulation of chromosomal DNA profile, cell survival, mutagenesis and gene conversion in X-irradiated yeast. *Indian J. Exp. Biol,* Vol. 34, (January), pp. 18-26.

Bala, M & Jain, V. (1994). Modulation of repair and fixation of UV-induced damage and its effect on mutagenesis in yeast. *Indian Journal of Experimental Biology,* Vol. 32, (December), pp. 860-864.

Blaheta, R.A., Franz, M., Auth, M.K., Wenisch, H.J., & Markus, B.H. (1991). A rapid non-radioactive fluorescence assay for the measurement of both cell number and proliferation. *J. Immunol. Methods,* Vol. 142, pp. 199-206.

Boothman, D.A., Meyers, M., Odegaard, E. & Wang, M. (1996). Altered G1 checkpoint control determines adaptive survival responses to ionizing radiation. *Mutat. Res,* Vol.358, No. 2 (November) pp.143-153.

Boreham, D.R. & Mitchel, R.E.J. (1991). DNA lesions that signal the induction of radioresistance and DNA repair in yeast. *Radiat. Res,* Vol. 128, No. 1 (October), pp. 19-28.

Boreham, D.R. & Mitchel, R.E.J. (1994). Regulation of heat and radiation stress response in yeast by hsp-104. *Radiat. Res.* Vol. 137, No. 2 (February), pp. 190-195.

Boreham, D.R., Dolling, J.A., Maves, S.R, Siwarungsun, N. & Mitchel, R.E.J. (2000). Dose rate effects for apoptosis and micronucleus formation in γ-irradiated human lymphocytes. *Radiat. Res,* 153, No. 5.1 (May), pp. 579-586.

Cai, L. & Liu, S.Z. (1990). Induction of cytogenetic adaptive response of somatic and germ cells in vivo and in vitro by low dose X-irradiation. *Int. J. Radiat. Biol.* Vol. 58, No. 1 (July) pp187-194.

Calabrese, E.J. & Baldwin, L.A. (2001). Hormesis: A generalizable and unifying hypothesis. *Crit. Rev. Toxicol,* Vol. 31, No. 4-5 (July), pp.353-424.

Carette, J., Lehnert, S. & Chow, T.Y. (2002). Implication of PBP74/mortalin/GRP75 in the radio-adaptive response. *Int. J. Radiat. Biol,* Vol. 78, No. 3 (March), pp 183-190.

Carson, C.T., Schwartz, R.A., Stracker, T.H., Lilley, C.E., Lee, D.V. & Weitzman, M.D. (2003). The Mre11 complex is required for ATM activation and the G2/M checkpoint. *EMBO J*, Vol. 22, No. 24 (December), pp. 6610- 6620.

Chaubey R.C., Bhilwade, H.N.Sonawane, V. R. & Rajagopalan R. (2006). Effect of low dose and low-dose rates of gamma radiation on DNA damage in peripheral blood leukocytes using comet assay *Int. J. of Low Radiation,* Vol. 2, No. ½, pp. 71-83.

Cohen, B.L. (1999). Validity of the linear threshold theory of radiation carcinogenesis at low doses. *Nuclear Energy,* Vol. 38, pp. 157-164.

Collis, S.J., Schwaninger, J.M., Ntambi, A.J., Keller T.W., Nelson, W.G., Dillehay, L.E. & Deweese, T.L. (2004). Evasion of early cellular response mechanisms following low level radiation induced DNA damage. *J. Biol. Chem,* 279, No. 48 (November), pp. 49624-49632.

Dasu, A. & Denekamp, J. (2002). Inducible repair and inducible radiosenstivity : a complex but predictable relationship? *Radiation Research.* Vol. 153, No. 3 (March) pp. 279-288.

Dwivedi, M., Singh, S., Sharma, A.K., Mathew, L. and Bala, M., (2001). Cytogenetis alterations associated with radio-adaptive response in proliferating and stationary phase cells of *Saccharomyces cerevisiae.* Proceedings of Biotechnocon-2001, first conference of Biotechnology Society of India. at V.P. Chest Institute, Delhi, Oct, 2001.

Dwivedi, M., Sehgal, N. & Bala M. (2008). The effects of low dose $^{60}$Co-gamma-radiation on radioresistence, mutagenesis, gene conversion, cell cycle and transcriptome profile in *Saccharomyces cerevisiae.Int. J. of Low Radiation,* Vol. 5, No. 4, pp 290-309.

Fenech, M. (1993). The cytokinesis-block micronucleus technique: a detailed description of the method and its application to genotoxicity studies in human populations. *Mutat. Res,* Vol. 285, No. 1 (January), pp. 35-44.

Franco, N., Lamartine, J., Frouin, V., Le Minter, P., Petat, C., Leplat, J.J., Libert, F., Gridol, X. & Martin, M.T. (2005). Low dose exposure to gamma rays induces specific gene regulations in normal human keratinocytes. *Radiation Research,* Vol. 163, No. 6 (June), pp. 623-635

Frappart, P.O., Tong, W.M., Demuth, I., Radovanovic I., Herceg, Z., Aguzzi, A., Digweed M. & Wang, Z.Q. (2005). An essential function for NBS1 in the prevention of ataxia and cerebellar defects. *Nature Medicine,* Vol. 11, No. 5 (May), pp. 538-544.

Gupta D, Arora R, Garg A P, Bala M, Goel H C. 2004. Modification of radiation-damage to mitochondrial system in vivo by Podophyllum hexandrum: mechanistic aspects. *Mol Cell Biochem* Vol. 266: pp. 65-77.

Hohl M, Kwon Y, Galván S.M., Xue X, Tous C, Aguilera A, Sung P, Petrini J.H.J. (2011) The Rad50 coiled-coil domain is indispensable for Mre11 complex functions. *Nature Structural & Molecular Biology* Volume: 18, pp. 1124-1131

Jager, De. M., Van N. J., Vangent, D.C., Dekker, C., Kanaar, R. & Wyman, C. (2001). Human Rad50/Mre11 is a flexible complex that

Kimata, Y., Kimata, Y.I., Shimizu, Y., Abe, H., Farcasanu, I.C. & Takeuchi, M. et al. (2003). Genetic evidence for a role of Bip/Kar2 that regulates Ire1 in response to accumulation of unfolded proteins, *Mol. Biol. Cell.* Vol. 14, No. 6 (June), pp. 2559-2569.

Luckey, T.D. (2008). Atomic bomb health benefits. *Dose Response,* Vol. 6, No. 4 (August), pp. 369-382.

Nambi, K.S.V. & Soman, S.D. (1990). Further observations on environmental radiation and cancer in India. *Health Phys.* Vol. 59, No. 3 (September), pp. 339-344.

Olivieri, G., Bodycot, J. & Wolff, S. (1984). Adaptive response of human lymphocytes to low concentration of radioactive thymidine. *Science* Vol. 223, No. 4636 (February), pp. 594-597.

Pandey, B.N., Sarma H.D., Shukla, D. & Misra, K. P. (2006). Low dose radiation induced modification of ROS and apoptosis in thymocytes of whole body irradiated mice, *Int. J. Low Radiation.* Vol. 2, No. ½ , pp. 111-118.

Sanderson, B.J. & Morley, A.A.(1986). Exposure of human lymphocytes to ionizing radiation reduces mutagenesis by subsequent ionizing radiation. *Mutat. Res.* Vol. 164, No. 6 (December), pp 347-351.

Sahara, T., Goda, T. And Ohgiya, S. (2002). Comprehensive expression analysis of time dependent genetic responses in yeast cells to low temperatures. *J Biol Chem,* Vol. 277, No.51 (December), pp. 50015-50021.

Sasaki, M.S., Ejima, Y., Tachibana, A., Yamada, T., Ishizaki, K., Shimizu, T. & Nomura, T. (2002). DNA damage response pathway in radio-adaptive response, *Mutat. Res,*Vol. 504, No. 1-2 (July), pp. 101-118.

Seo, H.R., Chung, H.Y., Lee, Y.J., Bae, S., Lee, S.J. & Lee, Y.S. (2006). p27Cip/Kip Is Involved in Hsp25 or Inducible Hsp70 Mediated Adaptive Response by Low Dose Radiation. *J. Radiat. Res.* Vol. 47, No. 1 (March), pp 83.90.

Shadley, J.D. & Wiencke J.K. (1989). Induction of the adaptive response by X-rays is dependent on radiation intensity. *Int. J. Radiat. Biol.* Vol. 56, No. 1 (July) pp 107-118.

Shadley, J.D. & Wolff, S. (1987). Very low doses of X-rays can cause human lymphocytes to become less susceptible to ionizing radiation. *Mutagenesis.* Vol.2, No.2 (March), pp. 95-96.

Sharma, A.K. & Bala, M., (2002). Beneficial effect of low level $^{60}$Co-$\gamma$ -radiation on yeast *Saccharomyces cerevisiae.* Vijnana Parishad Anusand. Patrika. Vol. 45, No. 1 (January), pp 77-82.

Smolka M. B., Albuquerque C. P., Chen S. H., Zhou H. (2007)"Proteome-wide identification of in vivo targets of DNAdamage checkpoint kinases." Proc. Natl. Acad. Sci. U.S.A. Vol. 104, pp. 10364-10369.

Stone D.E. & Craig, E.A. (1990) Self regulation of 70-kilodalton heat shock proteins in *Saccharomyces cerevisiae, Mol. Cell. Biol.* Vol. 10,No. 4 (April), pp. 1622-1632.

Stiff T., Reis C., Alderton G.K., Woodbine L., O'Driscoll M., Jeggo P.A. (2005) Nbs1 is required for ATR-dependent phosphorylation events. EMBO J. 24, pp. 199-208.

UNSCEAR (1994) United Nations Scientific Committee on the effects of Atomic Radiation report on "Adaptive response to radiation in cells and organisms, Sources and effects of ionizing radiation: Report to the General Assembly, with Scientific Annexes. Annex B. New York: United Nations.

UNSCEAR, (2000). Report to the General Assembly, with Scientific Annexes. Volume II: Effects. Annex I: Epidemiological Evaluation of Radiation-Induced Cancer No. E.00.IX.4. United Nations, New York.

Waltes R., Kalb R., Gatei M., Kijas A.W., Stumm M., Sobeck A., Wieland B., Varon R., Lerenthal Y., Lavin M.F., Schindler D., Doerk T. (2009) Human RAD50 deficiency in a Nijmegen breakage syndrome-like disorder. *Am. J. Hum. Genet.* Vol 84, pp. 605-616.

Zimmermann, F.K., Kern, R. & Rosenberger, H. (1975). A yeast strain for simultaneous detection of induced mitotic crossing over, mitotic gene conversion and reverse mutation. *Mutat. Res.* Vol. 28, No. 3 (June), pp. 381-388.

# 9

# Sterilization by Gamma Irradiation

Kátia Aparecida da Silva Aquino
*Federal University of Pernambuco-Department of Nuclear Energy*
*Brazil*

## 1. Introduction

Sterilization is defined as any process that effectively kills or eliminates almost all microorganisms like fungi, bacteria, viruses, spore forms. There are many different sterilization methods depending on the purpose of the sterilization and the material that will be sterilized. The choice of the sterilization method alters depending on materials and devices for giving no harm. These sterilization methods are mainly: dry heat sterilization, pressured vapor sterilization, ethylene oxide (EtO) sterilization, formaldehyde sterilization, gas plasma ($H_2O_2$ ) sterilization, peracetic acid sterilization, e-beam sterilization and gamma sterilization.

Gamma radiation sterilization and e-beam sterilization are mainly used for the sterilization of pharmaceuticals. Gamma radiation delivers a certain dose that can take time for a period of time from minutes to hours depending on the thickness and the volume of the product. E-beam irradiation can give the same dose in a few seconds but it can only give it to small products. Depending on their different mechanism of actions, these sterilization methods affect the pharmaceutical formulations in different ways. Thus, the sterilization method chosen must be compatible with the item to be sterilized to avoid damage.

To be effective, gamma or e-beam sterilization requires time, contact and temperature. The effectiveness of any method of sterilization is also dependent upon four other factors like the type of microorganism present. Some microorganisms are very difficult to kill. Others die easily the number of microorganisms present. It is much easier to kill one organism than many the amount and type of organic material that protects the microorganisms. Blood or tissue remaining on poorly cleaned instruments acts as a shield to microorganisms during the sterilization process, the number of cracks and crevices on an instrument that might harbor microorganisms. Microorganisms collect in, and are protected by, scratches, cracks and crevices such as the serrated jaws of tissue forceps.

Finally, here is no single sterilization process for all the pharmaceuticals and medical devices. It is hard to assess a perfect sterilization method because every method has some advantages and disadvantages. For this reason, sterilization process should be selected according to the chemical and physical properties of the product. It is fairly clear that different sterilization processes are used in hospital and in industry applications. While EtO or autoclave sterilization is used in hospitals, gamma radiation or e-beam sterilization is used in industry depending on the necessity of a developed institution. Superiority of radiation sterilization to EtO and other sterilization methods are known by all over the world. These factors facilitate to understand the relatively fast increase of the constitution of irradiation institutions. Thus, this chapter will discuss the use of sterilization by gamma radiation.

## 2. Radiation processing

Radiation processing refers to the use of radiation to change the properties of materials on an industrial scale. The term 'ionizing radiation' relates to all radiation capable of producing ionization cascades in matter. The energy range characteristic of ionizing radiation begins at about 1000 eV and reaches its upper limit at about 30 MeV. To avoid induced radioactivity, which may appear if the gamma ray energy is higher than 5 MeV or the energy of the fast electrons exceeds 10 MeV, it is prohibited to use for sterilization radiation characterized by energy higher than these values. On the other hand, the application of lower energy radiation (below 0.2 MeV) is not rational. Commercial gamma ray irradiation facilities are typically loaded with $^{60}Co$ of total activity from 0.3 to 3.0 MCi[1], while commercial e-beam facilities are equipped with one or two electron accelerators generating high power (10– 100 kW) beams of 8–10 MeV electrons.

When radiation passes through materials it breaks chemical bonds. Radiation processing has been used commercially for almost forty years. Gamma radiation from $^{60}Co$, electron beams and x-rays, are all used to sterilize the medical devices used in operations and other healthcare treatments. Implants, artificial joints, syringes, blood-bags, gowns, bottle teats for premature baby units and dressings are all sterilized using radiation. The surgical gloves are sterilized using gamma radiation from $^{60}Co$. Other industries that benefit from radiation processing include the food, pharmaceutical, cosmetic, horticultural, and automotive industries. In the horticultural industry, growing-mats, fleeces and pots may be reused after irradiation-reducing waste and cost and saving the environment from unnecessary waste. Similarly, commercial egg trays may be recycled after irradiation without risk of proliferating salmonella.

Gamma rays are formed with the self disintegration of Cobalt-60 ($^{60}Co$) or Cesium-137 ($^{137}Cs$) sources. Among thousands of gamma emitters only $^{137}Cs$ and $^{60}Co$ are indicated for radiation processing. The energy of gamma rays, as electromagnetic quantum waves, is similar to light, but with higher photon energy and shorter wavelength. The $^{60}Co$ radionuclide can be produced in a nuclear power reactor by the irradiation of $^{59}Co$ (metal), with fast neutrons. The radioactive isotope is formed by neutron capture as showed equation 1 (Laughlin, 1989).

$$_{27}Co^{59} + {_0}n^1 \rightarrow {_{27}}Co^{60} \tag{1}$$

The unstable nucleus of $^{60}Co$ emits photons of 1.17 and 1.33 MeV, decaying with a half-life of 5.2714 years to stable $^{60}Ni$ as shown the Figure 1 (Kaplan, 1955). The radioactive $^{60}Co$ source is composed of small pellets of cobalt that are loaded into stainless steel or zirconium alloy sealed tubes (pencil arrays).

Radiation is the unique source of energy which can initiate chemical reactions at any temperature, including ambient, under any pressure, in any phase (gas, liquid or solid),

[1] Ci (currie)=3.7 x $10^{10}$ Bq (becquerel)

without use of catalysts. Thus, radiation processing uses highly penetrating gamma radiation from sealed radiation sources travelling at almost the speed of light, to bombard and kill bacteria in products sealed inside their final packaging. In this way the irradiated product remains sterile until the packaging is removed. The energy carried by the gamma radiation is transferred to the product being irradiated by collisions between the radiation and the atoms of the product. In these collisions atoms lose their bound electrons in a process called ionization. It is this process that results in irreparable damage to the life sustaining chemistry of living organisms and the initiation of crosslinking chemistry or main chair scission in polymeric materials.

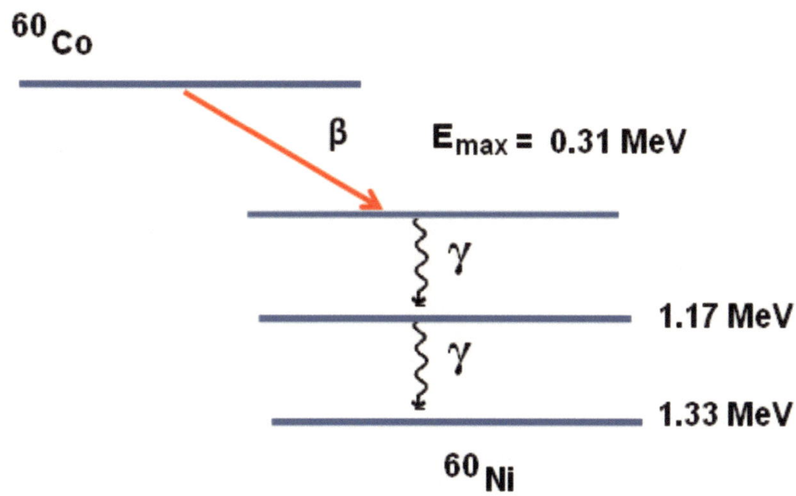


Fig. 1. Disintegration of $^{60}$Co

## 3. Gamma sterilization

### 3.1 General aspects

Gamma rays are generally used for the sterilization of gaseous, liquid, solid materials, homogeneous and heterogeneous systems and medical devices, such as syringes, needles, cannulas, etc. Gamma irradiation is a physical means of decontamination, because it kills bacteria by breaking down bacterial DNA, inhibiting bacterial division. Energy of gamma rays passes through hive equipment, disrupting the pathogens that cause contamination. These photon-induced changes at the molecular level cause the death of contaminating organisms or render such organisms incapable of reproduction. The gamma irradiation process does not create residuals or impart radioactivity in the processed hive equipment. Complete penetration can be achieved depending on the thickness of the material. It supplies energy saving and it needs no chemical or heat dependence. Depending on the radiation protection rules, the main radioactive source has to be shielded for the safety of the operators. Storage of is needed depending on emitting gamma rays continuously

The first aspect to consider when sterilizing with gamma is product tolerance to the radiation. During use of this type of radiation, high-energy photons bombard the product, causing electron displacement within. These reactions, in turn, generate free radicals, which aid in breaking chemical bonds. Disrupting microbial DNA renders any organisms that survive the process nonviable or unable.

Gamma radiation does have some significant advantages over other methods of producing sterile product. These benefits include: better assurance of product sterility than filtration and aseptic processing; no residue like EtO leaves behind; more penetrating than E-beam; low-temperature process and simple validation process.

Process validation may be defined as the documented procedure for obtaining, recording and interpreting the results required to establish that a process will consistently yield product complying with a predetermined specification. For sterilization, process validation is essential, since sterilization is one of those special processes for which efficacy cannot be verified by retrospective inspection and testing of the product. Process validation consists of: i. installation qualification of the facility; ii. operational qualification of the facility and iii. performance qualification of the facility (ISO 14937, 2000)

Radiation sterilization of medical products also is currently regulated by two standards, EN 552 (1994) and ISO 11137 (1995). These standards will be harmonized in the very near future into ISO 11137 (2006) part 1, part 2 and part 3. Currently, all three parts of ISO 11137 (2006) are at the Final Draft International Standard Stage (FDIS). These three documents are now published. All sterilization standards consider 'dose' as a key parameter in order to determine if a product is sterile. However, measurement of dose is not a trivial task and a commercial dosimetry system consists of dosimeters, readout equipment and procedure for its use. Dosimeters may be films, small plastic blocks, fluids or pellets where there is a known and reproducible response to radiation dose. The dosimetry system must be calibrated, and the calibration must be traceable to a national standard. ISO/ASTM standard 51261 gives guidelines for calibration procedures.

### 3.2 Effects of gamma rays on living organisms

Radiation effects on living organisms are mainly associated with the chemical changes but are also dependent on physical and physiological factors. Dose rate, dose distribution, radiation quality are the physical parameters. The most important physiological and environmental parameters are temperature, moisture content and oxygen concentration. The action of radiation on riving organisms can be divided into direct and indirect effects. Normally, the indirect effects occur as an important part of the total action of radiation on it. The Figure 2 shown that radiolytic products of water are mainly formed by indirect action on water molecules yielding radicals OH• , $e^-$ aq and H•. The action of the hydroxyl radical (OH•) must be responsible for an important part of the indirect effects. Drying or freezing of living organisms can reduce these indirect effects. If we consider pure water, each 100 eV of energy absorbed will generate: 2.7 radicals OH•, 2.6 e- aq, 0.6 radicals H•, 0.45 $H_2$ molecules and 0.7 molecules $H_2O_2$. (Borrely et al, 1998).

Several types of microorganism, mainly bacteria and, less frequently, moulds and yeasts, have been found on many medical devices and pharmaceuticals (Takehisa et al, 1998). Complete eradication of these microorganisms (sterilization) is essential to the safety of medical devices and pharmaceutical products. The sterilization process must be validated to verify that it effectively and reliably kills any microorganisms that may be present on the pre-sterilized product. Radiation sterilization, as a physical cold process, has been widely used in many developed and developing countries for the sterilization of health care products. Earlier, a minimum dose of 25 kGy was routinely applied for many medical devices, pharmaceutical products and biological tissues. Now, as recommended by the International Organization for Standardization (ISO), the sterilization dose must be set for each type of product depending on its bioburden. Generally, the determination of sterilization dose is the responsibility of the principal manufacturer of the medical product, who must have access to a well qualified microbiology laboratory.

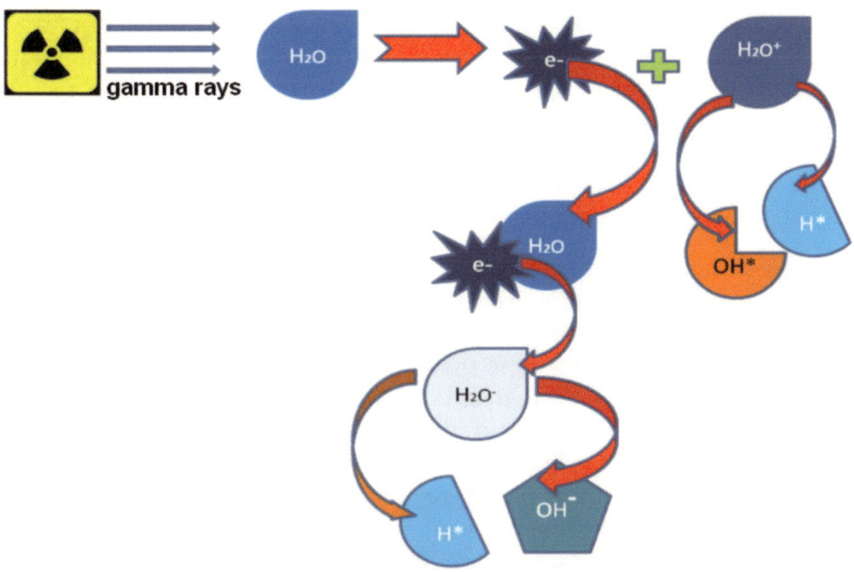


Fig. 2. Effect of gamma rays on water molecules

The lethal effect of ionizing radiation on microorganisms, as measured by the loss by cells of colony-forming ability in nutrient medium, has been the subject of detailed study. Much progress has been made towards identification of the mechanism of inactivation, but there still remains considerable doubt as to the nature of the critical lesions involved, although it seems certain that lethality is primarily the consequence of genetic damage. Many hypotheses have been proposed and tested regarding the mechanism of cell damage by radiation. Some scientists proposed the mechanism thought 'radiotoxins' that are the toxic substances produced in the irradiated cells responsible for lethal effect. Others proposed that radiation was directly damaging the cellular membranes. In addition, radiation effects on enzymes or on energy metabolism were postulated. The effect on the cytoplasmic membrane appears to play an additional role in some circumstances (Greez et al, 1983).

It is now universally accepted that the deoxyribonucleic acid (DNA) in the chromosomes represents the most critical 'target' for ionizing radiation because it is responsible for inhibition of cell division.

A DNA strand is composed of a series of nucleotides containing a purine (adenine, guanine) or a pyrimidine base (cytosine, thymine), a sugar (deoxyribose) bond to the base and a phosphate connected to the sugar. The nucleotides are joined by phosphodiester bonds between the sugar and the phosphate. DNA is composed of two complementary anti-parallel strands linked by hydrogen bonds between the bases. Thymine is complementary to adenine (two hydrogen bonds between them) whilst guanine is the complementary base to cytosine (linked by three hydrogen bonds). In the most frequent configuration, called B form, the two strands are twisted to form a right-handed double helix. Ionizing radiation can affect DNA either directly, by energy deposition in this critical target, or indirectly, by the interaction of radiation with other atoms or molecules in the cell or surrounding the cell like water. In particular, radiation interacts with water, leading to the formation of free radicals (see Figure 2) that can diffuse far enough to reach and damage DNA. It is worth mentioning that the OH• radical is most important; these radicals formed in the hydration layer around the DNA molecule are responsible for 90% of DNA damage. Consequently, in a living cell, the indirect effect is especially significant. In a general sense, the death of a microorganism is a consequence of the ionizing action of the high energy radiation. It is estimated that the irradiation of a living cell at one gray induces 1000 single strand breaks, 40 double strand breaks, 150 cross-links between DNA and proteins and 250 oxidations of thymine (ABCRI, 1992; Borrely et al, 1998) ).

Both prokaryotes (bacteria) and eukaryotes (moulds and yeasts) are capable of repairing many of the different DNA breaks (fractures). Living organisms have developed different strategies to recover from losses of genetic information caused by DNA damages. Damages to DNA alter its spatial configuration so that they can be detected by the cell. In the case of single strand breaks (Figure 3), the damaged DNA strand is excised and its complementary strand is used to restore it. Efficient and accurate repair of the damages can take place as long as the integrity of the complementary strand is maintained. Radiosensitivity is highly influenced by the capability of the strain to repair single-strand breaks. Strains that lack this ability are far more radiosensitive than the others (Tubiana et al., 1990; WHO, 1999). Double strand breaks are far more hazardous since they can lead to genome rearrangements. Two distinct mechanisms have been described for the repair of double strand breaks: non homologous end joining and recombination repair (Broomfield et al., 2001).

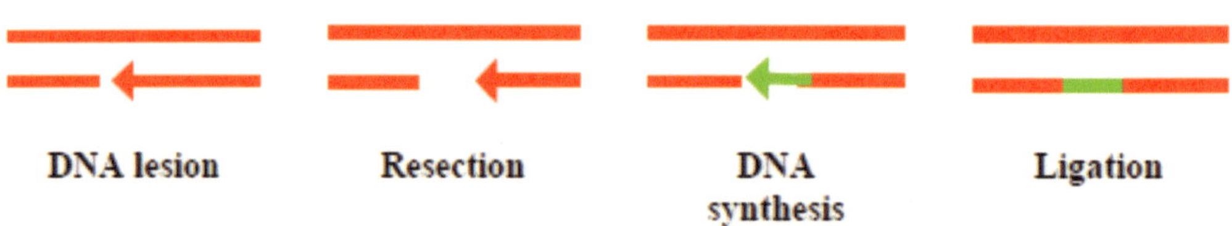


Fig. 3. Single strand breaks in DNA

1. For non homologous end joining, the free ends are joined by simple ligation which may result either to perfect reparation or to genetic mutation if sequences are not homologue.
2. Combinational repair (Figure 4) necessitates the presence of another copy of the genetic material within the cell since an identical DNA sequence is used as a template. This last mechanism cannot be achieved by all bacteria since some only possess one copy of genetic material per cell (Hansen, 1978; Kuzminov, 1999).

Apart from difficulties in location of the site of primary damage, there is still controversy as to whether the majority of radiation effects on biological systems are due directly to ionization or to the indirect action of the radiolysis products of water, or both. However, while the work on basic mechanisms continues, much is already known both qualitatively and quantitatively in relation to the radiation inactivation of microbial populations. Just as with heat resistance, there is considerable variability in radiation resistance between microbial species; in general, viruses are more radiation resistant than bacterial spores, which in turn are more resistant than vegetative organisms, yeasts and moulds. Moreover, the inactivation of microbial populations is considerably influenced by conditions of environment during irradiation-for example, gaseous composition, temperature, and nature of the suspending medium.

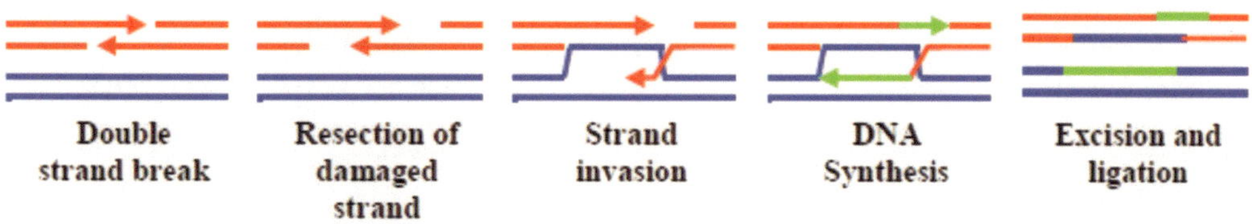


Fig. 4. Combinational repair of DNA double break

### 3.2.1 Decimal reduction dose

When a suspension of a microorganism is irradiated at incremental doses, the number of surviving cell forming colonies after each incremental dose may be used to construct a dose survival curve, as shown in Figure 5. The radiation resistance of a microorganism is measured by the so-called decimal reduction dose ($D_{10}$ value), which is defined as the radiation dose (kGy) required to reduce the number of that microorganism by 10-fold (one log cycle) or required to kill 90% of the total number (Whitby & Gelda, 1979). The $D_{10}$ value

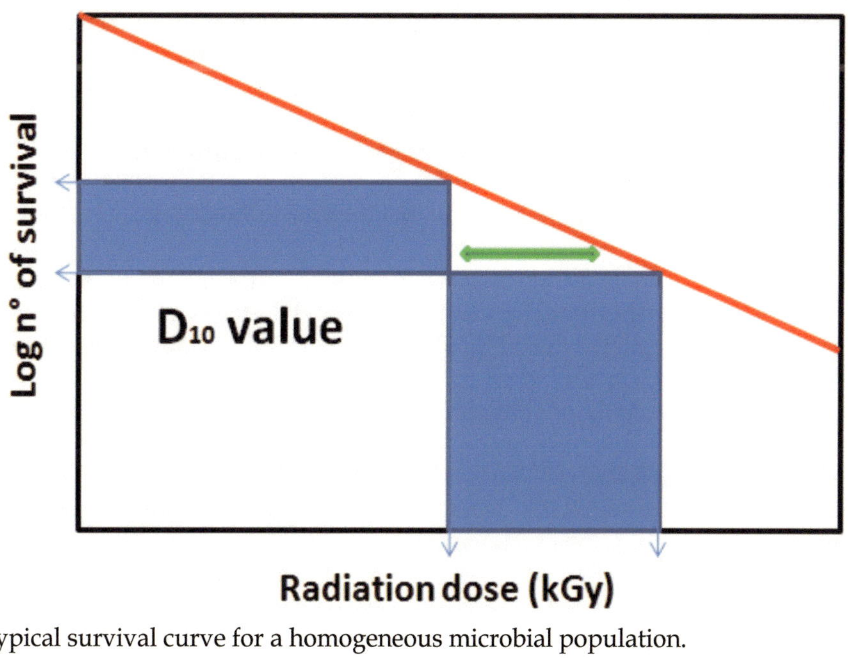


Fig. 5. Typical survival curve for a homogeneous microbial population.

can be measured graphically from the survival curve, as shown in Figure 5; the slope of the curve (mostly a straight line) is related to the $D_{10}$ value. With certain microorganisms, a 'shoulder' may appear in the low dose range before the linear slope starts. This 'shoulder' may be explained by multiple targets and/or certain repair processes being operative at low doses.

The decimal reduction dose is affected by irradiation conditions in which the microorganisms exist in dry or freezing, aerobic or anaerobic conditions. The $D_{10}$ value of some organisms (responsible for selected water-born diseases) irradiated in buffer solution is presented in Table 1.

| Microorganism | $D_{10}$ (kGy) | Desease | Reference |
|---|---|---|---|
| *Slamonella typhimurim* | 0.30 | Gastroenteritis | Borrely, 1998 |
| *Mycobacterium tuberculosis* | 0.30 | Tuberculosis | IAEA, 1975 |
| *Shigella dysenteriae* | 0.60 | Dysentery | IAEA, 1975 |
| *Vibrio cholerae* | 0.48 | Cholera | IAEA, 1975 |

Table 1. Decimal reduction dose ($D_{10}$) of some microorganisms

There are many factors affecting the resistance of microorganisms to ionizing radiation, thus influencing the shape of the survival curve. The most important factors are:

a. Size and structural arrangement of DNA in the microbial cell;
b. Compounds associated with the DNA in the cell, such as basic peptides, nucleoproteins, RNA, lipids, lipoproteins and metal ions. In different species of microorganisms, these substances may influence the indirect effects of radiation differently;
c. Oxygen: The presence of oxygen during the irradiation process increases the lethal effect on microorganisms. Under completely anaerobic conditions, the $D_{10}$ value of some vegetative bacteria increases by a factor of 2.5–4.7, in comparison with aerobic conditions;
d. Water content: Microorganisms are most resistant when irradiated in dry conditions. This is mainly due to the low number or absence of free radicals formed from water molecules by radiation, and thus the level of indirect effect on DNA is low or absent;
e. Temperature: Treatment at elevated temperature, generally in the sub-lethal range above 45°C, synergistically enhances the bactericidal effects of ionizing radiation on vegetative cells. Vegetative microorganisms are considerably more resistant to radiation at subfreezing temperatures than at ambient temperatures. This is attributed to a decrease in water activity at subfreezing temperatures. In the frozen state, moreover, the diffusion of radicals is very much restricted;
f. Medium: The composition of the medium surrounding the microorganism plays an important role in the microbiological effects. $D_{10}$ values for certain microorganisms can differ considerably in different media;
g. Post-irradiation conditions: Microorganisms that survive irradiation treatment will probably be more sensitive to environmental conditions (temperature, pH, nutrients, inhibitors, etc.) than the untreated cells.

In addition, it has been suggested that some pigments synthesized by microorganisms may play a role in their resistance towards ionizing radiation. For example, carotenoids synthesized by *Exiguobacterium acedicum* were found to be responsible for its radioresistance (Kim et al., 2007). Fungi that synthesize pigments such as *Curvularia geniculata* (melanin) or other *Dematiaceous fungi* that contain melanin and carotenoids have higher $D_{10}$ values (Saleh et al., 1988; Geis & Szaniszlo, 1984). These pigments appear to be involved in both photo- and radio-protection. It was also discovered that a higher amount of $Mn^{+2}$ in some radioresistant bacteria may partly explain their resistance due to the decrease of protein oxidation in presence of higher concentrations of $Mn^{+2}$ (Daly et al., 2007).

### 3.2.2 Sterilization dose

It can be defined as the absorbed energy per unit mass ([J.kg-1] = [Gy]). Survival fraction of the microorganisms is reversely proportional with the absorbed dose. Doses for sterilization should be chosen according to the initial bioburden, sterility assurance level (SAL) and the radiosensitivity of microorganisms. A sterility assurance level (SAL) is derived mathematically and it defines the probability of a viable microorganism being present on an individual product unit after sterilization. SAL is normally expressed as $10^{-n}$. SAL is generally set at the level of $10^{-6}$ microorganisms/ml or g for the injectable pharmaceuticals, ophtalmic ointment and ophtalmic drops and is $10^{-3}$ for some products like gloves that are used in the aseptic conditions. Generally for an effectively (F -value) of n = 8 is employed for sterilization of *Bacillus pumilus* for the standard dose of 25 kGy is equivalent to about eight times its $D_{10}$ (2.2-3 kGy).

The process of determining the sterilization dose is intended to establish the minimum dose necessary to achieve the required or desired sterility assurance level (SAL). Sterilization dose depends on: i. level of viable microorganisms on the product before the sterilization process (natural bioburden); ii. relative mix of various microorganisms with different $D_{10}$ values; iii. degree of sterility, i.e. sterility assurance level (SAL), required for that product. Because of this reason, the optimum sterilization dose is 25 kGy at the above level of bioburden (Takehisa et al, 1998).

On the other hand, the response of a microbial cell and hence its resistance to ionizing radiation depends of many factors like: i. nature and amount of direct damage produced within its vital target; ii. number, nature and longevity of radiation induced reactive chemical changes; iii. inherent ability of the cell to tolerate or correctly repair the damage and iv. influence of intra and extracellular environment on any of the above.

In general, bioburden on any product is made up of a mixture of various microbial species, each having its own unique $D_{10}$ value, depending on its resistance to radiation; these various species exist in different proportions. A standard distribution of resistances ($D_{10}$ values) has been agreed upon for the determination of sterilization dose based on Method 1 of ISO 11137 (1995). Thus, 65.487% of the microorganisms on a product has a $D_{10}$ value of 1.0 kGy, 22.493% of the microorganisms has a $D_{10}$ value of 1.5 kGy, etc. This is an average distribution based on significant amounts of data. It is not always that this distribution exists; it would depend on the conditions of manufacturing and subsequent processes. Method 1 of ISO 11137 (1995) is based on confirming that this distribution exists. From the reported survival data resulting from numerous investigations carried out on the effects of ionizing radiation on microorganisms, the following observations may be made:

1. Generally, bacterial spores are considered more radiation resistant than vegetative bacteria;
2. Among vegetative bacteria, gram-positive bacteria are more resistant than gram-negative bacteria;
3. *Vegetative cocci* are more resistant than vegetative bacilli;
4. Radiation sensitivity of moulds is of the same order as that of vegetative bacteria;
5. Yeasts are more resistant to radiation than moulds and vegetative bacteria;
6. Anaerobic and toxigenic Clostridium spores are more radiation resistant than the aerobic non-pathogenic Bacillus spores;
7. Radiation resistance of viruses is much higher than that of bacteria or even bacterial spores;
8. The majority of fungi have $D_{10}$ values between 100-500 Gy. *Dematiaceous fungi,* which are found in soils and rotten woods but normally not in pharmaceuticals, are highly radioresistant with $D_{10}$ values from 6 to 17 kGy. Yeast is more resistant than other fungi. *Candida albicans* for example was found to be quite radioresistant with $D_{10}$ of 1.1 to 2.3 kGy;
9. In general, it is observed that viruses are less sensitive towards ionizing radiation than bacteria and fungi. $D_{10}$ values for most viruses range from 3 to 5 kGy (Grieb et al., 2005), which is far more than bacteria. Radiation sensitivities of single stranded DNA viruses are higher than those of double stranded ones;
10. Viruses should not normally be found in pharmaceuticals, except in those originating from biotechnological processes. Biological products are submitted to specific guidelines (IAEA, 2004) and the use of higher irradiation doses may be validated for the elimination of viruses. Inactivation with a sufficient S.A.L. ($<10^{-9}$) of viruses such as HIV or hepatitis in grafts necessitates high doses from 60 to 100 kGy (Campbell & Li, 1999). Table 2 showed the radiosensivities of some micoorganisms at determined conditions.

### 3.2.3 Effect of temperature and additive on radiosensitivity of living organisms

Temperature plays a major role in the radiosensitivity of microorganisms. As temperature decreases, water radicals become less mobile. As a general rule, microorganisms are less radiosensitive when irradiated at low temperatures (Thayer & Boyd, 2001). For example, whilst sensitivity of spores from *Bacillus megaterium* was constant between -268 and -148°C, an increase in temperature to 20°C led to a 40% increase in sensitivity. Effect of temperature was observed to be similar for oxic and anoxic spores (Helfinstine et al., 2005).

The indirect effect is partially abolished by freezing the solution. The highest decrease in sensitivity is observed between 0 and -15°C. For example, $D_{10}$ value of *Escherichia* ***coli*** irradiated in meat increased from 0.41 kGy at +5°C to 0.62 kGy at -15°C. For *Staphylococcus* ***aureus***, $D_{10}$ at -76°C was 0.82 kGy instead of 0.48 kGy at +4°C (Sommers et al., 2002). Subfreezing temperatures offer less protection for spores than for vegetative species since they already have low moisture content. The irradiation of frozen aqueous solutions allowed minimizing the loss of active substance even for a 25 kGy dose. This approach seems to be the most promising method for terminal sterilization of aqueous solutions by ionizing radiations. The major radiolysis product was formed after the attack of the electron. Some of the radiolysis products detected were attributed to the attack of •OH,

| organism | classification | $D_{10}$ (kGy) | condition | reference |
|---|---|---|---|---|
| *Clostridium botulinum* spores | bacteria | 2.9 | 0°C, Phosphate Buffer | Grecz et al., 1965 |
| *Clostridium botulinum* spores | bacteria | 4.6 | Meat, 0°C | Grecz et al., 1965 |
| *Clostridium botulinum* spores | bacteria | 3.9 | Phosphate Buffer, -196°C | Grecz et al., 1965 |
| *Clostridium botulinum* spores | bacteria | 6.8 | Meat, -196°C | Grecz et al., 1965 |
| *Aspergillus flavus* | fungi | 0.60 | Aerated water, 20°C | Saleh et al., 1988 |
| *Aspergillus niger* | fungi | 0.42 | Aerated water, 20°C | Saleh et al., 1988 |
| *Cladosporium cladosporioides* | fungi | 0.03-0.25 | Aerated water, 20°C | Saleh et al., 1988 |
| *Curvularia geniculata* | fungi | 2.42-2.90 | Aerated water, 20°C | Saleh et al., 1988 |
| Coxsackievirus B-2 | viruses | 5.3 | Water, -90°C | Sullivan et al, 1973 |
| Coxsackievirus B-2 | viruses | 7.0 | Meat, 16°C | Sullivan et al, 1973 |
| Coxsackievirus B-2 | viruses | 8.1 | Meat, -90°C | Sullivan et al, 1973 |
| HIV | viruses | 8.8 | Bone, -78°C | Campbell and Li, 1999 |

Table 2. Radiosensivities of some micoorganisms

demonstrating the feasibility of a reaction between the •OH from ice radiolysis and the solute. A comparison was performed with irradiated frozen solutions of metoprolol, which has been studied in liquid aqueous solutions (Crucq et al, 2000). Degradation of metoprolol when irradiated in frozen solutions was negligible.

On the other hand, the evaluation of the radiosensitivity of bacteria as a function of the addition of radical scavengers is quite difficult since many experiments have been carried out either on isolated DNA, which does not take into account the effects within the cell. For experiments carried out on bacteria, the concentration of the scavenger within the cell was assumed to be equal to that of the extracellular media, which is generally not the case.

It was shown that the protection of bacteria against ionizing radiation in the presence of hydroxyl radical scavengers was highly dependent of the irradiation conditions (Billen, 1984). Scavengers are unable to prevent semi-direct effect due to the hydroxyl radicals from the bound water since the water lattice around DNA does not possess any solvent power (Korystov, 1992). Therefore, scavenging of the radicals from the bound water by an exogenous protector is almost impossible. It was observed that thiols are able to repair DNA damaged sites before a breakage occurs (ABCRI, 2001).

## 4. Gamma sterilization of human tissue grafts

Connective tissue allografts, such as bone, cartilage, tendons, ligaments, dura mater, skin, amnion, pericardium, heart valves and corneas, are widely used for reconstructive surgery in many clinical disciplines, including orthopaedics, traumatology, neurosurgery, cardiosurgery, plastic surgery, laryngology and ophthalmology. The grafts are prepared by specialized laboratories called 'tissue banks'. The risk of infectious disease transmission with tissue allografts is a major concern in tissue banking practice.

Microorganisms can be introduced into grafts during tissue procurement, processing, preservation and storage, but even if all these procedures are done under aseptic conditions, the possibility of bacterial, fungal and viral disease transmission of donor origin cannot be excluded. Bacterial, including tuberculosis, fungal, and viral infections, such as human immunodeficiency virus (HIV), hepatitis B and C (HBV, HCV), cytomegalo virus (CMV), as well as rabies and prion diseases, have been transmitted by tissue allografts. Thus, radiation sterilization of tissue grafts has been implemented in some tissue banks, and a dose of 25 kGy has been used in many of these tissue banks. The advantage of radiation sterilization is that it allows the processing of grafts, which have been previously sealed or tightly closed in special wrappings. Such procedures prevent any accidental recontamination during packing.

The problem is additionally complicated by the possible presence, in human tissues, of pathogenic viruses, such as the human immunodeficiency virus (HIV) (Daar et al, 1991), hepatitis viruses (HBV, HCV) (Conrad et al, 1995), cytomegalovirus or others. Data concerning the sensitivity of these viruses to ionizing radiation are scarce. This is mainly due to the fact that there are no suitable tests to study their inactivation, no appropriate animal models exist and no suitable method of in vitro culture of highly differentiated target cells (e.g. hepatocytes) for these viruses has yet been developed.

The wide range of $D_{10}$ values (4–8.3 kGy) determined for HIV and other viruses might be due to the influence of environmental conditions. Many factors can modify the sensitivity of pathogens microorganisms to ionizing radiation, including the temperature of irradiation. For example, the reduction of HIV virus was achieved with a dose of 50–100 kGy in frozen plasma (-80°C), and with 25 kGy at 15°C (Hiemstra et al, 1991). The $D_{10}$ value for HIV-1 irradiated at room temperature was 7.2 kGy, and 8.3 kGy at -80°C (Hernigou et al, 2000). The presence or absence of water and oxygen, and presence of radiation protectors are also factors can modify the sensibility of pathogens microorganisms. In the absence of water (for example, in dry air or lyophilized grafts) the resistance of pathogens increases. On the other hand, in the presence of water, an indirect effect of ionizing radiation predominates and the sensitivity of microorganisms increases. Oxygen enhances the damaging effect to microorganisms and further increases their sensitivity to radiation as discussed previously. Therefore, if lyophilization is used as a preservation procedure, it would be better to leave some amount of water in the tissue than attempt to remove as much water as possible. It should be noted that irradiation at low temperatures increases, while that at higher temperatures decreases the resistance of bacteria and viruses.

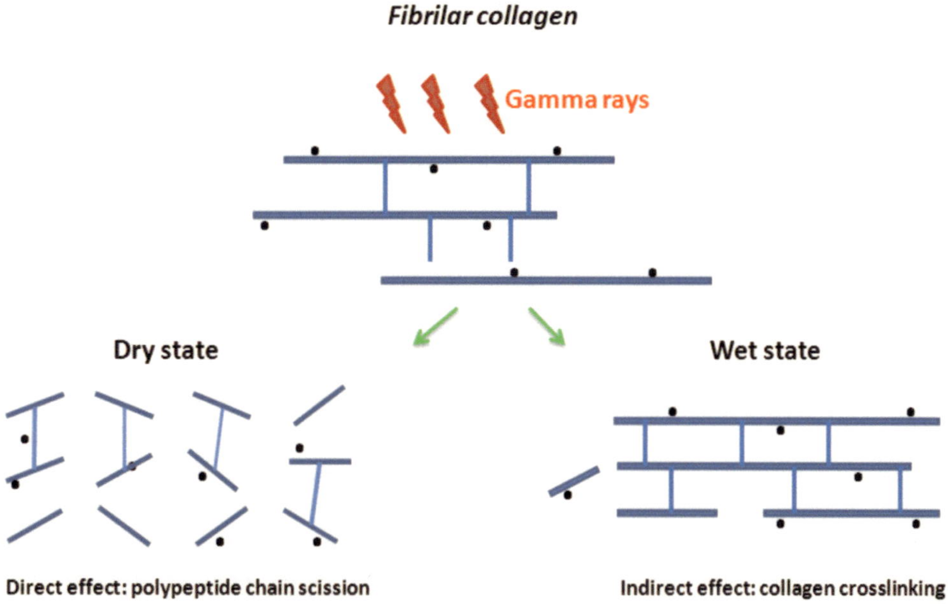


Fig. 6. Effects of gamma radiation on bone collagen molecules.

Collagen is a very variable protein, forming the basis of many connective and support tissues. It is a fibrous structural protein, with a distinctive structure. It has been postulated that polypeptide chain scissions (direct effect) predominate when collagen is irradiated in a dry state due to the direct effect of ionizing radiation, and this, in turn, dramatically increases collagen solubility in vitro and the rate of bone matrix resorption in vivo. It has been found, however, that a crosslinking reaction (indirect effect) appears during the irradiation of collagen in the presence of water (indirect effect), probably due to the action of highly reactive, short lived hydroxyl radicals (• OH) resulting from water radiolysis The Figure 6 shown the simplified scheme illustrating the direct and indirect effects of gamma irradiation on bone molecules.

## 5. Gamma sterilization of food

Food sterilization by gamma irradiation is the process of exposing food to ionizing radiation to destroy microorganisms, bacteria, viruses, or insects that might be present in the food. Irradiated food does not become radioactive, but in some cases there may be subtle chemical changes.

The treatment of solid food by ionizing radiation can provide an effect similar to heat pasteurization of liquids, such as milk. The use of the term "cold pasteurization" to describe irradiated foods is controversial, since pasteurization and irradiation are fundamentally different processes. Food irradiation is currently permitted by over 50 countries, and the volume of food treated is estimated to exceed 500,000 metric tons annually worldwide. (Farkas & Farkas, 2011).

By irradiating food, depending on the dose, some or all of the harmful bacteria and other pathogens present are killed. This prolongs the shelf-life of the food in cases where microbial spoilage is the limiting factor. Some foods, e.g., herbs and spices, are irradiated at sufficient doses (5 kGy) to reduce the microbial counts by several orders of magnitude; such ingredients do not carry over spoilage or pathogen microorganisms into the final product. It has also been shown that irradiation can delay the ripening of fruits or the sprouting of vegetables. Insect pests can be sterilized (be made incapable of proliferation) using irradiation at relatively low doses. The use of low-level irradiation as an alternative treatment to pesticides for fruits and vegetables that are considered hosts to a number of insect pests, including fruit flies and seed weevils. The table 3 showed some use of food irradiation.

Exposure to gamma irradiation doses below 10 kGy is effective in enhancing food safety through the inactivation of pathogenic microorganisms such as *Salmonella* and *Campylobacter* and in extending the shelf-life of the diet by eliminating the microorganisms responsible for food spoilage. Irradiation doses of between 20 to 25 kGy and between 20 to 30 kGy are used most frequently to treat diets intended for specific pathogen-free animals, whereas larger doses of 40 to 50 kGy are recommended for diets intended for gnotobiotic or germ-free animals, where absolute sterility is essential.

| **Type of food** | **Effect of Irradiation** |
|---|---|
| Meat, poultry | Destroys pathogenic fish organisms, such as Salmonella, Campylobacter and Trichinae |
| Perishable foods | Delays spoilage; retards mold growth; reduces number of microorganisms |
| Grain, fruit | Controls insect vegetables, infestation dehydrated fruit, spices and seasonings and reduces rehydration time |
| Onions, carrots, potatoes, garlic, ginger | Inhibits sprouting |
| Bananas, mangos,papayas, guavas, other non-citrus fruits | Delays ripening avocados, natural juices. |

Table 3. Food irradiation use

The effects of irradiation on the nutritive value of a product must be established before sterilization by radiation can become an important method for preserving food. The irradiation produces no greater nutrient loss than what occurs in other processing methods. Sample of a rat diet in which the protein 5, 10, 25, 35 and 70 kGy, and the effects on protein quality are given in Table 4. The results indicate no significant effect of irradiation on protein quality. Amino acid composition was similarly very little affected (Ley, 1969). By comparison of the different treatments (different radiation doses) and the control sample (not irradiated) of bean, it was observed that there was no significant alteration in the amino acid contents up to the maximum dose of 10 kGy. Even the more sensitive amino acids, such as the aromatic and basic, under the effect of gamma rays were kept intact in the samples. These results indicate that it is possible to use irradiation to reduce grain losses using different radiation doses without causing significant changes in the amino acid contents.

On the other hand, irradiation reduces the vitamin content of food, the effect of which may be indirect in that inadequate amounts of antioxidant vitamins (such as C, E, and β-carotene) may be available to counteract the effects of free radicals generated by normal cell metabolism. When food is irradiated, ionizing radiation reacts with water in the food, causing the release of electrons and the formation of highly reactive free radicals (see Figure 2). The free radicals interact with vitamins in ways that can alter and degrade their structure and/or activity (Murano, 1995). The extent to which vitamin loss occurs can vary based on a number of factors, including the type of food, temperature of irradiation, and availability of oxygen. Nonetheless, vitamin loss almost always increases with increasing doses of radiation. The destruction of vitamins continues beyond the time of irradiation. Therefore, when irradiated food is stored, it will experience greater vitamin loss than food that has not been irradiated. Cooking further accelerates vitamin destruction in irradiated food more than in non-irradiated food (Diehl, 1967).

| Dose (kGy) | True digestibility | Biological value | Net proteins utilization |
|---|---|---|---|
| 0 | 85.6 | 80.5 | 68.9 |
| 5 | 83.6 | 75.8 | 63.5 |
| 10 | 86.5 | 81.7 | 70.6 |
| 25 | 87.0 | 78.1 | 68.0 |
| 35 | 84.8 | 77.3 | 65.4 |
| 70 | 85.3 | 76.4 | 65.2 |

Table 4. Effect of gamma irradiation on the protein of rat diet

Vitamin C, vitamin B1, and, vitamin E are reduced in foods exposed to commercial levels of irradiation (1 kGy - 4.5 kGy). At the low doses of 0.3 to 0.75 kGy, food irradiation has been found to destroy up to 11% of vitamin C in fruit before storage, and up to 79% of vitamin C after three weeks of storage (Mitchell et al, 1992). Additionally, at the limit of its shelf life (270 days) irradiated mango pulp contains 57% less vitamin C than non-irradiated mango pulp at the limit of its shelf life (60 days). Whole grains, beans, and meat are important sources of thiamine (vitamin B1), which helps convert carbohydrates into energy. It is essential for heart, muscle, and nervous system function. Wheat flour irradiated at the low dose of 0.25 kGy lost up to 20 percent of thiamine initially and 62% after three months of storage.25 Beef irradiated at 3.0 kGy, which is below the legal limit, experienced a 19 % loss of thiamine. Oils, corn, nuts, seeds, and green vegetables are important sources of vitamin E, an antioxidant that protects body tissues and cells. It also may improve the immune system and help fight heart disease, cancer, Alzheimer's disease, and cataracts. Hazelnuts irradiated at 1.0 kGy lost 17% of vitamin E upon irradiation, and 58% of vitamin E after three months of storage and 30 minutes of baking. In addition, studies at higher levels of irradiation have demonstrated the destruction of vitamins A and K in food (Stevinson et al, 1959). The question of vitamin K in irradiated diets requires special considerations: i. it is known to be susceptible to destruction by y-irradiation (Ley, 1969); ii. it is synthesized by microbial action in the gut, and animals (particularly those that practice coprophagy) can satisfy part of their requirement by this means. Sterilized diets are usually only fed to specified-pathogen-free or gnotobiotic animals, i.e. those that have a limited gut microflora or none at all. Thus the organisms responsible for vitamin K synthesis are likely to be absent, and the animal's requirement for dietary vitamin K may be very much higher than that of its conventional counterpart. It is difficult, if not impossible to determine vitamin K chemically in animal diets because other components react as vitamin K to the assay procedure. Assessment of the vitamin K content of a diet must therefore depend on the response of the animals receiving it. The Table 5 gives the doses, which the some food vitamins lost.

| Food | Dose of sterilization (kGy) | Vitamins lost | reference |
|---|---|---|---|
| Mango | 10 | Vitamin C | Youssef et al., 2002 |
| Grapefruit | 10 | Vitamin C | Patil et al., 2004 |
| Pork | 10 | Thiamin | Fox et al., 1997 |
| Chicken | 30 | Vitamin E and Thiamin | Lakritz and Thayer, 1992 |
| Beef | 45 | Thiamin | Fox et al., 1995 |

Table 5. Vitamins lost after gamma irradiation of some food

A study of the vitamin contents of diets for guinea-pigs (RGP), chicks (SCM) and cats after irradiation at doses ranging from 20 to 50 kGy has been made (Coates et al., 1969). At doses of the order of 20 to 30 kGy, vitamin losses from the guinea-pig and chick diets were very small indeed, but a severe loss of vitamin A from the cat diet was observed after treatment at 25 kGy. The losses were such that they could have been compensated for by addition of about twice the usual supplement of the vitamins affected. Stability decreased markedly with increased moisture content of the diet.

Poly unsaturated fatty acids were reported to have beneficial effects on human health and also are susceptible to peroxidation damage (Haghparast et al., 2010). Therefore, stability of these components needs to be considered for the standardization of the radiation process (Erkan and Özden, 2007). Ionizing radiation causes the radiolysis of water which is present to a great extent in food. This generates free radicals (see Figure 2) all of which react with the food constituents. The most susceptible site for free radical attack in a lipid molecule is adjacent to the double bonds. The most affected lipids during irradiation are thus the polyunsaturated fatty acids that bear two or more double bonds (Brewer, 2009).

Study on chicken showed no significant difference in total saturated and unsaturated fatty acids between irradiated (1, 3, 6 kGy) and non-irradiated frozen chicken muscle (Rady et al.,1988), however Katta et al. (1991) found significant decrease in the amount of palmitic acid and increase in oleic acid as irradiation dose level increased (0.5-3 kGy) in chicken meat.

Changes in the palmitic (C16:0), oleic (C18:1) and linoleic (C18:2) fatty acids of soybeans at different radiation doses (1, 5, 10, 20, 40, 60, 80 and 100 kGy) were no found (Hafez et al.,1985). The irradiation at 10 kGy also changes the linoleic and linolenic acid contents of grass prawns. Irradiation caused a 16% decrease in linoleic acid content, whereas linolenic acid was not affected significantly (Hau and Liew, 1993).

The irradiation of fish no changes fatty acid compositions of two species of Australian marine fish irradiated at doses of up to 6 kGy (Armstrong et al., 1994), but chemical components of tilapia and Spanish mackerel has been reported (Al- Kahtani et al., 1996). Irradiation of tilapia at 1.5–10 kGy caused a decrease in myristic (C14:0), palmitic (C16:0) and palmitileic (C16:1) fatty acids. In the case of Spanish mackerel, palmitic (C16:0) and palmitileic (C16:1) fatty acids decreased when irradiated at 1.5–10 kGy. Contents of total saturated fatty acids in the muscle of non-irradiated sea bream was respectively lower than in 2.5 kGy irradiated sea bream and higher than in 5 kGy irradiated sea bream. There was significant difference in the content of total unsaturated fatty acids , mono unsaturated fatty acids between 2.5 kGy and 5kGy irradiated sea bream and no significant difference was determined in the content of unsaturated fatty acids, mono unsaturated fatty acids between non-irradiated and irradiated fish. On the other hand, the content of poly unsaturated fatty acids in the muscle of 5 kGy irradiated sea bream was significantly lower than in non-irradiated and 2.5 kGy irradiated sea bream (Erkan & Özden, 2007). All at same, the total saturated and total monounsaturated fatty acid contents were 27.97% and 24.72% for non-irradiated for sea bass, respectively. The amounts of these two fatty acids in irradiated samples increased to 28.18 and 25.75% for 2.5 kGy and 29.08 and 28.54% for 5 kGy. Significant difference also was found in the content of total unsaturated fatty acids, mono unsaturated fatty acids between 2.5 kGy (25.75%) and 5 kGy (28.54%) irradiated sea bass and between non-irradiated and irradiated fish. (Özden & Erkan, 2010).

Irradiated ground beef samples with 7 kGy had the highest total trans fatty acids, total monounsaturated and total unsaturated fatty acids than the other samples. Results showed an increase in trans fatty acids related to the increase on irradiation dose in ground beef and irradiation dose changed fatty acids composition especially trans fatty acids in ground beef (Yılmaz & Gecgel, 2007). Total saturated fatty acids and unsaturated fatty acids, mono unsaturated fatty acids of beef lipid increased with irradiation (1.13, 2.09 and 3.17 kGy), but ratios of unsaturated fatty acids, mono unsaturated fatty acids to saturated fatty acids did not change. Whilst, total poly unsaturated fatty acids reduced with irradiation, which resulted in poly unsaturated fatty acids to saturated fatty acids ratio decrease.

## 6. Gamma sterilization of medical devices

When radiation is used for the sterilization of medical devices, the compatibility of all of the components has to be considered. Ionizing radiation not only kills microorganisms but also affects material properties. Medical devices are made of many different materials, some of which are metals, but most are non-metals, such as formed polymers, composite structures and even ceramics. Radiation itself does not directly affect metals since sterilization energies are safely below any activation thresholds. Metals, such as those used in orthopaedic implants, are virtually unchanged by the radiation sterilization process. Nevertheless, it has to be kept in mind that some types of polymers when irradiated in contact with a metal can cause some corrosion of the metal or surface discolouration. This is generally caused from by products released by some polymers during irradiation.

Polymer devices subjected to irradiation sterilization will inevitably be affected by the radiation and the environment used during sterilization, and will experience changes in the polymer structure such as chain scission and crosslinking (Schnabel, 1981). For some polymers both processes coexist and either one may be predominant depending not only upon the chemical structure of the polymer, but also upon the conditions of irradiation is performed like temperature, environment, dose rate, etc. The crosslinking and main scissions that take place during irradiation may lead to sharp changes in physical properties of the polymers. These effects will lead to changes in the tensile strength, elongation at break and impact strength. The exact changes seen will depend both on the basic polymer and any additives used. The changes in mechanical properties may not be immediately apparent and there can be some time delay in their development. One visible side effect of irradiation sterilization is that many plastics will discolor or yellow as a result of the processing. Irradiated devices are completely safe to handle and can be released and used immediately after sterilization.

Many polymers are resistant to radiation at doses of up to around 25 kGy, the actual doses used will be higher than this to achieve sterilization, however complete sterilization and radiation damage of some magnitude will inevitably occur. The effect of radiation is cumulative and for items that must be repeatedly sterilized the total dosage can rise rapidly. For these items records need to be kept to insure that safe limits are not exceeded. Irradiation is very effective for fully packaged and sealed single-use items where only one radiation dose is required.

### 6.1 Effects of gamma sterilization in polymeric medical device

Poly(vinyl chloride), PVC, is a polymer widely used for radiosterilizable food packaging and medical devices. However when the polymer systems are submitted to sterilization by gamma radiation (25 kGy dose), their molecular structures undergo modification mainly as a result of main chain scission and crosslinking effects. For PVC both processes coexist and either one may be predominant depending upon the conditions (temperature, environment, dose rate, etc.) under which irradiation is performed. The crosslinking and main scissions that take place during irradiation may lead to sharp changes in physical properties of the PVC (Vinhas et al, 2004).

During the interaction of gamma radiation with PVC, the reactions shown in Figure 7 can take place (Bacarro et al., 2003). This interaction gives rise to macroradicals deriving from C-Cl bond scission reactions (reaction I). The chlorine radical continues the reaction by way of a form center reaction in which HCl is formed and acts as a catalyst (reaction II). The A, B or C macroradicals recombine with each other forming networks due the restricted mobility of the macroradicals in the solid state (reaction III). It was reported, which crosslinking effect is predominant for PVC irradiated at lower doses (Silva et al, 2008). Oxidation reactions of macroradicals A, B or C (reaction IV), interaction of radical A with neighboring double bonds and other macroradicals from the impurities or from direct action of gamma radiation also can play an important role on crosslinking effect of PVC irradiated at lower radiation dose. However in presence of air the polymeric radicals A, B and C react with oxygen from air producing the peroxyl macroradical (reaction V). This radical formed can them undergo further reactions leading to main chain scission. This effect is predominant when the PVC molecule is irradiated at higher doses. Thus in the sterilization dose the commercial PVC undergoes the main chain scission (Ferreira et al, 2008).

Poly(methyl methacrylate), PMMA, also is used in manufacturing of medical supplies that can be sterilized by gamma irradiation at dose of 25 kGy and used in absorbed dose measurements in intense radiation fields. In general, polymer radicals are responsible for changes in the physical properties of PMMA. In particular, gamma irradiation of PMMA causes main scission and hydrogen abstraction from an α-methyl or methylene group. The extent of formation of each of the derivatives resulting from irradiation depends on the physical state of PMMA (Schnabel, 1981). The great majority of authors have reported that scission results from a macroradical that itself is radiolysis product of a lateral bond as shown in the Figure 8 (reaction I ) (Guillet, 1985). The volatile products like $HCOOCH_3$, CO, $CO_2$, $HCOCH_3$ and $CH_4$, can be accounted for by the subsequent reactions of the carbomethoxy radical (B radical). The formation of C radical is the basic reason for the radiation-induced degradation of PMMA. Under air atmosphere the C radical undergoes the chain oxidation process forming the peroxyl free radical (D). Once D radical is formed in PMMA, it can abstract hydrogen from PMMA chains to form hydroperoxide. The hydroperoxide decomposes slowly but steadily at room temperature to generate new oxidative products, which induce further degradation. In addition, it is believed that the free radical A, peroxyl radical (B) and the hydroperoxides are the main substances, which induce the changes in PMMA properties when it is gamma irradiated (Schnabel, 1981).

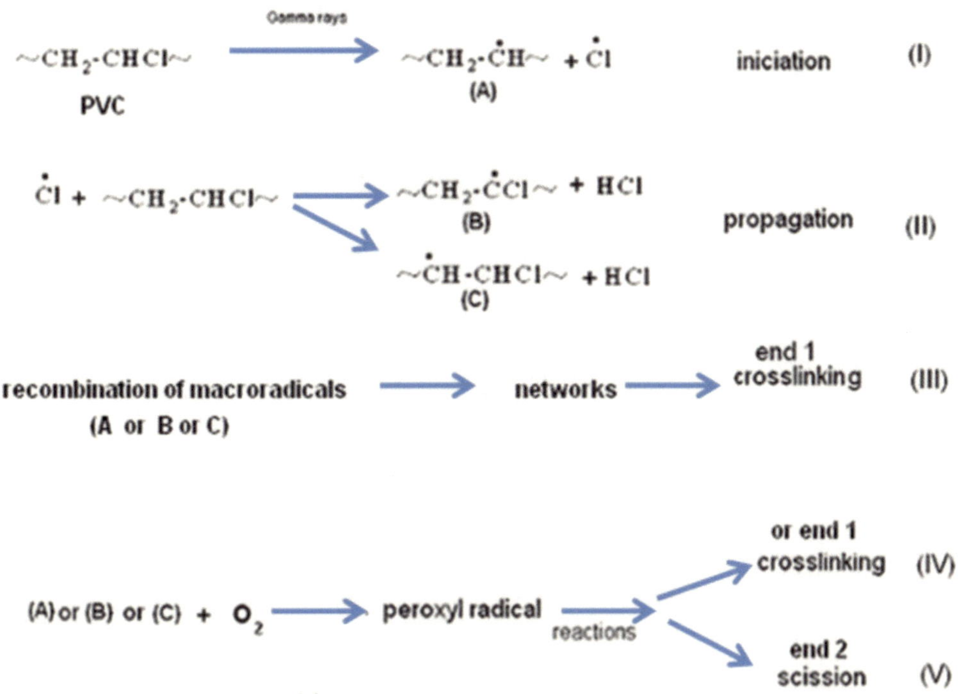


Fig. 7. Effects of gamma irradiation on PVC molecule

Polycarbonate (PC) fills an important niche as one of the most popular engineering resins in the medical device market. Bisphenol-A polycarbonate has been commercially available since the 1960s, and its use in medical devices dates from approximately that time. Possessing a broad range of physical properties that enable it to replace glass or metal in many products, polycarbonate offers an unusual combination of strength, rigidity, and toughness that helps prevent potentially life-threatening material failures. In addition, it provides glasslike clarity, a critical characteristic for clinical and diagnostic a setting in which visibility of tissues, blood, and other fluids is required because biocompatibility is essential for any material used in direct or indirect contact with patients (Freitag et al., 1988).

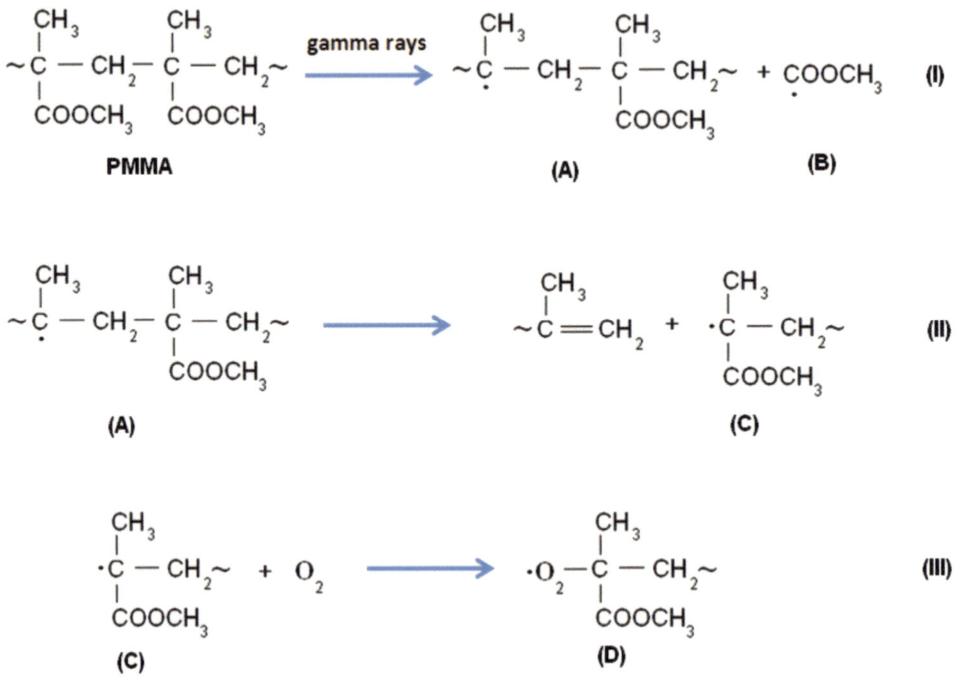


Fig. 8. Radiolytic degradation of PMMA

The radiation-induced main chain scissions on PC occur in the carbonate groups, causing the evolution of carbon monoxide, carbon dioxide and hydrogen. The radiolysis of PC produces phenoxy and phenyl polymeric radicals that cause yellowness of the polymer. However, it has been reported in the literature that the crosslinking effect predominates at small doses, whereas at higher doses the main chain scission is more pronounced (Araujo et al, 1998).

Polyurethane (PU) is widely used in various medical devices because of its biocompatibility, and has some reports concerning its physicochemical stability and biological safety. However, among substances which were produced by degradation of PU, it was reported that a carcinogen, 4,4'-methylenedianiline (MDA), was produced from PU sterilized by gamma irradiation. On the other hand, a modified PU was produced and called thermosetting PU. In the case of thermosetting PU used in medical devices such as potting material in artificial dialysis devices, plasma separators, etc., the production of MDA upon sterilization showed a reverse tendency to non modified PU (Shintani, 1992). Their components and characteristics used in PU fabrication are much different, however their influences on the production of MDA by sterilization have not been sufficiently clarified.

As shown in Figure 9, it was suggested that the mechanism of MDA production might be the cleavage at urethane linkage successive to the terminalamino group, by radiation or hydrolysis (Shintani, 1992). Since more hydrophilic components were detected in the current experiment, we speculate the major cleavage portion will be at urethane linkage, thus producing MDA. The possibility of the cleavage at benzene-CH, linkage will not be significant due to no aniline or p-toluidine production.

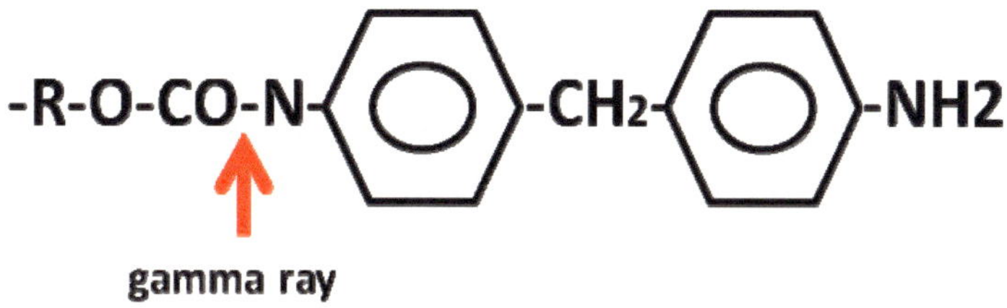


Fig. 9. Proposed mechanism of MDA formation by gamma irradiation of PU

Ultrahigh molecular weight polyethylene (UHMWPE) possesses a unique structure and properties which have resulted in its having been the most widely used material for replacing damaged or diseased cartilage in total joint replacements for the last 35 year. UHMWPE is a linear (non-branching) semi-crystalline polymer which can be described as a two phase composite of crystalline and amorphous phases. The two resins of UHMWPE that are currently used in orthopaedics are GUR 1020 (3.5 million g /mol) and GUR 1050 (5.5-6 million g /mol). Orthopaedic components machined from UHMWPE are typically sterilized by irradiation with 25 kGy of $^{60}Co$ gamma rays (Goldman et al, 1998). Such strong ionizing radiation is likely to have a detrimental effect upon the microstructure, such as entanglement density and tie molecules that give UHMWPE its needed properties for total joint replacement applications. The high-energy photons, such as gamma rays, can generate free radicals in polymers (P) through homolytic bond cleavage (reaction 1 in Figure 10). These radicals have been shown to have long lifetimes, especially those generated in the crystalline regions of the polymer where they can diffuse at low mobility into the amorphous regions of the polymer, and can therefore continue to undergo chemical reactions for many months and beyond. This time-dependent free-radical reaction mechanism poses serious concern for the radiation degradation of polymers, especially in the presence of oxygen as is observed in the reactions showed in Figure 10 (reactions 2 and 3), which has a high difusional mobility and is very reactive with the radicals. Hydroperoxides also are formed as the first product of oxidation and upon their decomposition free radicals are re-generated (reaction 4 in Figure 10). Every molecule of hydroperoxide produced subsequently undergoes radiolysis to generate an alkoxy radical which both provides new initiating radicals and at the same time produces carbonyl compounds (reaction 5 in Figure 10). Thus, the process is autocatalytic and can lead to the further formation ketones, alcohols, esters, and carboxylic acids in the polyethyelene chains. Therefore, as long as there is an oxygen source, the cycle can continue and the number of oxidation products will increase without any further irradiation (Schanbel, 1981). This process is known as post-irradiation aging and has been shown to occur in implants that were gamma sterilized in air and packaged in air-permeable packaging. Changes in physical, chemical and mechanical properties of UHMWPE as a consequence of oxidative degradation (Costa & P. Bracco, 2004). Property changes include an increase in percent crystallinity, an increase in density (an indirect measure of oxidation), an increase in elastic modulus, and a decrease in elongation to failure.

As the evidence of the clinical consequences of oxidative degradation of UHMWPE total joint replacement components increased, the orthopaedic implant manufacturers began to study and then to employ alternative sterilization methods such as gamma radiation sterilization in inert environment (e.g. argon, nitrogen, vacuum) packaging as a means to minimize oxidation during shelf aging. For UHMWPE, cross-linking dominates when the polymer is irradiated in nitrogen, while chain scission dominates when the material is irradiated in air. This is due to the fact that oxygen is extremely reactive with the free radicals produced by irradiation, forming peroxides which can break down and lead to further radical production, so that the total number of free radicals generated and the total extent of chain scission, are greatly increased. It should be noted that, without some additional manufacturing step to extinguish any remaining entrapped free radicals, oxidation will occur upon exposure to oxygen (such as during in vivo use). For gamma sterilization in an inert environment to be successful, it must be combined with barrier packaging to prevent access of atmospheric oxygen to the UHMWPE during shelf storage. Thus, barrier packaging is expected to effectively reduce the risk of oxidative degradation of UHMWPE during shelf storage (Rimnac & Kurtz, 2005)

$$P \xrightarrow{\text{gamma rays}} P\cdot \qquad (1)$$

$$P + O_2 \longrightarrow PO_2\cdot \qquad (2)$$

$$PO_2\cdot + PO_2\cdot \longrightarrow POOP + O_2 \qquad (3)$$

$$PO_2\cdot + P'H \longrightarrow POOH + P'\cdot \qquad (4)$$

$$POOP' \xrightarrow{\text{gamma rays}} P\text{-}CHO + P'O\cdot \qquad (5)$$

Fig. 10. Polyolefins oxidation caused by gamma sterilization

Polypropylene (PP) is one of the most widely used plastics for packaging applications. Polypropylene is one of the most popular polymers in the manufacturing of medical disposables, since it exhibits high transparency, good mechanical properties, low cost and chemical inertness over other polymers. In a continuously increasing part of this market, especially in the pharmaceutical area, but also in food packaging and especially in the manufacturing of syringes, security lenses, surgical clothing, etc. Medical instruments employed in the diagnosis or treatment of a patient, especially those that can penetrate the protective, barrier of the skin, must be completely exempt of germs.

Changes in polymer properties were observed when PP medical devices are sterilized by gamma irradiation undergoing oxidative degradation if sterilized in air. Oxidation of PP is usually relatively easy to detect owing to the strong absorption by the carbonyl group in the FT-IR spectrum as is showed in Figure 11. Polypropylene has a relatively simple spectrum with few peaks at the carbonyl position. The integrated absorption of the C=O band centered about 1720 $cm^{-1}$ has been assumed to give a quantitative evaluation of the radiation induced oxidation. Since the PE is a polyolefin the carbonyl group is obtained in reaction 5 of the scheme in Figure 10. Oxidation tends to start at tertiary carbon atoms because the free radicals formed here are more stable and longer lasting, making them more susceptible to attack by oxygen. The carbonyl group can be further oxidized to break the chain, this weakens the material by lowering its molecular weight, and cracks start to grow in the regions affected.

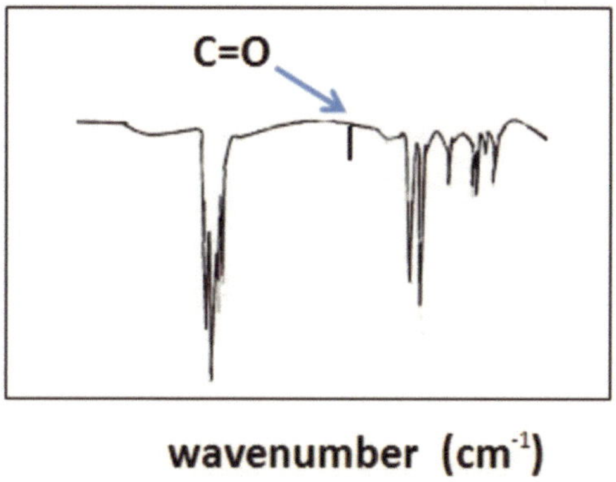


Fig. 11. FT-IR spectrum of PP exposed to gamma sterilization in air

The changes in the PP molecule by gamma sterilization are associated with the changes in crystallinity and morphology of the polymer. The correlations between the changes in both morphology and crystallinity with other properties during irradiation are important to explain the mechanism that lead to crystallinity change. Some studies investigated the response of PP to γ-radiation and relate the crystallinity and morphological changes to corresponding changes in other properties such as mechanical properties, viscosity, melting temperature, etc. Kushal et al. (1995) relate the drop in the melting temperature, viscosity and mechanical properties versus the increases in crystallinity during γ-irradiation to the breakdown of crystallites with a concomitant formation of smaller crystalline entities.

The extent of chain scission and crosslinking of PP is dependent on the γ-irradiation dose but not the initial starting morphology (Zhang and Cameron 1999). Using WAXD (Wide angle X-ray diffraction) and DSC (Differential scanning calorimetry) techniques, Alariqi et al. (2006) found change in the degree of crystallinity, which caused by γ-irradiation, depends on the γ-irradiation dose ( see Table 6) and Kostoski and Stojanovic (1995) found the increase in crystallinity of oriented isotactic polypropylene with low absorbed doses of γ-radiation, up to 200 kGy. They have also found that the peak melting temperature decreased with absorbed dose. The results were explained in terms of the scission of the tie molecules followed by the growth of new thin crystal lamellae, as well as to the fact that irradiation produces defects in the polymer structure which decrease its thermal stability. However, the number of chain scission increased with decreasing the dose rate. From, lowering molecular weight, increased chain scissions, increased crystallinity, it can be understood that the rise in crystallinity is due to re-crystallization of shorter chains which are produced by the chain scission of tie molecules forming new perfect crystallites leading to an increase in crystallinity. On the other hand, the decrease in crystallinity was attributed to the formation of crosslinking. Krestev et al. (1986) have found that part of monoclinic α-phase of PP is converted into triclinic γ-phase during gamma irradiation. It was reported that the formation of γ-phase was not due to the crystallization of low molecular fraction but to the high internal pressure caused by the crosslinking.

| Irradiation dose (kGy) | Degree of crystallinity (%) | |
|---|---|---|
| | WAXD | DSC |
| 0 | 38.5 | 36.3 |
| 10 | 48.0 | 42.9 |
| 25 | 33.2 | 32.6 |

Table 6. Effect os gamma sterilization on crystallinity of Polypropylene

Polyisoprene, especially in the form of natural rubber latex, is widely used in prophylactic medical disposables, such as gloves and condoms, and found to be an effective barrier. Because of its unsaturation, natural rubber and many other elastomers will slightly crosslink when exposed to radiation sterilization conditions. Such crosslinking will not detract from the overall extensibility or elongation of these rubber devices. Natural rubber formulations, as well as formulations based on other elastomers, can also be used as gasketing materials in devices. Although isobutylene is well known to scission when exposed to radiation, a halogenated copolymer of isobutylene and isoprene, commonly brominated butyl rubber (BIIR), can be formulated to exhibit radiation response when used in the tyre industry. Having been previously crosslinked with a zinc oxide system, BIIR can withstand the radiation exposure required for sterilization. Such elastomeric materials form the sealed caps on injectable drugs, being able to reseal themselves after having been penetrated by the needle of a syringe.

Silicone rubber is widely used in medical applications, where sterilized is an essential requirement for all medical tools and devices that contact the body or bodily fluid and medical components must be sterilized frequently by gamma irradiation. Gamma radiation is known to induce changes in the molecular architecture of silicone rubber, resulting in an increase in molecular weight and a decrease in elasticity. This effect is also observed in samples previously subjected to post-cure treatments. Radicals are generated by chain scission and/or methyl or hydrogen abstraction (see Figure 12) and are subsequently terminated via oxidation reactions or coupled to form longer chain branches. Although these two mechanisms compete against each other, crosslinking reactions dominate in silicone materials; higher dosages of gamma radiation and longer treatment cycles have been shown to result in higher crosslink densities (Traeger & Castonguar, 1966). An increase in polymer-filler interfacial interactions through crosslinking reactions is also observed.

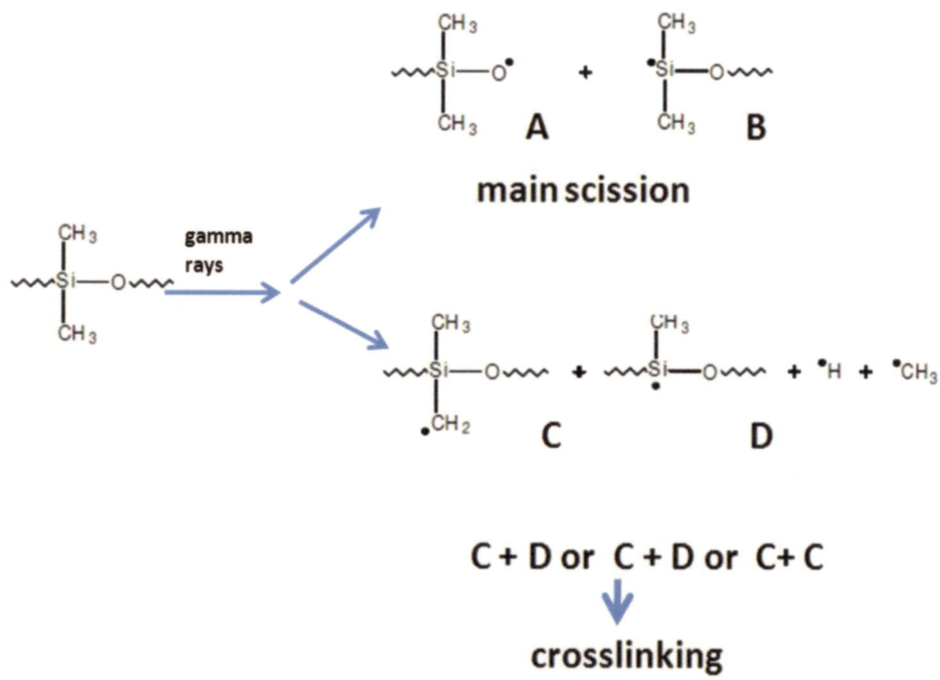


Fig. 12. Effects of sterilization by gamma irradiation in silicone molecule

### 6.2 Action of stabilizers in polymeric medical device exposed to gamma sterilization

With the development of space science, the stability of polymeric materials against radiation has been drawing the attention of scientists. Polymers which contain aromatic groups are well known to have relatively good radiation stability, but are also very expensive. The practical solution of these protection tasks are connected to specific chemical agents, well engineered polymer additives, elaborated mainly for the stabilization of general purpose polymers. The radiation stabilizers, called "antirads" represent only a modest, but flourishing fraction of that thermo-oxidative- and UV stabilizers.

The reason behind the parallel technical development of conventional and radiation stabilizers is related to the fact, that the UV degradation and thermo-oxidative degradation as well as radiation degradation of polymers are all similar chain reactions. As such, these processes consist of several steps of: chain initiation, chain propagation, chain branching and chain termination. The scheme according to which these reactions proceed on a H containing polymer chain P is seen in Figure 6. In spite of the differences in fine details the task is similar in all the three main (thermooxidative, UV and radiation) degradation processes, namely to control and/or diminish the danger of deterioration of properties either by preventing chain initiation, and/or stopping chain propagation.

Additives may promote radiolytic stabilization on properties of polymers thought two primary mechanisms: a) scavenging of excited-state energy (quenching), and b) scavenging of paramagnetic species (free radicals, secondary electrons). Also the incorporation of additives, plasticizing type, act as "mobilizer" on polymer chains. Additives and stabilizers are commonly included in small amounts (less than 1%) in commercial polymer products to aid in processing, stabilize the material and impart particular properties to the product. In controlling the route of those oxidative chain reactions, there are two main types of antioxidant stabilizers:

- Primary or chain-breaking antioxidants interfere with the chain propagation step. That step is the main carrier of the oxidative degradation.
- Secondary or preventive antioxidants destroy hydro-peroxide groups, responsible for chain initiation and chain branching.

Typical primary antioxidants, interfering with the chain-carrying radicals are the orthodisubstituted phenols, alkylphenols, hydroxyphenyl propionates and hydroxybenzyl compounds. Irganox 1010 (see structure in Table 7) is one of the most important additive protecting PE and PP in radiation sterilization. It is important to note, that such stabilizers are never used alone. Secondary antioxidants represent an even greater group of sophisticated organic molecules: aromatic amines, organic sulfur compounds (typically thiobisphenols and thioethers) as well as phosfites and sterically hindered amines. These two latter type of compounds are successfully applied in the radiation-stabilization of PP (Williams et al., 1977). The Table 7 showed some commercial antioxidant structures.

Clearly, the radiation-protection stabilizer systems should fulfill a whole series of other requirements such as chemical, physical and toxicological safety. Take for example the blood-taking and transfusion sets, made out of plasticized PVC, radiation sterilized and then stored (standing by) for years, later filled with chemically stabilized blood, and cooled and stored again. During all these procedures the protected polymeric material should be stable, should not loose its elasticity, and in the last steps there are strict limitations on traces of all chemicals extractable by the blood.

In relation to radiation stability improvement, discoloration of PP homopolymer was eliminated by the incorporation of the light protector in the samples prepared by injection moulding. Samples of PP homopolymer with different additives also were prepared by compression moulding were irradiated to 25 kGy. The commercial additives used were Tinuvin 622, Irganox B225 (blend of Irganox 1010 and Irganox 168), Irganox PS800 and Irganox 1010. The structures of these commercial additives were showed in Table 7. Elongation to break was measured after irradiation and at 12 months of aging at room temperature. Previous experiments with samples prepared in the same way without additives had shown a strong effect of post irradiation degradation and this effect could be anticipated by the effect of a higher applied dose. The effect of additives was significant as all of additived samples could be considered functional after aging. Addition of antioxidants improved mechanical stability in samples prepared by compression molding and by injection moulding, but they had a negative effect in discoloration (Gonzales & Docters, 1999)

Hindered Amine Light Stabilizer (HALS) is among the more extensively used additives for protecting polymers against degradation by the combined effect of light, temperature, and atmospheric oxygen. The protection of the polymer from the light by these compounds takes place via a mechanism involving photo-oxidation of the amines to nitroxyl radicals (Lucarini et al, 1996). Nitroxyl radical is capable of scavenging the radicals through a reaction called Denison cycle. On the other hand, very little information on this additive for

| Structure | Comercial name | Classification of antioxidant |
|---|---|---|
| 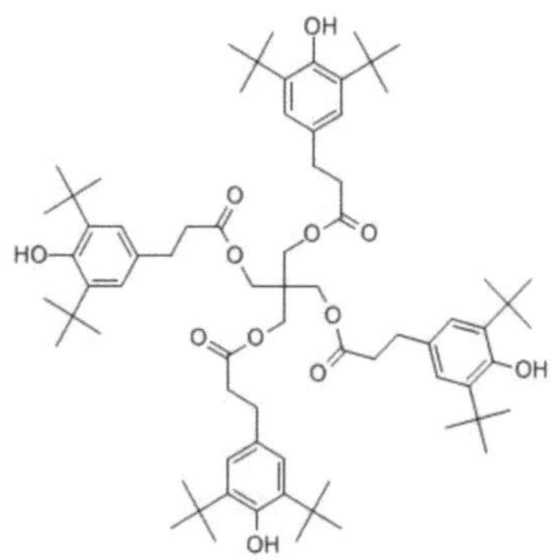 | Irganox 1010 | primary |
| 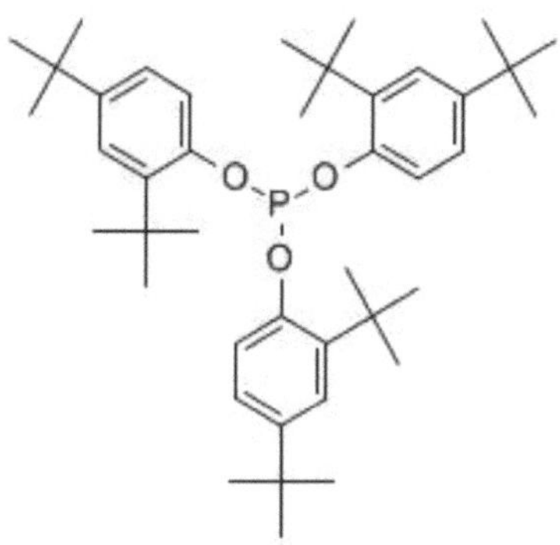 | Irganox 168 | secondary |
| 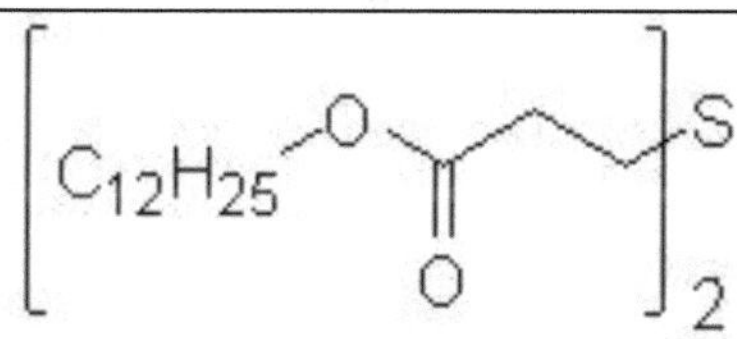 | Irganox PS 800 | secondary |
| 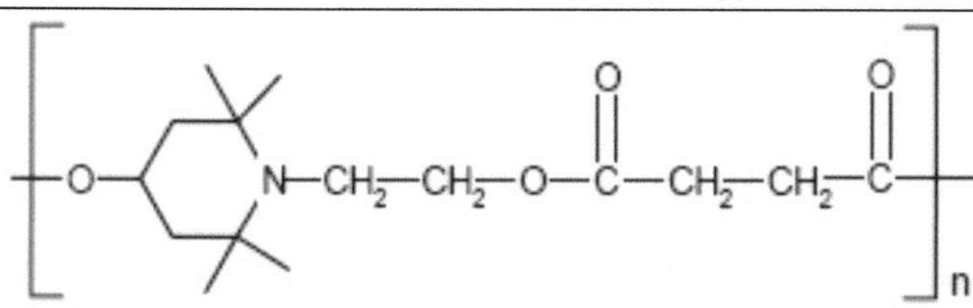 | Tinuvin 622 | secondary |

Table 7. Some commercial additives used in the stabilization of polymer gamma sterilized

radiolytic stabilization of polymers has been published. The efficiency of a certain additive in the stabilization of polymer molecules against radiation may be evaluated by measuring the effect of this additive on the free radical population after irradiation, as well as on its rate of decay. The efficiency of HALS additives depends on their molecular weight, structure, solubility and concentration in the polymer matrix. Conversion of amines into nitroxyl radicals following the reaction with peroxyl radicals leads to relatively stable intermediate species. Regeneration of the nitroxyl radical limits the consumption of HALS during degradation allowing the use of these additives in low concentration. Tinuvin 622 (see structure in Table 7) is a macromolecular HALS, which exhibits a quite high thermal stability. This additive starts to decompose around 400ºC with intramolecular ester group rearrangement. Final decomposition events occur at 900ºC and include nitrile and hydrogen cyanide formation (Lucarini et al., 1996)

Samples of PMMA irradiated at 30 kGy and containing Tinuvin 622 showed more resistance to radiation damage. Tinuvin 622 also induces a faster evolution of radicals produced on PMMA radiolysis, which can result in the inhibition of free radical damage. Above 30 kGy, both PMMA without (PMMA-control) PMMA with Tinuvin 622 (PMMA-622) undergo significant changes in the yellowness index with increase of absorbed dose due to conjugated center that absorbs light in the visible range. After 63 days of storage at 30 kGy dose, the yellowness index measured was 2.78 and 0.17 to PMMA-control and PMMA-622, respectively. These results showed that the Tinuvin 622 is a good alternative for stabilizing the PMMA against gamma irradiation damage in sterilization processes with low cost (Aquino et al., 2010)

The scheme in Figure 13 is generally accepted to explain the aspects of the chemistry mechanism of HALS action to inhibit polymer photo-oxidation. This scheme was used to guide a strategy to assess Tinuvin 622 action in radiolytic stabilization of PMMA (Aquino & Araujo, 2008). According to this scheme, the tetramethylpiperidine moiety, which is the basic structure of HALS, is initially oxidized to produce a nitroxyl radical by gamma irradiation. The nitroxyl radical acts as a scavenger of the radical originating from the irradiation of polymer chain substrate to form an alkylated aminoether. From the aminoether, the nitroxyl radical is regenerated through quenching another peroxyradical produced by oxidation of the polymer chain. Thus the nitroxyl radicals could regenerate many times through the chain reaction before their depletion.

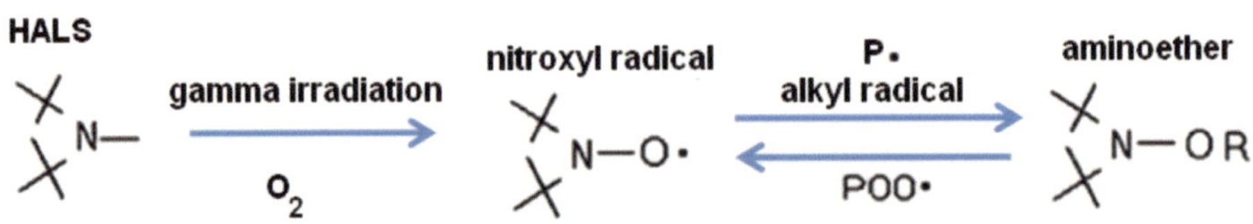


Fig. 13. Typical photo-stabilizing action of HALS in the polymer system

The single Electron Spin Resonance (ESR) spectrum was obtained for Tinuvin 622 sample irradiated at 100 kGy and was attributed to nitroxyl radical (Aquino et al., 2010). The chemistry of HALS had been widely documented for UV irradiation. As the Tinuvin 622 is a tertiary HALS a sequence of reactions starting with amine ionization and a-aminoalkyl radicals are formed. These radicals rapidly react with oxygen and fragment under the elimination of formaldehyde to nitroxyl radicals. Thus, similar stabilization mechanism is attributed to Tinuvin 622 when the PMMA is submitted to gamma irradiation. The gamma rays can break covalent bonds in PMMA molecule to directly produce the free radicals as was shown in Figure 8 (II). The gamma rays can also produce excited states in PMMA which undergo further reactions to produce the A radical (Figure 8) indirectly. Thus, there are two ways for Tinuvin 622 to decrease the main scission effect of gamma-irradiated PMMA. One way is to directly inhibit the formation of A radical by quencher mechanism when Tinuvin 622 may be to absorb the energy of the excited molecules in PMMA via an intermolecular energy transfer and possibly convert the absorbed energy. The other way is based on ESR results, the A radical to be scavenged by a nitroxyl radical and an alkyloxyamine is formed. The alkyloxyamine scavenges a peroxyl radical in a second step, in which the nitroxyl radical is regenerated (Aquino & Araujo, 2008).

Vinhas et al. (2004) also reported radio-protective action of a common photo-oxidative stabilizer, HALS, in PVC films plasticized with DEHP (di-2-ethylhexyl phthalate). The HALS additive is believed to interrupt oxidative propagation reaction by scavenging of chlorine radical formed in PVC radiolysis.

On the other hand polymer blends are an attractive route to formation of a novel material. Polystyrene, PS, contains aromatic groups that increase radiation resistance and stabilizes the excited species formed by irradiation. The presence of PS in PVC/PS blends could be an interesting route to PVC radiolytic stabilization. The analysis of Figure 14 revealed that at 0-15 kGy the main effect of gamma irradiation on PVC is crossllinking and at 25-100 kGy the main chain effect is predominant. However, the PS in the blend system inhibits crosslinking in the lower irradiation dose range (0–15 kGy) and less chain scission occurs in PVC/PS film than in PVC film. At a sterilization dose (25 kGy) were found a decrease of 65% (95/05) and 47% (90/10) in scissions per original molecule of PVC (Silva et al, 2008). The PS molecule acts as an additive and the aromatic groups of PS structure absorb the excitation energy and a lower bond cleavage yield is noted. This in turn causes a decrease in the formation of free radicals, which are responsible for scission degradation reaction. The mechanisms of main scission and crosslinking of PVC have been showed in Figure 7.

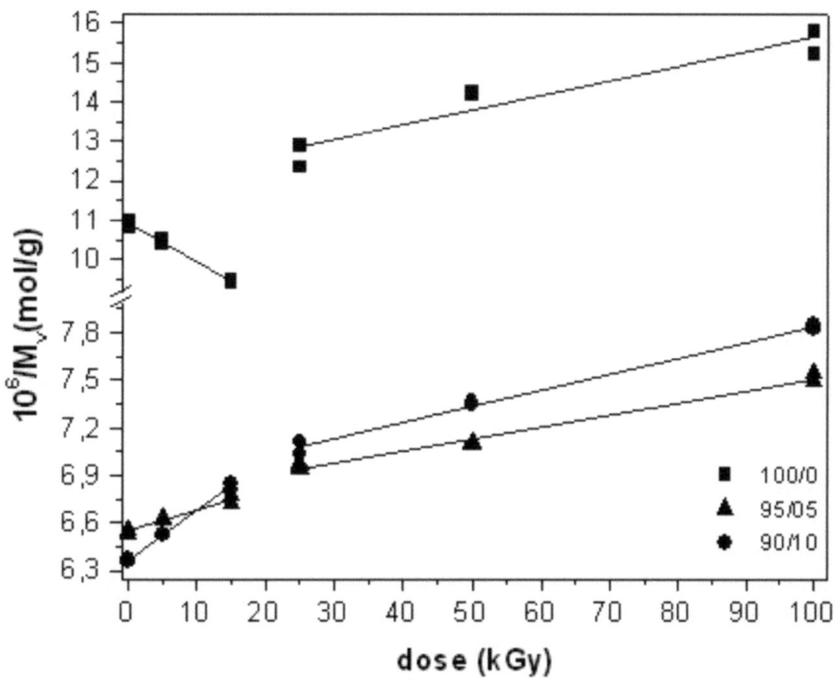


Fig. 14. Reciprocal of $M_v$ as a function of the irradiation dose of PVC and PVC/PS blends

In addition, the preparation of polymer films containing disperse nanoparticles has a great interest. The importance of these nanocomposites is due the mechanical, electrical, thermal, optical, electrochemical, catalytic properties that will differ markedly from that of the component materials. For example, the synthesis of $Sb_2S_3$ nanoparticles by sonochemical route under ambient air from solution containing antimony chloride as metal source and thioacetamide as a sulfur source produced amorphous powder with monodiperse nanospheres, whose diameters were calculated in the range of 300-500 nm. Films of PVC with $Sb_2S_3$ (PVC/Sb) nanoparticles were exposed to gamma irradiation at sterilization dose and the effects of the nanoparticles on the viscosity average molar mass ($M_v$) of sterilized PVC were studied. The results revealed less chain scissions occur in PVC/Sb films at 0.30 wt% concentration. At sterilization dose (25 kGy) was calculated a decrease of 67% in scissions per original molecule of PVC. No information about use of $Sb_2S_3$ in the radiolytic stabilization of polymers has been published and consequently the mechanism of radiolytic stabilization effect of these nanoparticles is not clear. However, some probable reactions may be going on under gamma irradiation.

## 7. Conclusion

Sterilization is defined as any process that effectively kills or eliminates almost all microorganisms like fungi, bacteria, viruses, spore forms. Gamma radiation sterilization are mainly used for the sterilization of pharmaceuticals. Depending on their different mechanism of actions, this sterilization method affects the pharmaceutical formulations in different ways. Thus, the sterilization method chosen must be compatible with the item to be sterilized to avoid damage.

Radiation processing has been used commercially for almost forty years. Gamma radiation from cobalt-60 is used to sterilize the medical devices used in operations and other healthcare treatments. Implants, artificial joints, syringes, blood-bags, gowns, bottle teats for premature baby units and dressings are all sterilized using radiation. Gamma irradiation is a physical means of decontamination, because it kills bacteria by breaking down bacterial DNA, inhibiting bacterial division.

The radiation resistance of a microorganism is measured by the so-called decimal reduction dose ($D_{10}$ value), which is defined as the radiation dose (kGy) required to kill 90% of the total number. Survival fraction of the microorganisms is reversely proportional with the absorbed dose. Doses for sterilization should be chosen according to the initial bioburden, sterility assurance level (SAL) and the radiosensitivity of microorganisms. Temperature plays a major role in the radiosensitivity of microorganisms. As a general rule, microorganisms are less radiosensitive when irradiated at low temperatures

On the other hand, radiation sterilization of tissue grafts has been implemented in some tissue banks, and a dose of 25 kGy has been used in many of these tissue banks. The advantage of radiation sterilization is that it allows the processing of grafts, which have been previously sealed or tightly closed in special wrappings. Such procedures prevent any accidental recontamination during packing. Food also can is sterilized by gamma irradiation and the process exposing food to ionizing radiation to destroy microorganisms, bacteria, viruses, or insects that might be present in the food. Irradiated food does not become radioactive, but in some cases there may be subtle chemical changes. The use of low-level irradiation as an alternative treatment to pesticides for fruits and vegetables that are considered hosts to a number of insect pests including fruit flies and seed weevils. The irradiation produces no greater nutrient loss than what occurs in other processing methods. However, the irradiation reduces the vitamin content of food, the effect of which may be indirect in that inadequate amounts of antioxidant vitamins (such as C, E, and β-carotene) may be available to counteract the effects of free radicals generated by normal cell metabolism. In addition, the most affected lipids during irradiation are thus the polyunsaturated fatty acids that bear two or more double bonds.

When radiation is used for the sterilization of medical devices, the compatibility of all of the components has to be considered. Ionizing radiation not only kills microorganisms but also affects material properties. When the polymer systems are submitted to sterilization by gamma radiation (25 kGy dose), their molecular structures undergo modification mainly as a result of main chain scission and crosslinking effects. Both processes coexist and either one may be predominant depending not only upon the chemical structure of the polymer, but also upon the conditions like temperature, environment, dose rate, etc., under which irradiation is performed.

The protection of polymers against sterilization dose requires efficient additives preventing and/or stopping chain reaction type oxidative degradation. Primary and secondary antioxidants work well here in synergy. Polymer blend and nanoparticles also may be used in radiolytic stabilization of polymer used in medical devices. Commercial raw materials are available for radiation-sterilizable medical devices made of polyolefins and other thermoplastics. Similarly, polymer compounds of suitable formulae are offered commercially for high-dose applications in nuclear installations.

## 8. References

Action biologique et chimique des rayonnements ionisants (ABCRI). (2001). B.Tilquin (Ed.), 115 pp. Paris, France

Al-Kahtani, H. A.;Abu-Tarboush, M. H.; Bajaber, A. S.; Atia, M. ;Abou-Arab, A. A.& El-Mojaddidi, M. A. (1996). Chemical changes after irradiation and post-irradiation storage in Tilapia and Spanish mackerel. *Journal of Food Science*, Vol. 61, pp. 729-733

Alariqi, S.A.S.; Pratheep, A. K.; Rao, B. S. M. & Singh, R. P. (2006). Biodegradation of γ-sterilised biomedical polyolefins under composting and fungal culture environments. *Polymer Degradation Stability*, Vol. 91, N° 6, pp. 1105-1116.

American Society for Testing and Materials. (2006). Standard Guide for Selection and Calibration of Dosimetry Systems for Radiation Processing, ISO/ASTM 51261, Annual Book of ASTM Standards, ASTM, p.p 970-988, USA, New York,

Aquino, K. A. S. & Araujo, E. S. (2008). Effects of a Hindered Amine Stabilizer (HAS) on Radiolytic and Thermal Stability of Poly(methyl methacrylate). *Journal of Applied Polymer Science*, Vol. 110, N° 1, pp. 401-401

Aquino, K. A. S.; Araujo, E. S & Guedes, S. M. (2010). Influence of a Hindered Amine Stabilizer on optical and mechanical properties of poly (methyl methacrylate) exposed to gamma irradiation. *Journal of Applied Polymer Science*, Vol.116, N° 2, pp. 748-753

Araujo, E. S.; Khoury, H. J. & Silveira, S. V. (1998). Effects of gamma-irradiation on some properties ofdurolon polycarbonate. *Radiation Physics and Chemistry*, Vol. 53, pp. 79-84

Armstrong, S. G.; Wyllie, S. G. & Leach, D. N. (1994) Effects of preservation by gamma irradiation on the nutritional quality of Australian fish. *Food Chemistry*, Vol.50, pp. 351-357.

Baccaro, S.; Brunella, V.; Cecília, A. & Costa, L. (2003). $\gamma$ irradiation of poly(vinyl choride) for medical applications. *Nuclear Instruments and Methods in Physics Research*, Vol. 208, pp. 195-198

Brewer, M. S. (2009). Irradiation effects on meat flavor: A review. *Meat Science*, Vol. 81, N° 1, pp. 1-14

Billen, D. (1984). The role of hydroxyl radical scavengers in preventing DNA strand breaks induced by X irradiation of toluene treated *E. coli*. *Radiation Research*, Vol.97, pp. 626-629

Borrely, S. I.; Cruz, A. C.; Del Mastro, N. L.; Sampa, M. H. O. & Somessari, E. S. (1998). Radiation processing of sewage and sudge. A review. *Progress in Nuclear Energy*, Vol. 33, pp. 3-21

Broomfield, S.; Hryciw, T. & Xiao, W. (2001). DNA postreplication repair and mutagenesis in *Saccharomyces cerevisiae*. *Mutation Research*, Vol. 486, pp. 167-184

Cadet, J.; Douki, T.; Gasparutto, D.; Gromova, M;, Pouger, J. P.; Ravanat, J. L.; Romieu, A. & Sauvaigo S. (1999). Radiation-induced damage to DNA: mechanistic aspects and measurement of base lesions. *Nuclear Instruments and Methods. in Physics Research section B*, Vol. 151, pp. 1-7

Campbell, D.G. & Li P. (1999). Sterilization of HIV with irradiation: relevance to HIV infected bone allografts. *NZ Journal of Surgery*, Vol. 69, pp. 517-521

Coates, M. E.; Ford, J. E.; Gregory, M. E & Thompson, S.Y. (1969). Effects of gamma-irradiation on the vitamin content of diets for laboratory animals. *Laboratory Animals*, Vol. 3, pp. 39-49

Costa, L. & Bracco, P. (2004) Mechanisms of crosslinking and oxidative degradation of UHMWPE, In: *The UHMWPE Handbook*, S. Kurtz, (Ed), 345-347, Elsevier, Academic Press, Boston

CONRAD, E. U.; Ernest, U. David, R.; Gretch, M. D.; Kathyn, R.; Obermeyer, B. S.; Margery, S.; Moogk, M. S.; Merlyn, S. M. D.; Jefrey, J.; Wilson, B. S. & Michael, S. (1995). The transmission of hepatitis C virus through tissue transplantation, *Journal of Bone and Joint Surgery*, Vol. 77-A, pp. 214–224

Crucq A. S.; Slegers C.; Deridder V. & Tilquin B. (2000). Radiosensitivity study of cefazolin sodium. *Talanta*, Vol. 52, pp. 873-877

DAAR, E. S.; MOUDGIL, T.; MEYER, R. D.& HO, D. D. (1991). Transient levels of viremia in patients in primary human immunodeficiency virus type 1 infection. *New England Journal of Medicine*. Vol. 324, pp. 961–964

Daly M. J.; Gaidamakova, E. K.; Matrosova, V.Y.;Vasilenko, A.; Zhai, M.; Leapman, R. D.; Lai, B.; Ravel, B. S.W., Kemner, K. M. & Fredrickson J.K. (2007). Protein oxidation implicated as the primary determinant of bacterial radioresistance. *PLOS Biology*,Vol. 5, pp. 769-779

Diehl, J.H. (1967). Combined effects of irradiation, storage, and cooking on the vitamin E and B1 levels of foods. *Food Irradiation*, Vol. 10, N° 2-7,pp. 1967

EN 552. (1994). *Sterilization of medical devices – Validation and routine control of sterilization by irradiation.* Brussels CEN

Erkan, N. & Özden, Ö. (2007). The changes of fatty acid and amino acid compositions in sea bream (*Sparus aurata*) during irradiation process. *Radiation Physics and Chemistry,*Vol 76, N° 10, pp. 1636-1641

Farkas, J. & Farkas C. M. (2011) History and future of food irradiation, *Trends Food Science & Technology*, Vol. 22, N° 2-3, pp.121-126

Freitag, D.; Grigo, U.; Mueller, P. (1988). *Polycarbonates.* In: Encyclopedia of Polymer Science and Engineering, Vol. 11, 2nd ed, 648-718, New York, Wiley

Fox, J. B. J.; Lakritz, L. & Thayer, D. W. (1997). Thiamin, riboflavin and α-tocopherol retention in processed and stored irradiated pork. *Journal of Food Science,* Vol. 62, N°5, pp. 1022-1025

Fox, J. B. J.; Lakritz, L.; Hampson, J.; Richardson, F.; Ward, K & Thayer, D. W. (1995). Gamma irradiation effects on thiamin and riboflavin in beef, lamb, pork, and turkey. *Journal of Food Science,* Vol. 60, N° 3, pp. 596-598

Geis, P. A. & Szaniszlo, P. J. (1984). Carotenoid pigments of the dermatiaceous fungus *Wangiella dermatitidis. Mycologia,* Vol. 76, pp. 268-273

Goldman, M.; Gronsky, R. & Pruitt, L. (1998). The influence of sterilization technique and ageing on the structure and morphology of medical-grade ultrahigh molecular weight polyethylene. *Journal of Materials Science: Materials in Medicine,* Vol. 9, pp. 207- 212

Gonzalez, M E. & Docters, A. S. (1999). Evaluation of stability of polymeric insulation materials in radiation fields and development of radiation stable PVC and polypropylene for medical devices, In: *Stability and stabilization of polymers under irradiation,* IAEA tecdoc- 1062 (Ed), 97-110, Vienna, Austria

Guillet, J. (1985). *Polymer photophysics and photochemistry,* Cambridge University Press, New York, USA

Grecz, N.; Snyder, O. P.; Walker A. A. & Anellis A. (1965). Effect of temperature of liquid nitrogen on radiation resistance of spores of *Clostridium botulinum. Applied Microbiology,* Vol. 13, pp. 527–536

Greez, N.; Rowley, D.B. & MATSUYAMA, A. (1983). The action of radiation on bacteria and viruses, In: *Preservation of Food by Ionizing Radiation,* Vol. 2, Josepson, E. S. & Peterson, M. S. (Eds), 167, CRS Press, Boca Raton FL, USA

Grieb, T. A.; Forng, R. Y.; Stafford, R. E., Lin, J.; Almeida, J.; Bogdansky, S.; Ronholdt, C.; Drohan, W. N. & Burgess, W. H. (2005). Effective use of optimized, high dose (50kGy) gamma irradiation for pathogen inactivation of human bone allographs. *Biomaterials,* Vol. 26, pp. 2033-2042

Hafez, Y. S.; Mohamed, A. I.; Singh, G. & Hewedy, F. M. (1985). Effects of gamma irradiaton on proteins and fatty acids of soybean. *Journal of food science,* Vol. 50, pp. 1271-1274

Haghparast, S.; Kashiri, H.; Shabanpour, B. & Pahlavani, M. H. (2010). Antioxidant properties of sodium acetate, sodium citrate and sodium lactate on lipid oxidation in rainbow trout (*Oncorhynchus mykiss*) sticks during refrigerated storage (4°C). *Iranian Journal of Fisheries Sciences,* Vol. 9, pp. 73-86

Hansen M.T. (1978). Multiplicity of genome equivalents in the radiation-resistant bacterium *Micrococcus radiodurans. Journal of Bacteriology,* Vol. 134, pp. 71-75

Hau, L. B. & Liew, M. S. (1993). Effects of gamma irradiation and cooking on vitamin B6 and B12 in grass prawns (*Penaeus monodon*). *Radiation Physics and Chemistry*, Vol. 42, pp. 297-300

Helfinstine S. L.; Vargas-Aburto C.; Uribe R.M. & Woolverton C. J. (2005). Inactivation of *Bacillus* endospores in envelopes by electron beam irradiation. *Applied and Environmental Microbiology*, Vol. 71, pp. 7029-7032

Hernigou, P.; Gras, G.; Marinello, G. & Dormont, D. (2000). Influence of irradiation on the risk of transmission of HIV in bone grafts obtained from appropriately screened donors and followed by radiation sterilization, *Cell Tissue Bank*, Vol.1 4, pp. 279–289

Hiemstra, H.; Tersmette, M.; Vos, A. H. V.; Over, J.; Van Berkel, M. P. & Bree, H. (1991). Inactivation of HIV by gamma radiation and its effect on plasma and coagulation factors, *Transfusion*, Vol. 31, N° 1, pp. 32–39

International Atomic Energy Agency (IAEA). (2004). *International standards for tissue banks*, IAEA, Vienna, Austria

International Atomic Energy Agency (IAEA). (1975). Radiation for a clean environment, *Proceedings of International Symposium on the use of high-level radiation in waste treatment,* Munich, Germany, March17-21, 1975

International Organization for Standardization. (1995). *Sterilization of healthcare products – Requirements for validation and routine control – Radiation sterilization,* ISO 11137, ISO, Geneva, Switzerland

International Organization for Standardization. (2006). *Sterilization of Health Care Products – Radiation – Part 1: Requirements for the Development, Validation and Routine Control of a Sterilization Process for Medical Products,* ISO 11137-1, ISO, Geneva, Switzerland

International Organization for Standardization. (2006). *Sterilization of Health Care Products – Radiation – Part 2: Establishing the Sterilization Dose,* ISO 11137-2, ISO, Geneva, Switzerland.

International Organization for Standardization. (2006). *Sterilization of Health Care Products – Radiation – Part 3: Guidance on Dosimetric Aspects,* ISO 11137-3, ISO, Geneva, Switzerland

International Organization for Standardization. (2000). *Sterilization of Medical Devices - General Requirements for Characterization of a Sterilizing Agent and the Development, Validation and Routine Control of a Sterilization Process for Medical Devices,* ISO-14937(E), ISO, Geneva, Switzerland

Kaplan, I. (1955). *Nuclear Physics.* Addison-Wesley Pub. Co., Boston, USA

Katta, S. R.; Rao, D. R.; Sunki, G. R. & Chawan, C. B. (1991). Effects of gamma irradiation of whole chicken carcasses on bacterial loads and fatty acids. *Journal of food science*, Vol. 56, pp. 371-372

Kim, D.; Song, H.; Lim, S.; Yun. H. & Chung J. (2007). Effects of gamma irradiation on the radiation-resistant bacteria and polyphenols oxidase activity in fresh kale juice. *Radiation Physics and Chemistry, Vol.* 76, pp. 1213-1217

Kostoski, D & Stojanovic Z. (1995) Radiation-induced crystallinity changes and melting behavior of drawn isotactic polypropylene. *Polymer Degradation and Stability*, Vol. 47, pp. 353-356

Korystov, Y. N. (1992). Contribution of the direct and indirect effects of ionizing radiation to reproductive cell death. *Radiation Research.* Vol. 129, pp. 228-234

Krestev, V.; Dobreva, B.; Atanasov A. & Nedkov E. (1986). *Morphology of polymers*, Walter De Gruyter, Berlin, Germany

Kushal, S. & Praveen, K. (1995). Influence of gamma-irradiation on structural and mechanical properties of polypropylene yarn. *Journal of Applied Polymer Science*, Vol. 55, N° 6, pp. 857–863

Kuzminov, A. (1999). Recombinational Repair of DNA Damage in Escherichia coli and Bacteriophage λ. *Microbiology and molecular biology reviews*, Vol. 63, pp. 751-813

Laughlin, W. L.; Boyd, A. W.; Chadwich, K. H.; Donald, J. C. & Miler, A. (1989). Dosimetry for Radiation Processing, Taylor & Francis Ed, New York, USA

Lakritz, L. & Thayer, D.W. (1992). Effect of ionizing radiation on unesterified tocopherols in fresh chicken breast muscle. *Meat Science*, Vol. 32, pp. 257- 265

Ley, F. J. (1969). Sterilization of laboratory animal diets using gamma radiation. *Laboratory Animals*. Vol. 3, pp. 201-254

Lucarini, M.; Pedulli, G. F.; Borzatta, V. & Lelli, N. (1996). The determination of nitroxide radical distributions in polymers by EPR imaging. *Polymer Degradation and Stability*, Vol. 53, pp. 9-17

Mitchell, G. E.; McLauchlan, R. L.; Isaacs, A. R. & Nottingham, S. M. (1992). Effect of low dose irradiation on composition of tropical fruits and vegetables. *Journal of Food Composition and Analysis*, Vol. 5, pp. 291-311

Muranno, E. & Haves, D. J. (1995). *Food Irradiation: A Sourcebook*, Iowa State University: Blackwell Pub Professional, Ames, USA

Özden, Ö. & Erkan, N. (2010). Impacts of gamma radiation on nutritional components of minimal processed cultured sea bass (*Dicentrarchus labrax*). *Iranian Journal of Fisheries Sciences*, Vol. 9, N° 2, pp. 265-278

Patil, B. S.; Vanamala, J. & hallman, G. (2004). Irradiation and storage influence on bioactive components and quality of early and late season 'Rio Red' grapefruit (Citrus paradisi Macf.). *Postharvest Biology and Technology*, Vol. 34, N° 1, pp. 53-64

Rady, A. H.; Maxwell, J.; Wierbicki, E. & Phillips, J. G. (1988). Effect of gamma radiation at various temperatures and packaging conditions on chicken tissues. I. Fatty acid profiles of neutral and polar lipids separated from muscle irradiation at -20° C. *Journal of physical chemistry*, Vol. 31, pp. 195-202

Rimnac, C. M. & Kurtz, S. M. (2005). Ionizing radiation and orthopaedic prostheses. *Nuclear Instruments and Methods in Physics Research B*, Vol. 236, pp. 30–37

Saleh, Y.G.; Mayo, M. S. & Ahearn D. G. (1988). Resistance of some common fungi to gamma irradiation. *Applied Environmental Microbiology*. Vol. 54, pp. 2134-2135.

Schnabel, W. (1981). *Polymer Degradation-Principles and Practical Applications*, Macmillan Publishing Co, New York, USA

Silva, F. F.; Aquino, K. A. S. & Araujo, E. S. (2008). Effects of gamma irradiation on poly(vinyl chloride)/polystyrene blends: Investigation of radiolytic stabilization and miscibility of the mixture. *Polymer Degradation and Stability*, Vol. 93, pp. 2199-2003

Shintani H. (1992) Gamma-radiation and autoclave sterilization of thermoplastic and thermosetting polyurethane. *Journal of Radiation Sterilization*, Vol. 1, pp. ll-13

Sommers C. H.; Niemira B. A.; Tunick M. & Boyd G. (2002). Effect of temperature on the radiation resistance of virulent *Yersinia enterocolitica*. *Meat Science*, Vol. 61, pp. 323-328.

Stevenson, M. H. (1994). Nutritional and other implications of irradiating meat. *Proceedings of the Nutrition Society*, Vol. 53. pp. 317-325

Sullivan R.; Scarpino P. V.; Fassolitis A. C.; Larkin E. P. & Peeler J. T. (1973). Gamma Radiation Inactivation of Coxsackievirus B-2. *Applied Microbiology*, Vol. 26, pp. 14-17

Takehisa, M., , Shintani, H.; Sekiguchi, M.; Koshikawa ,T.; Oonishi, T.; Tsuge, M.; Sou, K.; Yamase, Y.; Kinoshita, S.; Tsukamoto, H.; Endo, T.; Yashima, K.; Nagai, M.; Ishigaki, K.; Sato, Y. & Whitby, J. L. (1998). Radiation resistance of the bioburden from medical devices, Radiation Physics and Chemistry. Vol. 52, pp. 21–27

Thayer, D. W. & Boyd G. (2001). Effect of irradiation temperature on inactivation of *Escherichia coli* O157:H7 and *Staphylococcus aureus*. *Journal of Food Protection*, Vol. 64, pp. 1624-1626

Tubiana, M.; Dutreix, J. & Wambersie A. (1990). *Introduction to radiobiology*, Taylor and Francis, New York, USA

Traeger, R. K. & Castonguay, T. T. (1996). Effect of γ-radiation on the dynamic mechanical properties of silicone rubbers. *Journal of Applied Polymer Science*, Vol. 10, N° 4, pp. 535-550

Vinhas, G. M.; Souto-Maior, R. M.; Almeida, Y. M. B. & Neto. B. B. (2004).. Radiolytic degradation of poly(vinyl chloride) systems. *Polymer Degradation and Stability*, Vol. 86, pp. 431-436

Whitby, J. L. & Gelda, A. K. (1979). Use of incremental doses of cobalt 60 radiation as a means to determining radiation sterilization dose. *Journal of Parenteral Drug Association*, Vol. 33, pp. 144–155

World Health Organization (WHO). (1999). *High-dose irradiation: wholesomeness of food irradiated with doses above 10kGy*, WHO, Technical Report, Geneva, Switzerland

WILLIAMS, J. L. (1977). Radiation stability of polypropylene. *Radiation Physics and Chemistry*, Vol. 9, pp. 444- 454

Yılmaz, I. & Gecgel, U. (2007). Effects of gamma irradiation on *trans* fatty acid composition in ground beef. *Food Control*, Vol. 18, N° 6, pp. 635–638

Youssef, B. M.;Asker, A. A; El-Samahy, S. K. & Swailam, H. M. (2002). Combined effect of steaming and gamma irradiation on the quality of mango pulp stored at refrigerated temperature. *Food Research International*, Vol. 35, N° 1, pp. 1-13

Zhang, X. C. & Cameron, R. E. J. (1999). The morphology of irradiated isotactic polypropylene. Journal of Applied Polymer Science, Vol. 74, N° 9, pp. 2234-2242

# 10

# Induction of Genetic Variability with Gamma Radiation and Its Applications in Improvement of Horsegram

K. N. Dhumal[1] and S. N. Bolbhat[2]
[1]*Department of Botany, University of Pune, Pune (M.S.),*
[2]*Dada Patil Mahavidyalaya Karjat, Dist- A. Nagar (M.S.),*
*India*

## 1. Introduction

Induced mutation is one of the best alternatives for the improvement of horsegram as it can help to regenerate and restore the variability, which is generally lost in the process of adaptation to various stresses. Genetic variability is the most essential prerequisite for any successful crop improvement programme as it provides spectrum of variants for the effective selection, which can be achieved through the processes of hybridization, recombination, mutation and selection.

Genetic variability has been exhausted in horsegram due to natural selection and hence conventional breeding methods are not fruitful (Wani and Anis, 2001). Legumes generally loose different alleles for high productivity, seed quality, pest and disease resistance during the processes of adaptation to environmental stress.

Gamma sources are used to irradiate a wide range of plant materials, like seeds, whole plants, plant parts, flowers, anthers, pollen grains and single cell cultures or protoplasts. Radiations have been used successfully to induce useful mutations for plant breeding. The lower doses/ concentrations of the mutagenic treatments could enhance the biochemical components, which are used for improved economic characters (Muthusamy *et al.*, 2003). Gamma radiation can induce useful as well as harmful effects on crops so there is need to predict the most beneficial dose for improvement of specific traits of crop plants (Jamil and Khan, 2002).

Improvement in yield and productivity of pulses is the need of the hour, but for this marginal land, aberrant rainfall, non availability of improved seeds, less or no input and poor crop management are the main constraints. Amongst pulses horsegram (*Macrotyloma uniflorum* (Lam.) Verdc) is highly neglected in India and hence require more emphasis on its improvement as it has nutritional, medicinal and fodder value. In Maharashtra, during the year 2008-2009, horsegram was cultivated on 0.466 lakh ha with annual production of 0.3232 lakh tones. The average yield per hactar was 693.56 Kg.

Horsegram is drought tolerant and having good nitrogen fixing ability, but receives a low priority in cropping system, soil types etc. It is grown in *kharif* and *rabbi* seasons, as main crop, or as a mixed crop. It is cultivated in areas with annual rainfall 300-600 mm, but does not tolerate flooding or water logging. The favourable average temperature is 18 to 27°C, and adapted to a wide range of well-drained soils.

The use of dry seeds of horsegram as human food is limited due to its poor cooking quality, presence of high level of enzyme inhibitors and heamagglutinin activities (Ray 1969). The seeds are rich in tannins and polyphenols compared to the other legumes (Kadam and Salunkhe 1985). Antinutrients like phytates, tannins and oxalic acid reduce the availability of iron.

### 1.1 Experimental layout

The authentic seeds of horsegram (*Macrotyloma uniflorum* (Lam.) Verdc) cultivar Dapoli Kulthi-1 were procured from Head, Department of Botany, College of Agriculture, Dr. Balasaheb Savant Konkan Agricultural University, Dapoli, Dist-Ratnagiri (M.S.) India.

The field experiments were conducted on the experimental field at Department of Botany, University of Pune, Pune 411 007 (M.S.). The crop of horsegram Dapoli Kulti-1 was grown in *Kharif* season under uniform conditions. All the experiments were carried out in triplicate, following RBD design. The size of each plot was 3.75m X 2.75m and each plot had 225 plants. The distance between two rows and two plants was 30 X 15 cm and the distance between two adjacent plots was one meter.

A total of 42 combination treatments in $M_1$ generation with untreated dry seeds, which was used as control. Final 675 seeds of each treatment were used to raise $M_1$ generation. The $M_2$, $M_3$, $M_4$ and mutants were raised in next rainy season.

## 2. Results

Results obtained in the present investigation are discussed in brief.

### 2.1 $M_2$ generation

#### 2.1.1 Chlorophyll mutations

The chlorophyll mutants were scored from 7 to 10 days after sowing. The different types of chlorophyll mutants such as albina, xantha, chlorina and viridis were reported in all mutagenic treatments.

Amongst all the treatments used gamma radiations had induced the highest chlorophyll mutation frequency, followed by combination treatments. Prakash and Halaswamy, (2006) and Manigopa-Chakraborty *et al.*, (2005) and Bolbhat and Dhumal (2009) have also reported induction of chlorophyll mutation in horsegram with GR and their combination.

The frequency of total chlorophyll mutants varied for single as well as combination of gamma radiation and EMS. In gamma radiation the average percentage of chlorophyll mutation frequency was 1.47% which was slightly higher than combination treatments (1.23%). The highest chlorophyll mutation frequency was recorded in 200Gy (1.70%). The values of chlorophyll mutation frequency were ranging from 1.09% to1.70%. Amongst all types of chlorophyll mutations, albina and chlorina types showed higher percentage

(0.64 and 0.58%). The percentage of xantha and viridis was 0.41 and 0.28 respectively. The average number of chlorophyll mutations was 24.25, while the average frequency was 1.47%.

The combination treatments showed wide range of total percentage of chlorophyll mutations. The range varied from 0.74% to 1.72%. Highest chlorophyll mutation frequency (1.72%) was noted in 100Gy + 0.3%EMS.

Reddy and Annadurai (1992) claimed that chlorophyll mutation can be used as an index for evaluating the mutagenic action of different mutagens. It is also important for assessing the potency of mutagen and also can be used as an indicator of factor mutations. Chlorophyll mutations are used as a dependable index for evaluating the genetic effects of mutagens.

Albina, xantha, chlorina and viridis were found to be the most abundant type of chlorophyll mutants induced by GR and combination treatments in horsegram Bolbhat and Dhumal (2009). Manjaya *et al.*, (2007) and Tambe *et al.*, (2010) attributed this genes concerned with the development and expression of chlorophyll pigments.

### 2.2 Viable mutations

The mutations affecting gross morphological changes in plant habit, leaf and pod morphology, and maturity were scored as viable mutations. These mutants were characterized and named on the basis of specific characters constantly observed in them throughout the course of investigation. Viable mutants and their characteristic features are given in Table-2. Effect of mutagens on the frequency and spectrum of different types of viable mutations in $M_2$ generation is illustrated in Table-1.

### 2.3 Frequency and spectrum of viable mutations

All the treatments used have induced the widest spectrum of viable mutations. The range of viable mutations in gamma radiation and GR + EMS (Table-1) was 0.61 to 2.16% and 0.21 to 1.81% respectively. The highest percentage of frequency of mutations noted for plant habit, leaf and pod morphology and maturity, due to various treatments of gamma radiation was 0.77%, 0.66%, 0.58% and 0.30% respectively. These results also indicated that the percentage of plant habit mutations was maximum as compared to others. In combination treatments the percentage of frequency for plant habit, leaf, pod and maturity type mutations, was 0.76%, 0.63%, 0.49% and 0.28% respectively. The highest percentage of mutations was obtained in gamma radiation (2.16%) as compared to combination treatment (1.81%).

The widest spectrum and frequency of viable mutations may be due to differential mode of action of the mutagens on different base sequences in various genes. The results indicated that the variety used for study was sensitive to mutagenic treatments.

### 2.4 Plant habit mutations

Tall and gigas mutants obtained in the present investigation showed vigorous growth (Plate-2). According to Weber and Gottschalk (1973) and Blonstein and Gale (1984) the

| Mutants | Gamma rays (Gy) | | | | | | | |
|---|---|---|---|---|---|---|---|---|
| | 100 | | 200 | | 300 | | 400 | |
| Total plants studied | 1560 | | 1710 | | 1675 | | 1632 | |
| | No. | % | No. | % | No. | % | No. | % |
| **Plant habit** | | | | | | | | |
| Gigas | 2.00 | 0.13±0.01 | 5.00 | 0.29±0.02 | 2.00 | 0.12±0.01 | 0.00 | 0.00±0.00 |
| Tall | 2.00 | 0.13±0.01 | 0.00 | 0.00±0.00 | 3.00 | 0.18±0.01 | 0.00 | 0.00±0.00 |
| Dwarf | 2.00 | 0.13±0.01 | 2.00 | 0.12±0.01 | 1.00 | 0.06±0.00 | 1.00 | 0.06±0.00 |
| Compact | 2.00 | 0.13±0.01 | 1.00 | 0.06±0.00 | 2.00 | 0.12±0.01 | 1.00 | 0.06±0.00 |
| Bouquet | 0.00 | 0.00±0.00 | 0.00 | 0.00±0.00 | 0.00 | 0.00±0.00 | 0.00 | 0.00±0.00 |
| Erect | 2.00 | 0.13±0.00 | 0.00 | 0.00±0.00 | 1.00 | 0.06±0.00 | 0.00 | 0.00±0.00 |
| Tendrilar | 0.00 | 0.00±0.00 | 2.00 | 0.12±0.01 | 0.00 | 0.00±0.00 | 0.00 | 0.00±0.00 |
| Spreading | 2.00 | 0.13±0.01 | 2.00 | 0.12±0.00 | 1.00 | 0.06±0.00 | 2.00 | 0.12±0.00 |
| Total | 12.00 | 0.77±0.04 | 12.00 | 0.70±0.04 | 10.00 | 0.60±0.03 | 4.00 | 0.25±0.01 |
| **Leaf** | | | | | | | | |
| Gigas | 2.00 | 0.13±0.01 | 4.00 | 0.23±0.01 | 2.00 | 0.12±0.01 | 0.00 | 0.00±0.00 |
| Broad | 1.00 | 0.06±0.00 | 0.00 | 0.00±0.00 | 1.00 | 0.06±0.00 | 0.00 | 0.00±0.00 |
| Narrow | 0.00 | 0.00±0.00 | 2.00 | 0.12±0.00 | 1.00 | 0.06±0.00 | 0.00 | 0.00±0.00 |
| Small | 2.00 | 0.13±0.01 | 1.00 | 0.06±0.00 | 2.00 | 0.12±0.01 | 0.00 | 0.00±0.00 |
| Stalked | 0.00 | 0.00±0.00 | 0.00 | 0.00±0.00 | 2.00 | 0.12±0.00 | 1.00 | 0.06±0.00 |
| Close pinnae | 2.00 | 0.13±0.01 | 0.00 | 0.00±0.00 | 1.00 | 0.06±0.00 | 0.00 | 0.00±0.00 |
| Curly | 1.00 | 0.06±0.00 | 1.00 | 0.06±0.00 | 2.00 | 0.12±0.01 | 3.00 | 0.18±0.01 |
| Long | 0.00 | 0.00±0.00 | 0.00 | 0.00±0.00 | 0.00 | 0.00±0.00 | 0.00 | 0.00±0.00 |
| Total | 8.00 | 0.51±0.02 | 8.00 | 0.47±0.01 | 11.00 | 0.66±0.02 | 4.00 | 0.25±0.01 |
| **Maturity** | | | | | | | | |
| Early | 0.00 | 0.00±0.00 | 0.00 | 0.00±0.00 | 1.00 | 0.06±0.00 | 0.00 | 0.00±0.00 |
| Late | 4.68 | 0.30±0.02 | 3.42 | 0.20±0.01 | 1.68 | 0.10±0.01 | 0.00 | 0.00±0.00 |
| Total | 4.68 | 0.30±0.02 | 3.42 | 0.20±0.01 | 2.68 | 0.16±0.01 | 0.00 | 0.00±0.00 |
| **Pod** | | | | | | | | |
| Gigas | 2.00 | 0.13±0.01 | 4.00 | 0.23±0.01 | 2.00 | 0.12±0.00 | 0.00 | 0.00±0.00 |
| Long | 1.00 | 0.06±0.00 | 0.00 | 0.00±0.00 | 2.00 | 0.12±0.01 | 0.00 | 0.00±0.00 |
| Broad | 1.00 | 0.06±0.00 | 0.00 | 0.00±0.00 | 0.00 | 0.00±0.00 | 1.00 | 0.06±0.00 |
| Narrow | 3.00 | 0.19±0.01 | 0.00 | 0.00±0.00 | 2.00 | 0.12±0.00 | 0.00 | 0.00±0.00 |
| Small | 2.00 | 0.13±0.01 | 2.00 | 0.12±0.01 | 2.00 | 0.12±0.01 | 1.00 | 0.06±0.00 |
| Total | 9.00 | 0.58±0.02 | 6.00 | 0.35±0.01 | 8.00 | 0.48±0.01 | 2.00 | 0.12±0.00 |
| **High yielding** | 0.00 | 0.00±0.00 | 0.00 | 0.00±0.00 | 2.00 | 0.12±0.01 | 0.00 | 0.00±0.00 |
| Total freq. | | 2.16±0.10 | | 1.72±0.09 | | 1.99±0.09 | | 0.61±0.03 |

Table 1. Effect of gamma radiation on frequency and spectrum of viable mutations in $M_2$ generation of horsegram cv. Dapoli Kulthi-1.

| Mutants | Characters | Treatments |
|---|---|---|
| **Plant habit mutants** | | |
| **Gigas** | These mutants were vigorous, upright and tall with larger, thick, close pinnae and thick profuse branching at the base. Pods were large containing bold seeds. | 100, 200 & 300Gy, and various combination treatments. |
| **Tall** | These mutants were vigorous, tall, with medium and thick leaves, normal flowers and pods. | 100 and 300Gy and various combination treatments. |
| **Dwarf** | Plant height of these mutants ranged from 15 to 20cm and had profuse branching at the base which formed a dense umbrella like canopy. | All GR treatments and combinations. |
| **Compact** | These mutants were characterized with dwarf, profuse and compact branches. The branching was more at the base, giving rise to dense, interwoven secondary branches, which ultimately made the mutant compact. | All GR treatments and combination treatments. |
| **Bouquet** | These mutants had profuse branching at the base, which remained very close to each other forming a bunch. The canopy of secondary branches and leaves together gave an appearance of bouquet. | Combination treatments |
| **Erect** | The mutant was erect and tall with shy branching and light green pinnae. | 100 and 300Gy and combination treatments. |
| **Tendrilar** | These types of mutants were very weak, slender, branched or unbranched with very few leaflets. The distal portion of stem was tendrilar and had twining tendency. | 200Gy and combination treatments. |
| **Spreading** | These mutants were creeping on the ground with terminal branches (30 to 60 cm length). | GR and combination treatments. |
| **Leaf mutants** | | |
| **Super gigas** | These mutants had extremely large, thick, dark green leaflets with very prominent midrib, thick and semi erect stem and sparse branching. | 100Gy + 0.2%EMS, 300Gy + 0.5%EMS. |
| **Gigas** | These mutants were vigorous, upright tall with large thick pinnae, profuse branching. Pods were large with bold seeds. | 200Gy (0.23%) and combination treatments |
| **Broad** | These mutants were vigorous, upright, dwarf with large, thick, close pinnae. | Combination treatments |
| **Small** | These mutants were associated with dwarfness, small leaflets and light green colour with profuse branching at the base. | GR and 400Gy + 0.4%EMS. |
| **Curly** | This mutant had curly leaflets with elongated petiole. The leaflets were wedge shaped with shorter leaf lamina, curling towards centre. The plants were erect with pale green foliage | GR and various combination treatments. |
| **Long** | These mutants were dwarf having narrow and very long leaflets. | Combination treatments. |

| Mutants | Characters | Treatments |
|---|---|---|
| **Plant habit mutants** | | |
| **Maturity mutants** | | |
| **Early** | These mutants were associated with dwarfness and showing early flowering within a short span of 30-33 days in comparison with the flowering duration of 53-57 days in control. The total duration of the crop in such mutant was 65-70 days against 100-110 days in control. | 300Gy, 400Gy + 0.5%EMS and various combination treatments |
| **Late** | These mutants were with broad and thick dark green coloured foliage, tall and bearing late flowers as compare to control. Flowering in these mutants was achieved in 60 to 65 DAS against 53 to 57 DAS in control. The total duration of the crop in such mutant was 115-120 DAS against 100-110 DAS in control. | GR and combination treatments |
| **Pod mutants** | | |
| **Gigas** | These mutants were associated with gigas plant type, with very large pods containing 5-6 bold seeds per pod. | GR and combination treatments. |
| **Long** | The plants were normal in appearance with comparatively much longer pods, containing 7 to 8 seeds per pod. | 100 and 300Gy and combination treatments |
| **Broad** | The mutants had broad pods containing 5-6 bold seeds per pod. | 100 and 400Gy and combination treatments |
| **Small** | It was associated with dwarf plant type mutation having very small pods containing 2-3 seeds per pod. | GR and combinations treatments. |
| **High yielding** | These mutants possess high yield contributing characters viz. number of pods, thousand seed weight and total seed yield per plant. | GR and combinations treatments. |

Table 2. Viable mutants and their characteristic features

increase in plant height was due to the changes in the internodal length and increase in cell number and cell length or both.

### 2.5 Leaf mutations

Different types of leaf mutations e.g. super gigas, gigas, broad, narrow, small, tiny, stalked, close pinnae, curly and long were observed in $M_2$ generation of horsegram with GR and GR + EMS treatments.

Tara and Dnyansagar (1979) stated that the changes in shapes of the leaves were due to chromosomal aberrations, induced by chemical mutagens and ionizing radiation. The leaf mutations obtained in horsegram may be ascribed to above cited reasons, which may be useful as gene markers in conventional breeding. These may be useful for understanding the genetic control of leaf formation and regulation of their size, shape and form.

(a) Field preparation

(b) Seedling stage

(c) 30 days crop

(d) 45 days crop

(e) 60 days crop

(f) Mutant screening

(g) Maturity stage

(h) Harvesting stage

Plate 1. $M_2$ generation in field

Plate 2. Plant habit mutants.

### 2.6 Maturity mutations

#### 2.6.1 Early and late

The early mutants were recorded in $M_2$ with GR and GR + EMS. These mutants showed rapid growth and early maturity. Several workers like Dalvi (1990), Rudraswamy *et al.*, (2006) and Bolbhat and Dhumal (2010) reported early and late mutants in horsegram.

In present investigation the early mutants of horsegram show pod maturity within 50-55 DAS in the gamma radiation and combination treatments. The agronomic traits like early flowering and pod maturity have bean always given paramount importance, while planning the breeding strategies. Gottschalk and Wolff (1983) explained the early mutants could be very much useful for genecological studies. The earliness was mainly achieved through rapid growth, during early stages of ontogeny and initiation of first inflorescence. Early maturity in the mutants may be due to the physiological, biochemical, enzymological and hormonal changes induced by the mutagens.

The late mutants were noted in $M_2$ generation of horsegram with gamma radiation and GR + EMS. The main reason attributed to the late maturity were inadequate production of flowering hormones, physiological disturbances, enhanced production of a floral inhibitors and reduced ability to respond to the floral stimulus in the shoot apex (Beveridge and Murfet, 1996). According to Zakri and Jalani (1998) late or early maturity has agronomic significance as these mutants suit for the specific requirement of breeding strategy. The lateness in maturing is worthwhile for prolonging the vegetative phase and allowing the development of a strong sink, which may help to enhance the yield. In addition the period from flowering to maturity should also be long enough, for better seed filling. The late mutants were noted in horsegram with the treatments of gamma radiation and their combinations.

### 2.7 Pod mutations

Pod mutations such as long, large, narrow and small were recorded in $M_2$ generation (Plate-3). These mutants were also reported in horsegram Bolbhat and Dhumal (2010).

### 2.8 High yielding mutants

The high yielding mutants obtained in horsegram due to treatments of GR and GR + EMS showed increased number of pods and grain yield per plant over control.

### 2.9 Quantitative characters in $M_2$ and $M_3$ generations

Gamma radiations and GR + EMS proved to be very effective to induce variability in quantitative traits like plant height, primary branches per plant, days required for first flowering and first pod maturity, number of pods per plant, pod length, number of seeds per pod, 1000 seed weight and yield per plant in $M_2$ and $M_3$ generations (Table-3 and 4 ).

### 2.10 Quantitative traits

#### 2.10.1 Plant height

The treatments of gamma radiations and combinations were effective for reducing the plant height (Table-3) and the maximum reduction was noted in 100Gy.

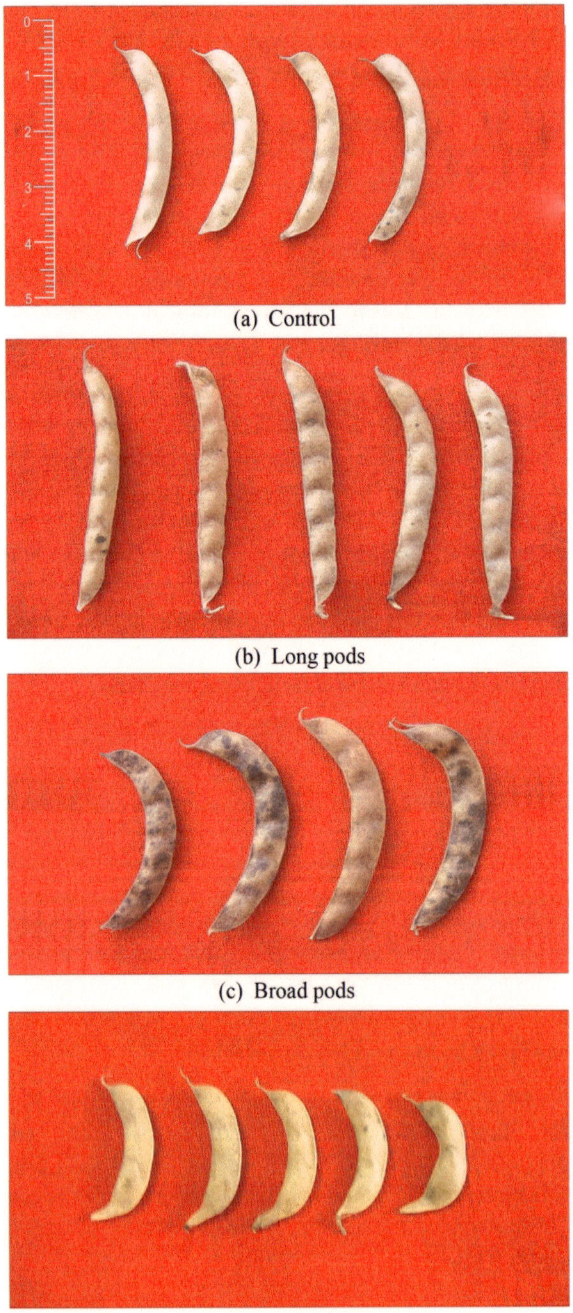

(a) Control

(b) Long pods

(c) Broad pods

(d) Short pods

Plate 3. Pod mutants.

| Treat | Plant height (cm) | Pri. br./ plant | DAS for first flowering | DAS for first pod maturity | No. of pods/ plant | No. of seeds/ pod | 1000 seed Wt. (g) | Seed yield/ plant (g) |
|---|---|---|---|---|---|---|---|---|
| Control | 46.70±1.87 | 6.30±0.25 | 53.30±2.13 | 85.20±3.41 | 71.20±2.85 | 7.20±0.29 | 26.80±1.07 | 14.54±0.58 |
| 100Gy | 35.40±1.77 | 6.66±0.33 | 53.07±2.65 | 81.70±4.09 | 67.70±3.39 | 6.66±0.33 | 28.30±1.42 | 13.36±0.67 |
| 200 | 39.40±2.76 | 6.20±0.43 | 53.70±3.76 | 78.90±5.52 | 59.50±4.17 | 6.33±0.44 | 28.20±1.97 | 11.22±0.79 |
| 300 | 41.30±1.24 | 7.66±0.23 | 54.80±1.64 | 84.60±2.54 | 83.80±2.51 | 7.10±0.21 | 27.50±0.83 | 16.76±0.50 |
| 400 | 40.10±2.41 | 6.10±0.37 | 54.70±3.28 | 84.40±5.06 | 71.30±4.28 | 6.33±0.38 | 26.30±1.58 | 12.87±0.77 |
| 100Gy + 0.2%EMS | 28.70±1.72 | 6.10±0.37 | 49.40±2.96 | 79.20±4.75 | 38.40±2.30 | 6.20±0.37 | 25.60±1.54 | 8.83±0.53 |
| 100 + 0.3 | 29.40±1.18 | 8.40±0.34 | 47.60±1.90 | 77.30±3.09 | 49.70±1.99 | 6.10±0.24 | 26.50±1.06 | 7.64±0.31 |
| 100 + 0.4 | 26.50±1.86 | 5.60±0.39 | 47.30±3.31 | 79.70±5.58 | 37.80±2.65 | 6.30±0.44 | 26.20±1.83 | 8.46±0.59 |
| 100 + 0.5 | 31.20±0.94 | 5.10±0.15 | 47.50±1.43 | 78.60±2.36 | 41.40±1.24 | 6.00±0.18 | 27.50±0.83 | 8.29±0.25 |
| 200 + 0.2 | 33.20±1.66 | 5.90±0.30 | 49.30±2.47 | 77.80±3.89 | 38.70±1.94 | 6.50±0.33 | 27.60±1.38 | 8.87±0.44 |
| 200 + 0.3 | 25.70±0.77 | 6.10±0.18 | 49.40±1.48 | 79.30±2.38 | 41.30±1.24 | 6.20±0.19 | 26.60±0.80 | 9.12±0.27 |
| 200 + 0.4 | 27.50±1.65 | 6.20±0.37 | 47.60±2.86 | 79.80±4.79 | 40.70±2.44 | 6.00±0.36 | 27.80±1.67 | 9.76±0.59 |
| 200 + 0.5 | 30.20±2.11 | 7.10±0.50 | 49.20±3.44 | 78.50±5.50 | 39.80±2.79 | 6.50±0.46 | 26.30±1.84 | 8.90±0.62 |
| 300 + 0.2 | 34.30±1.37 | 6.50±0.26 | 54.90±2.20 | 79.40±3.18 | 54.80±2.19 | 7.10±0.28 | 26.40±1.06 | 13.10±0.52 |
| 300 + 0.3 | 31.20±1.56 | 7.10±0.35 | 54.40±2.72 | 79.60±3.98 | 50.70±2.54 | 7.00±0.35 | 26.60±1.33 | 11.53±0.58 |
| 300 + 0.4 | 29.10±2.04 | 5.20±0.36 | 54.30±3.80 | 78.70±5.51 | 33.50±2.35 | 7.10±0.50 | 30.50±2.14 | 9.98±0.70 |
| 300 + 0.5 | 35.30±1.06 | 9.10±0.27 | 50.80±1.52 | 80.60±2.42 | 71.70±2.15 | 7.10±0.21 | 29.80±0.89 | 16.01±0.48 |
| 400 + 0.2 | 42.50±2.13 | 7.800.39± | 50.40±2.52 | 76.80±3.84 | 71.40±3.57 | 6.50±0.33 | 28.70±1.44 | 14.44±0.72 |
| 400 + 0.3 | 33.20±1.99 | 6.60±0.40 | 56.30±3.38 | 77.60±4.66 | 70.40±4.22 | 7.10±0.43 | 25.80±1.55 | 12.69±0.76 |
| 400 + 0.4 | 40.90±1.64 | 6.70±0.27 | 40.80±1.63 | 57.70±2.71 | 52.80±2.11 | 6.20±0.25 | 27.70±1.11 | 10.38±0.42 |
| 400 + 0.5 | 47.50±1.43 | 8.30±0.25 | 40.60±1.22 | 66.80±2.00 | 74.40±2.23 | 6.20±0.19 | 27.70±0.83 | 14.10±0.42 |
| SEM± | 1.42 | 0.27 | 2.14 | 3.30 | 2.25 | 0.27 | 1.41 | 0.47 |
| F-value | 42.03 | 29.99 | 8.26 | 3.64 | 95.80 | 4.82 | 2.45 | 69.09 |
| P-value | 0.00 | 0.00 | 0.00 | 0.00 | 0.00 | 0.00 | 0.01 | 0.00 |
| LSD $_{0.05}$ | 2.78 | 0.53 | 4.19 | 6.47 | 4.41 | 0.53 | 2.76 | 0.92 |

Table 3. Micromutations in $M_2$ generation of horsegram cv. Dapoli Kulthi-1.

### 2.11 Number of primary branches per plant

Data obtained in $M_2$ generation on number of primary branches per plant (Table-3) indicated that the mean values of this parameter showed positive and negative influence. Maximum number of primary branches per plant was recorded in 300Gy (7.66) over control.

All the treatments in $M_3$ showed increase in number of primary branches per plant. Maximum number was recorded in 100Gy (12.58) and 200Gy + 0.2 %EMS (11.25) over control (10.99). Dalvi (1990) also noted similar trend with physical as well as chemical mutagens in horsegram.

### 2.12 Number of days required for first flowering

GR and GR + EMS treatments have induced the variability in number of days required for first flowering in $M_2$ generation. However some treatments were stimulatory and others were inhibitory for inducing the flowering. The minimum number of days required for first flowering were 50 DAS in 300Gy and 42 DAS in 400Gy + 0.4%EMS.

Dalvi (1990) and Rudraswami *et al.*, (2006) also noted similar results in horsegram with different mutagens.

### 2.13 Number of days required for first pod maturity

The data recorded in (Table-3) indicated that all the treatments of GR and combinations had succeeded in reducing the number of days required for first pod maturity as compared to control. The combination treatments were highly significant. The average minimum number of days (57.70 DAS) required for first pod maturity was noted in 400Gy + 0.4%EMS. The data obtained for $M_3$ generation for this trait was on par with of $M_2$ generation (Table-4). All the treatments of GR + EMS, caused reduction in number of days required for first pod maturity than control. The results of Nawale (2004) supported the above findings.

| Treat | Plant height (cm) | Pri. br./ plant | DAS for 1st flowering | DAS for 1st pod maturity | No. of pods/ plant | No. of seeds/ pod | 1000 seed Wt. (g) | Seed yield/ plant (g) |
|---|---|---|---|---|---|---|---|---|
| Control | 50.99±2.04 | 10.99±0.44 | 54.00±2.16 | 83.00±3.32 | 100.55±4.02 | 6.84±0.27 | 25.07±1.00 | 16.58±0.66 |
| 100Gy | 49.73±2.49 | 12.58±0.63 | 53.67±2.68 | 82.44±4.12 | 106.18±5.31 | 6.83±0.34 | 25.31±1.27 | 18.62±0.93 |
| 200 | 44.57±3.12 | 11.49±0.80 | 54.72±3.83 | 84.45±5.91 | 110.04±7.70 | 6.84±0.48 | 26.05±1.82 | 19.14±1.34 |
| 300 | 49.85±1.50 | 11.98±0.36 | 50.98±1.53 | 79.17±2.38 | 110.32±3.31 | 6.85±0.21 | 25.93±0.78 | 19.60±0.59 |
| 400 | 48.75±2.93 | 10.03±0.60 | 54.77±3.29 | 85.78±5.15 | 109.62±6.58 | 6.88±0.41 | 26.70±1.60 | 18.90±1.13 |
| 100Gy + 0.2%EMS | 45.10±2.71 | 6.78±0.41 | 54.00±3.24 | 83.67±5.02 | 86.88±5.21 | 6.55±0.39 | 24.80±1.49 | 14.12±0.85 |
| 100 + 0.3 | 48.50±1.94 | 6.39±0.26 | 52.63±2.11 | 81.31±3.25 | 100.11±4.00 | 6.53±0.26 | 24.14±0.97 | 16.35±0.65 |
| 100 + 0.4 | 49.27±3.45 | 7.30±0.51 | 54.86±3.84 | 84.16±5.89 | 90.25±6.32 | 6.62±0.46 | 24.56±1.72 | 14.95±1.05 |
| 100 + 0.5 | 46.11±1.38 | 8.48±0.25 | 51.44±1.54 | 79.18±2.38 | 93.33±2.80 | 6.42±0.19 | 22.84±0.69 | 15.27±0.46 |
| 200 + 0.2 | 50.61±2.53 | 11.25±0.56 | 57.34±2.87 | 88.66±4.43 | 85.22±4.26 | 7.08±0.35 | 25.53±1.28 | 13.83±0.69 |
| 200 + 0.3 | 45.81±1.37 | 11.06±0.33 | 52.94±1.59 | 81.86±2.46 | 93.42±2.80 | 6.54±0.20 | 23.85±0.72 | 15.71±0.47 |
| 200 + 0.4 | 50.42±3.03 | 10.44±0.63 | 58.12±3.49 | 89.82±5.39 | 97.28±5.84 | 7.16±0.43 | 27.31±1.64 | 16.96±1.02 |
| 200 + 0.5 | 47.07±3.29 | 9.68±0.68 | 53.23±3.73 | 82.65±5.79 | 86.53±6.06 | 6.54±0.46 | 25.09±1.76 | 14.71±1.03 |
| 300 + 0.2 | 50.94±2.04 | 9.64±0.39 | 55.64±2.23 | 85.75±3.43 | 83.49±3.34 | 6.52±0.26 | 25.81±1.03 | 13.68±0.55 |
| 300 + 0.3 | 50.78±2.54 | 9.11±0.46 | 54.67±2.73 | 84.00±4.20 | 87.33±4.37 | 6.18±0.31 | 24.28±1.21 | 13.52±0.68 |
| 300 + 0.4 | 47.00±3.29 | 8.67±0.61 | 45.19±3.16 | 76.23±5.34 | 87.18±6.10 | 6.84±0.48 | 24.27±1.70 | 13.40±0.94 |
| 300 + 0.5 | 46.03±1.38 | 9.87±0.30 | 46.71±1.40 | 78.44±2.35 | 92.39±2.77 | 6.42±0.19 | 24.57±0.74 | 14.23±0.43 |
| 400 + 0.2 | 44.10±2.21 | 9.67±0.48 | 43.29±2.16 | 72.35±3.62 | 102.18±5.11 | 5.82±0.29 | 23.45±1.17 | 13.30± |
| 400 + 0.3 | 41.84±2.51 | 10.51±0.63 | 49.88±2.99 | 76.90±4.61 | 98.42±5.91 | 6.74±0.40 | 25.65±1.54 | 15.59± |
| 400 + 0.4 | 49.12±1.96 | 9.22±0.37 | 42.34±1.69 | 71.14±2.85 | 96.08±3.84 | 6.18±0.25 | 24.40±0.98 | 15.20± |
| 400 + 0.5 | 48.91±1.47 | 11.01±0.33 | 48.52±1.46 | 74.68±2.24 | 85.53±2.57 | 6.78±0.20 | 25.52±85.53 | 15.18±0.46 |
| SEM± | 1.99 | 0.41 | 2.19 | 3.42 | 3.95 | 0.29 | 1.05 | 0.66 |
| F-value | 3.44 | 32.25 | 7.99 | 4.20 | 9.60 | 2.57 | 2.07 | 18.14 |
| P-value | 0.00 | 0.00 | 0.00 | 0.00 | 0.00 | 0.00 | 0.02 | 0.00 |
| $LSD_{0.05}$ | 3.90 | 0.80 | 4.29 | 6.70 | 7.74 | 0.57 | 2.06 | 1.29 |

Table 4. Micromutations in $M_3$ generation of horsegram cv. Dapoli Kulthi-1.

### 2.14 Number of pods per plant

The data recorded in (Table-3) revealed that the some of GR and GR + EMS treatments had stimulatory as well as inhibitory effect. In $M_2$ generation maximum number of pods per plant (83.80) were noted in 300Gy than control (71.20). The minimum number of pods per plant (59.50) were recorded at 200Gy. However all the combination treatments have caused reduction in number of pods per plant. The trend in variation of pod number observed in $M_3$ generation was similar to that of $M_2$ generation (Table-4).The results of Dalvi (1990) for horsegram were in agreement with the present study.

### 2.15 Total number of seeds per pod

Data on total number of seeds per pod in $M_2$ and $M_3$ progeny showed non significant change as compared to control.

### 2.16 1000- seed weight

Results recorded on 1000-seed weight (Table-3) indicated that all the treatments of GR were non significant but the combination treatments such as 300Gy + 0.4%EMS and 300Gy + 0.5%EMS (30.50g and 29.80g) had shown considerable increase in 1000 seed weight over control. The results of $M_3$ generation were on par with $M_2$ generation (Table-4).

### 2.17 Seed yield per plant

Mean values for seed yield per plant decreased in treatments of GR and GR + EMS as compared to controls (Table-3). The maximum seed yield (16.76g) was noted in 300Gy and minimum (11.22g) in 200Gy as compared to control (14.54g). The combination treatment 300Gy + 0.5%EMS had induced maximum increase in seed yield per plant (16.01g) over control (14.54g). But all other treatments had caused reduction in seed yield per plant. In $M_3$ generation seed yield per plant was increased in all gamma radiation treatments, but it decreased in combinations as compared to control (Table-4). Maximum total seed yield was recorded in 300Gy (19.60g) and in 200Gy + 0.4%EMS (16.96g) as compared to control (16.58g). Hakande (1992) reported wider variability in yield due to mutagenic treatments in winged bean, which was attributed to pollen sterility and genetical as well as physiological alterations caused by mutagens. Previous studies indicated that both additive and non-additive genes contribute to yield. Luthra *et al.*, (1979), Reddy and Sree Ramulu (1982) also supported the above view.

### 2.18 Harvest index

Mean values for seed yield per plant, biological yield and harvest index decreased with few exceptions in all mutagen treated populations as compared to their controls (Fig-1). In gamma radiation maximum seed yield (16.76g) as well as biological yield (41.55g) were recorded in 300Gy. The highest value of harvest index (40.81) was reported in 400Gy. In combination treatments seed yield 16.01g (300Gy + 0.5%EMS), biological yield 37.44g (300Gy + 0.5%EMS) and harvest index 44.77 (400Gy + 0.5%EMS) were recorded.

$M_3$ progeny showed increase in seed yield, decrease in biological yield and increase in harvest index as compare to $M_2$ (Fig-1). Jain (1975) claimed that high value of harvest index and dry matter production contributes to yield. The genotype with ability for converting larger part of dry matter in to economic yield is highly preferable (Donald, 1962). According to Vaghela *et al.*, (2009) the biological yield per plant and harvest index were found to be the most valuable traits for formulating the selection criteria to improve seed yield in chickpea.

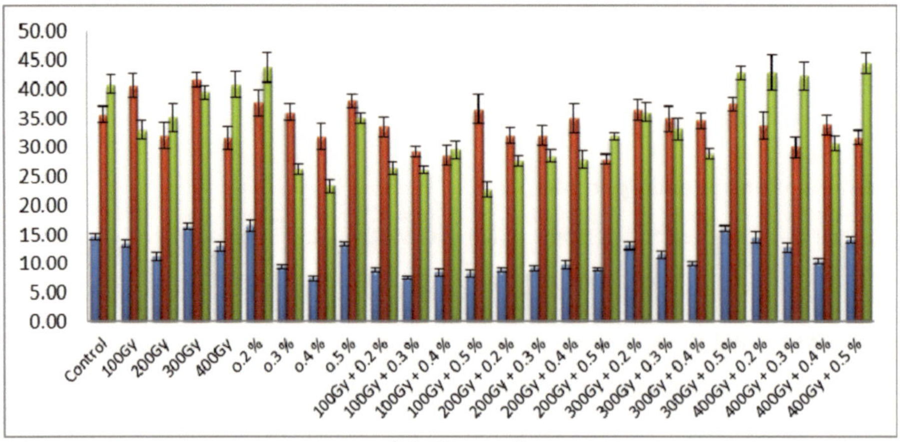


a. $M_2$ Generation

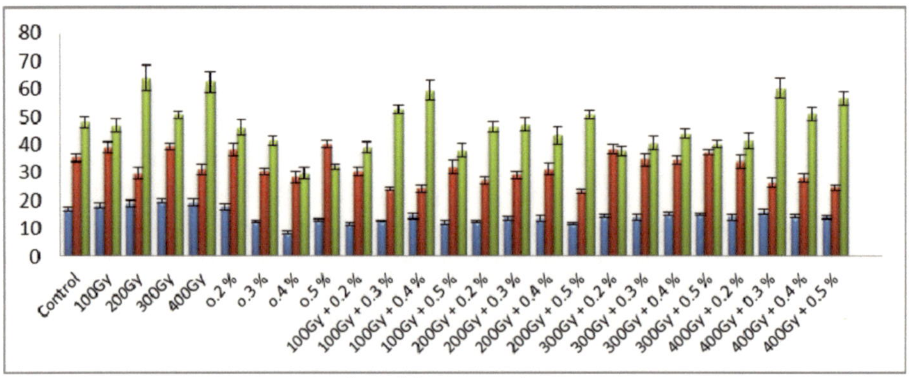


b. $M_3$ generation

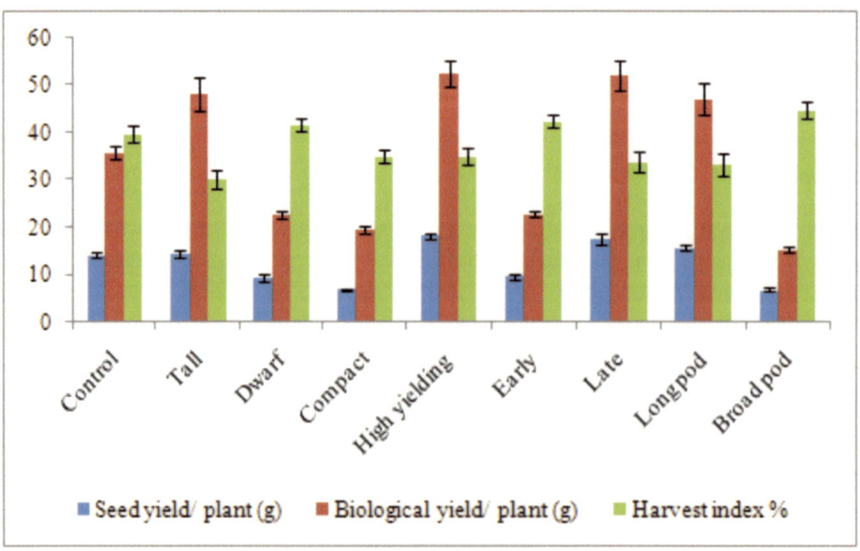


c. (c) $M_3$ Mutants

Fig. 1. Mutagenic effect on seed yield, biological yield and harvest index

The relationship of biological yield with economic yield or grain yield may help to prodict the performance and yield efficiency of the genotypes.

### 2.19 Mutants in $M_3$

Mean values for seed yield per plant, biological yield and harvest index were positively or negatively influenced in all mutants as compared to controls (Fig-1). Maximum seed and biological yield was recorded in tall, high yielding, late and long pod mutants over control. The highest seed yield (18.29g) and biological yield (52.40g) in high yielding mutant and harvest index (44.72%) in broad pod mutant was recorded. In present investigation the rate of dry matter production in horsegram mutants was significantly correlated with seed yield (Fig-1). Work of Sahane *et al.,* (1995) in horsegram was inconformity with present study.

### 2.20 Morphological and yield attributes of $M_3$ mutants

Desirable mutants such as tall, dwarf, early, late, long and broad pod as well as high yielding were isolated from $M_2$ and $M_3$ generations. More than seven true breeding macromutants (Table-5) were characterized on the basis of morphological and yield attributes, which may be useful for future breeding programme in horsegram yield improvement.

| Mutants | Plant height (cm) | Pri. br./ plant | DAS for 1st flowering | DAS for 1st pod maturity | No. of pods/ plant | No. of seeds/ pod | 1000 seed Wt. (g) | Seed yield/ plant (g) |
|---|---|---|---|---|---|---|---|---|
| Control | 56.70±2.27 | 6.30±0.25 | 61.70±2.47 | 102.20±4.09 | 71.20±2.85 | 6.90±0.28 | 26.80±1.07 | 14.15±0.57 |
| Tall | 71.50±3.58 | 7.30±0.37 | 63.60±3.18 | 105.40±5.27 | 82.40±4.12 | 6.40±0.32 | 27.20±1.36 | 14.54±0.73 |
| Dwarf | 16.30±1.14 | 4.20±0.29 | 58.30±4.08 | 97.70±6.84 | 47.00±3.29 | 7.00±0.49 | 25.80±1.81 | 9.49±0.66 |
| Early | 22.80±1.37 | 5.40±0.32 | 30.60±1.84 | 65.80±3.95 | 55.00±3.30 | 6.20±0.37 | 27.10±1.63 | 9.64±0.58 |
| Late | 60.90±3.65 | 8.10±0.49 | 65.40±3.92 | 110.20±6.61 | 85.30±5.12 | 7.20±0.43 | 28.30±1.70 | 17.58±1.05 |
| Long pod | 59.10±2.36 | 7.10±0.28 | 64.30±2.57 | 106.10±4.24 | 80.20±3.21 | 7.30±0.29 | 26.20±1.05 | 15.74±0.63 |
| Broad pod | 24.60±1.72 | 6.20±0.43 | 60.40±4.23 | 97.50±6.83 | 33.40±2.34 | 6.40±0.45 | 27.80±1.95 | 6.84±0.48 |
| High yielding | 59.40±1.78 | 8.10±0.24 | 62.80±1.88 | 109.30±3.28 | 87.20±2.62 | 7.20±0.22 | 28.50±0.86 | 18.29±0.55 |
| SEM± | 1.86 | 0.27 | 2.48 | 4.17 | 2.67 | 0.29 | 1.58 | 0.52 |
| F-value | 267.16 | 43.22 | 37.50 | 20.70 | 141.32 | 4.67 | 1.21 | 143.49 |
| P-value | 0.00 | 0.00 | 0.00 | 0.00 | 0.00 | 0.00 | 0.35 | 0.00 |
| $LSD_{0.05}$ | 3.91 | 0.57 | 5.21 | 8.76 | 5.61 | 0.61 | 3.32 | 1.09 |

Data are means of three replicates ± standard deviation. Significant difference due to treatments was assessed by Fisher's LSD as a post-hoc test.

Table 5. Morphological and yield attributes of $M_3$ mutants of horsegram cv. Dapoli Kulthi-1.

### 2.21 Tall

These mutants were vigorous, tall (71.50cm), with medium and thick leaves with delayed flowers and pod formation compared to control. There was significant increase in primary branches, pods, 1000 seed weight and seed yield per plant (Plate- 4).

(a) Control (b) Tall

(c) Dwarf (d) Compact

(e) Early (f) Late

(g) Long Pods (h) High yeilding

Plate 4. Stable mutants (M3 generation)

### 2.22 Dwarf

Plant height of these mutants ranged from 15 to 20cm and had profuse branching at the base, which formed a dense umbrella like canopy (Plate- 4 ). There was significant reduction in number of days required for first flowering and pod maturity period.

The induced dwarfness is desirable agronomic trait, because plant density per unit area, will be very high. It will help for fast translocation of metabolites from source to sink (Auti, 2005). Such mutants are valuable for investigating plant gene function and developing new crop variety with lodging resistance (Wei *et al.*, 2008).

### 2.23 Early

Early mutants were associated with dwarfness and showing early flowering within a short span of 30-33 days in comparison with the flowering duration of 53-57 days in control. The total duration of the crop was 65-70 days against 100-110 days in control (Plate- 4). Such type of mutants were recorded earlier in horsegram (Dalvi, 1990, Rudraswamy *et al.*, 2006 and Bolbhat and Dhumal (2010). These mutants are highly desirable to reduce the crop duration. Short duration variety of horsegram will play a key role in avoiding drought and water stress faced by rainfed agriculture.

### 2.24 Late

These mutants were with broad and thick dark green foliage, tall and bearing late flowers as compared to control. Flowering was achieved in 60 to 65 DAS against 53 to 57 DAS in control. Plant height, primary branches per plant, pods per plant, pod length, seeds per pod, 1000 seed weight and seed yield per plant was improved over control (Plate- 4).

According Auti (2005) late maturity in these mutants was due to genetic damage caused by the mutagens. Same may be the reason for horsegram late mutants.

### 2.25 Long pod

The plants were normal in appearance with comparatively longer pods, containing seven to eight seeds per pod. (Plate- 4). Plant height, primary branches per plant, pods per plant, pod length, seeds per pod and seed yield per plant were significantly increased over control.

These mutants showed considerable increase in all quantitative traits, contributing to increase in yield. This type of mutants were reported earlier in mungbean (Auti and Apparao, 2009), urdbean (Sagade, 2008 and Gahlot *et al.*, 2008), cowpea (Girija and Dhanavel, 2009) and horsegram (Bolbhat and Dhumal, 2010).

### 2.26 Broad pod

These mutants had broad pods containing 6-7 bold seeds per pod (Plate- 4). Significant decrease was noted in days required for first flowering 60.40 DAS (control 61.70DAS) and maturity period 97.50DAS (control 102.20DAS). There was significant increase in 1000 seed weight 27.80g (control 26.80g).

### 2.27 High yielding

These mutants showed superior yield attributes such as length and number of pods, number of seeds, 1000 seed weight and yield per plant as compared to control (Plate- 4). Auti (2005) stated that some unknown mechanism must be inducing the expression of yield controlling genes, which are responsible to increase the yield. Same may be the reason for getting high yielding mutants in horsegram. These mutants can break the constrains of low yield of horsegram, in rainfed agriculture.

### 2.28 Antinutrients in seeds

#### 2.28.1 Polyphenols

The seeds of all viable mutants showed reduced which will be of immense importance for the feature development of desirable varieties of horsegram with improved nutritional quality.

### 2.29 Phytate

The results shown in Table- 6 revealed that all the mutants except dwarf and compact showed decrease in phytate content. Phytic acid decides the nutritional quality of cereals and seeds of legume as it is a strong antinutritional factor. It is also implicated in the "hard-to-cook" phenomenon of legumes (Stanley, 1985). Hence development of new cultivars with low phytate is the prime need of time. In present study tall, high yielding, early, late and long pod mutants showed reduced phytic acid contents (Table- 6). All such mutants can be exploited in future hybridization programme of horsegram aimed for improving nutrient quality.

| Mutants | Polyphenols (mg $100g^{-1}$) | Phytate (mg $100g^{-1}$) | Trypsin inhibitors (TIU) |
|---|---|---|---|
| Control | 442.25±17.69 | 2639.71±105.59 | 185.81±7.43 |
| Tall | 420.12±21.01 | 2565.51±128.28 | 110.43±5.52 |
| Dwarf | 447.14±31.30 | 2647.94±185.36 | 104.19±7.29 |
| Compact | 408.55±12.26 | 2643.83±79.31 | 120.71±3.62 |
| Early | 396.26±23.78 | 2425.56±145.53 | 98.75±5.93 |
| Late | 412.21±16.49 | 2544.99±101.80 | 126.96± |
| Long pod | 411.37±28.80 | 2372.45±166.07 | 136.32±9.54 |
| Broad pod | 409.07±12.27 | 2641.59±79.25 | 129.98±3.90 |
| High yielding | 402.26±24.14 | 2289.66±137.38 | 138.65±8.32 |
| SEM± | 17.81 | 106.29 | 5.36 |
| F-Value | 1.88 | 3.24 | 46.18 |
| P-Value | 0.13 | 0.02 | 0.00 |
| $LSD_{0.05}$ | 37.42 | 223.32 | 11.26 |

Data are means of three replicates ± standard deviation. Significant difference due to treatments was assessed by Fisher's LSD as a post-hoc test.

Table 6. Nutrient quality in seeds of $M_3$ mutants of horsegram cv. Dapoli Kulthi 1

### 2.30 Trypsin inhibitor

All the mutants showed decreased levels of trypsin inhibitor over control. Lowest values of TI were found in early mutant (98.75 TIU). The values of TI in different mutants in order as given below dwarf > tall >compact >late >long pod >high yielding.

The legume seeds like horsegram contain trypsin and chemotrypsin, which are the main anti-nutritional factors (Chavan and Hejgaard, 1981) as they decreases the digestibility of protein and cause pancreatic hypertrophy (Manjaya *et al.*, 2007). The decrease in the levels of trypsin inhibitor as a result of the treatments of mutagens was reported by Harsulkar (1994). Kim *et al.*, (2010) reported increase as well as decrease in trypsin inhibitor in mutants induced by gamma radiation. The mutants of horsegram with low TI have a great role in vegetarian diet of Indians.

## 3. Conclusion

In conclusion GR and GR+EMS had induced sufficient genetic variability in horsegram cv. Dapoli Kulthi-1. The agronomically and nutritionally superior mutants will be the promising material for plant breeders in feature. Such cultivars will surve the purpose of protein malnutrition in vegetarian diets and will also economically benefit the resource poor farmers of rainfed area.

The high yielding mutants will play a major role to break the yield constraints in horsegram, as a result of this the farmers can be attracted towards the cultivation of this low input, nitrogen fixer legume crop and there by it will help to improve the economic status of farmers from rainfed, drought prone areas of Maharashtra States. The early mutants will be of \ great help to reduce the crop duration and thereby help to avoid droughts in late stage of crop. The mutants with low TI, phytate and phenols will be of revolutionary importance, as they will be free from all antinutritional factors.

The stabilization of desirable mutants, through multilocation trails is in progress on farmers' field at Karjat (Dist-Ahmednagar), Research farms of Dr. B.S.S.K.A.U., Dapoli (Dist-Ratnagiri) and at Department of Botany, University of Pune (MS). The efforts are also being made to release the most desirable mutants of horsegram, Dapoli Kulthi-1.

## 4. References

Auti, S.G. (2005). Mutational Studies in mung (*Vigna radiata* (L.) Wilczek). Ph.D. Thesis. University of Pune, Pune (MS), India.

Auti, S.G. and Apparao, B.J. (2009). Induced mutagenesis in mungbean (*Vigna radiata* (L.) Wilczek). *In*: Induced Plant Mutations in the Genomics Era. Food and Agricultural Organization of the United Nations. Shu, Q.Y. (Ed.). pp. 107-110. Italy, Rome.

Beveridge, C. and Murfet, I. (1996). The gigas mutant in pea is deficient in the floral stimulus. *Physiol. Plant*. 96: 637-645.

Blonstein, A.D. and Gale, M.D. (1984). Cell size and cell number in dwarf barley and semidearf cereal mutants and their use in cross breeding II (Teidse 407) FAO/IAEA. Vienna. pp. 19-29.

Bolbhat, S.N. and Dhumal, K.N. (2009). Induced macromutations in horsegram (*Macrotyloma uniflorum* (Lam.) Verdc). *Legume Res.* 32 (4): 278-281.

Bolbhat, S.N. and Dhumal, K.N. (2010). Desirable mutants for pod and maturity characteristics in $M_2$ generation of horsegram (*Macrotyloma uniflorum* (Lam.) Verdc). *Res. on Crops* 11 (2): 437-440.

Chavan, J.K. and Hejgaard, J. (1981). Detection and partial characterization of subtilisin inhibitors in legume seeds by isoelectric focusing. *J. Sci. Food Agric.* 32: 857-859.

Dalvi, V.V. (1990). Gamma rays induced mutagenesis in horsegram (*Macrotyloma uniflorum* (Lam.) Verdc). M.Sc. dissertation. Dr. B. S. K. K. Vidyapeeth, Dapoli (MS), India.

Donald, C.M. (1962). In search of yield. *J. Aust. Inst. Agric. Sci.* 28 (3): 171-178.

Gahlot, D.R., Vatsa, V.K. and Kumar, D. (2008). Some mutants for pod and maturity characteristics in $M_2$ generation of urdbean (*Vigna mungo* (L.) Hepper). *Legume Res.* 31 (4): 272-275.

Girija, M. and Dhanavel, D. (2009). Mutagenic effectiveness and efficiency of gamma rays, EMS and their combined treatments in cowpea (*Vigna unguiculata* (L.) Walp). *Global J. Mol Sci.* 4 (2): 68-75.

Gottschalk, W. and Wolff G. (1983). The alteration of flowering and ripening times, pp.75-84. In: Induced Mutations in plant Breeding. Springer Verlag. Berlin Heidelberg New York, Tokyo.

Hakande, T.P. (1992). Cytological studies in *Psophocarpus tetragonolobus* L.D.C. Ph.D. Thesis, Marathwada University, Aurangabad (MS) India.

Harsulkar, A.M. (1994). Studies on the mutagenic effects of pesticides in barley. Ph.D. thesis. Dr. Babasaheb Ambedkar Marathwada University, Aurangabad (MS), India.

Jain, H.K. (1971). New type in pulses. Ind. FMG. 21 (8): 9-10

Jamil, M. and Khan, U.Q. (2002). Study of genetic variation in yield components of wheat cultivar Bukhtwar-92 as induced by gamma radiation. *Asian J. of Plant Sci.* 5 (1): 579-580.

Kadam, S.S. and Salunkhe, D.K. (1985). Nutritional composition, processing, and utilization of horsegram and moth bean. CRC Rev. *Food Sci. Nutri.* 22: 1–26.

Kim, D.S., Lee, K.J., Kim, J.B., Kim, S.H., Song, J.Y., Seo, Y. W., Lee, B.M. and Kang, S.Y. (2010). Identification of Kunitz trypsin inhibitor mutations using SNAP markers in soybean mutant lines. *Theor. Appl. Genet.* 121 (4): 751-760.

Luthra, O.P., Arora, N.D., Singh, R.K. and Chaudhary, B.D. (1979). Genetics analysis for metric traits in mungbean (*Vigna radiata* L. Wilczek). Haryana Agricultural University Journal of Research. 9: 19-24.

Manigopa-Chakraborty, Ghosh, J., Singh, D. N., Virk, D.S. and Prasad, S.C. (2005). Selection in $M_2$ generation of horsegram (*Macrotyloma uniflorum*) through participatory plant breeding. *J. Arid-Legumes.* 2 (1):1-4.

Manjaya, J.G. and Nandanwar, R. S. (2007). Genetic improvement of soybean variety JS 80-21 through induced mutations. *Plant Mutation Reports*. 1 (3):36-40.

Muthusamy, A., Vasanth, K. and Jayabalan, N. (2003). Response of physiological and biochemical components in *Gossypium hirsutum* L. to mutagens. *J. Nuclear Agric. Biol.* 32 (1): 44-51.

Nawale, S.R. (2004). Studies on induced mutagenesis in cowpea (*Vigna unguiculata* (L.) Walp.) M.Sc. dissertation, Dr.B.S.K.K.Vidyapeeth, Dapoli (MS), India.

Prakash, B.G. and Halaswamy, K.M. (2006). Chemical mutagenesis and their effectiveness under M<ovid: sub>2/ovid: sub> generation in horsegram (*Macrotyloma uniflorum* (Lam.) Verdc). *J. Arid-Legumes*. 3 (1):11-14.

Ray, P.K. (1969). Toxic factor(s) in row horsegram (*Dolichos biflorus*). *J. food Sci.* 6: 207-211.

Reddy, P.R.R. and C. Sree Ramulu. (1982). Heterosis and combining ability for yield and its components in greengram (*Vigna radiata* (L.) Wilczek). *Genetica Agraria*. 36: 297-308.

Reddy, V.R.K. and Annadurai, M. (1992). Induction of chlorophyll mutants in lentil. Res. JPI Environ 8 (1-2): 59-69.

Rudraswami, P., Vishwanatha, K.P. and Gireesh, C. (2006). Mutation studies in horsegram (*Macrotyloma uniflorum* (Lam.) Verdc). BARC, LSS-2006, Mumbai (MS), India. pp-88-89.

Sagade, (2008). Genetic improvement of urdbean (*Vigna mungo* L. Hepper) through mutation breeding. Ph.D. Thesis. University of Pune, Pune (MS), India.

Sahane, D.V., Dhonukshe, B.L. and Navale, P.A. (1995). Relative dry matter efficiency and harvest index in relation to grain yield of horsegram. *J. Maharashtra Agric. Univ.* 20 (1):136-137.

Stanley, D.W., Aguilera, J.M. (1985). A review of textural defects in cooked reconstituted legumes: the influence of structure and composition. *J. Food Biochem.* 9: 277–323.

Tambe, A.B., Pachore, M.V., Giri, S.P., Andhale, B.S.and Apparao, B.J. (2010). Induced Chlorophyll Mutations in Soybean *Glycine max* (L.) Merrill. *Asian J. Exp. Biol. Sci. Spl.*:142-145.

Tara, J.L. and Dnyansagar, V.R. (1979). Effect of gamma rays and EMS on growth and branching in *Turnera ulmifolia. J. Cytol. Genet.* 14: 118-123.

Vaghela M.D., Poshiya, V.K., Savaliya, J.J., Davada, B.K. and Mungra, K.D. (2009). Studies on character association and path analysis for seed yield and its components in chickpea (*Cicer arietinum* L.) *Legume Res.* 32 (4): 245-249.

Wani, A.A. and M. Anis. (2001). Gamma rays induced bold seeded high yielding mutant in chickpea. *Mutation Breeding Newsletter*. 45: 20-21.

Weber, E. and Gottschalk, W. (1973). DieBeziehungen Zuischen Zellagrobe and internodienuange tri starhleindyzei earten *Pisum* mutant. *Beitr Bio Pfl.* 49: 101-126.

Wei, F.H., You, F., Li, A. and Qing, Y.S. (2008). A revisit of mutation induction by gamma rays in rice (*Oriza sativa* L.): implications of microsatellite markers for quality control. *Mol Breeding* 22: 281-288.

Zakri, A.H., and Jalani, B.S. (1998). Improvement of soybean through mutation breeding. Improvement of grain legume production using induced mutations. Proceedings of a workshop Pullman Washington, USA,1-5 July 1986.pp 451-461.Vienna, Austria, International Atomic Energy Agency.

# 11

# Ultrastructure Alterations in the Red Palm Weevil Antennal Sensilla Induced By Gamma Irradiation

Eman A. Mahmoud, Hussein F. Mohamed* and Samira E.M. El-Naggar
*Biological Applications Department, Nuclear Research Center,*
*Atomic Energy Authority, Abo-Zaabal, Cairo,*
*Egypt*

## 1. Introduction

Palm beetles (Coleoptera: Curculionidae) are harmful pests for cultivated palm - oil, date or coconut palms (Bedford, 1980, Giblin-Davis *et al,* 1996). The red palm weevil (RPW) *Rhynchophorous ferrugineus* (Oliv.) is the most important pest of the date palm in the Middle-East (Abraham & Pillai, 1998). The insect was discovered in Saudi Arabia in the mid-1980s (Gush, 1997 & Faleiro *et al.*, 1999). Since then it has spread over most of the date palm areas (Bokhari & Abuzuhira 1992 & Vidyasagar *et al.*, 2000), due to the transfer of infested offshoots and palm trees.

It was introduced accidentally into the Gulf states from Pakistan in 1985, from whence it spread throughout the Arabian Peninsula. It was first reported in Egypt in 1992. It infested the governorates Ismailia and Sharkia, an area with an estimated one million palm trees (Salama & Abdel Aziz, 2001). lthough they have been known as major pests for a long time, efficient and acceptable methods of controlling them are still lacking in many cases. In addition, the selection of plants for oviposition is determined by the physical nature of their surface and by chemical factors which are detected after contact (Fenomore, 1980, Chadha & Roome, 1980). As such, the identification of olfactory and contact chemoreceptors in insects has received the attention of many workers (Albert, 1980, Bland, 1984 & Salama *et al.*, 1987).

Antennae in insect are organs of taste, smell and stimulation (Wiggles worth, 1972). Other authors (Stubbs, 1980, Nakamuta, 1985 & Obata, 1986) have suggested a possible role of sensory stimuli (olfactory, gustatory or mechanical) in prey, or mate detection. The antennae also play kinetic roles and normally keep the nervous system in a state of tone in which it responds to stimuli of all kinds. Antennae of insects vary greatly in length, overall size, size of the individual segments, segmentation, setation and other aspects with the structures being closely related to their function ( Srivastava & Omkar , 2003). Notably, some authors studied the ultrastructurally effects of gamma rays on the features of normal antennal sensilla of *R. Ferrugineus* as Oland *et al.*, (1988); El-kholy & Mikhaiel (2008) and El- Akhdar & Afia, 2009.

Despite the importance of sense organs on the antennae of red palm weevil (for oviposition, feeding and mating) there is little information about it. Therefore, the objective of current study is to describe the distribution, types of sensillae in *Rhynchophorus ferrugineus* using scanning electron microscopy, and estimate the effects of gamma radiation on them. This information may be useful for controlling the pests.

## 2. Materials and methods

### 2.1 Insect rearing technique

Red palm weevil, *R. ferruginous* adults were obtained from cocoons collected from infested date palm trees in the Salhia district (Sharkiya governorate, Egypt). A large number of cocoons were put into oblong opaque -white plastic boxes ( 21 x10.5x 7 cm in length, width and high respectively) covered with tight-fitting and perforated lids. Cocoons were kept in wet toweling until adult emergence.The newly emerged adults were sexually differentiated. Newly laid eggs from the adult culture were collected using a 0.2 mm brush and inserted into round plastic Petri dishes (50 mm diameter x 20 mm high) containing tissue paper soaket in Benzoate solution, which was renewed daily. The Petri dish had a screw cap and uniformly spaced perforations (2 mm diameter) around the surface of the box including lid and bottom.

The hatched larvae were maintained in plastic cups (250 ml capacity) containing 5g of artificial diet as described by Rahalkar *et al.,* (1985). The plastic cups were covered with muslin cloth and kept at controlled laboratory conditions (27°C and 85% RH). Daily inspection was carried out until eggs hatched and the larval and pupal stages were completed. After emergence, adult weevils were transferred into wire screen cages containing pieces of suger cane as food.

### 2.2 Irradiated treatments

Adults weevils, 2 weeks after emergence, irradiated with 15, 20 Gy of gamma radiation using gamma cell ($^{60}$Cobalt -Source), installed in the Middle Eastern Regional Radioisotope Center for the Arab Countries, Cairo, Egypt, with the dose rate of 1 Gy / 1.95 sec.

### 2.3 Examination with the scanning electron microscope (SEM)

unirradiated and the irradiated adult weevils were freeze by liquid nitrogen then dried in the chamber of the scanning electron -microscope, SEM (Jeol-JSM-5600 LV in SEM Unit, Central Laboratory for Elemental Analysis, Inshas, Egypt) in the low vacuum mode. Then the micrographs were taken, this technique called low vacuum scanning electron microscope freeze drying (L V-SEM)(Gasser *et al.,*2008) .This technique resulted in the presence of few small particles in white color represents ice during freeze drying technique of the specimens in low vacuum SEM on the micrographs

## 3. Results

The antenna of the red palm weevil measures about 0.9 cm in length (geniculate antenna). It arises from the elongated anta cava in front of the anterior margin of the compound eye and at the base of the rostrum. It consists of three segments, (scape, pedicel, and flagellum) differing in size and shape (Fig.1).

### 3.1 The scape

It is the longest segment; it measures more than one-third of the length of the antenna. Proximally, it has a hook fitting in the antennal groove, while distally it is somewhat thickened (Fig.1).

### 3.2 Pedicel

The pedicel has a small triangular base that fits in a comparatively large cavity at the distal end of the scape (Fig.1). Two subtypes of sensilla were distinguished

### 3.3 The flagellum

It consists of six segments; the first five segments known as the funicle (resemble the pedicel in form), while the sixth flagellar segment is known as the club (Fig.1). Six subtypes of sensillae were recognized on the funicle, S. trichodea 1, 2, 3, S. chaetica 2 and S. coeloconic 1, 3. The characteristic morphological features of sensilla are shown in figures (2D, 3A).

### 3.4 The club

It is the last segment of flagellum (similar to funnel in shape), longer and broader than other antennal segments (Fig.1). The light microscopic observation revealed that the sensilla are densely and homogenously packed on this area. Also, the electronic microscopic study revealed the presence of four types (seven subtypes) of sensilla highly crowded over the whole of this segment.

Fig. 1. General morphology of adult antenna *Rhynchophorus Ferrugineus* showing: Scape (S), Pedicel (P), Funicle (F) and Club (C). [Scale bar indicate 1 mm]

### 3.5 The different types of sensilla

#### 3.5.1 Sensilla coeloconica

Sensilla coeloconica 1 (S.C1) were located in a large and low lumen on both sides of the scape (Fig.2A). S.C1 was distinguished by their finger-like appearance, which characterizes it from the other sensilla. On high magnification, the sensilla appears to be formed of multiple straight tubules directed towards the antennal shaft giving a pyramidal appearance, opening at the apex with many ridges on the surface wall of the sensilla.. Each one arises from a narrow socket; the base of sensilla may be attached to the socket in some parts. The sensilla measures about 16.8 μm in length and about 11.3 μm in diameter (Fig.2A). Two subtypes of these sensilla were distinguished, on the pedicel, the first, Sensilla coeloconica 1 was scattered in a little numbers and have been described previously, the second, Sensilla coeloconica 2 which emerges from a well-defined socket. The fingers are long and tapering apically, some arise from sockets with tilting towards the antennal shaft, and have triangular shape. They have veins ornamented vertically on the surface of the sensilla wall. These sensilla ranged from about 30.8 to 41.3 μm in length, and about 15.8 μm in diameter (Fig.2B). Few numbers of S.C1 are distributed on the $3^{rd}$, $4^{th}$ and 5 $^{th}$ segments of funnels on the flagellum while, S.C 3 are found on the $3^{rd}$ segment of the funnel. The finger-like processes of S. coeloconica 3 are long and fused into a blunt tip. Number of them is tapering apically, extends with the same diameter and has an elongated triangular appearance. They grow out from wide sockets and extend in a straight line on the antennal shaft (Fig.3A). These sensilla measured about 18 μm in length, and about 5 μm in diameter.

#### 3.5.2 Sensilla basiconica

Sensilla basiconica 1 are found on the ventral side between the scape and the pedicle in one row. It is a hair with a simple bulky base and a wide socket, and this hair ending fans out into two curved parts in line with the curve of the whole sensilla. These sensilla ranged from about 12.5 to 18.8 μm in length, and ranged about 5.0 to 6.5 μm in diameter (Fig.2C). Sensilla basiconica 2 are spread in a very few number on the surface of the club. They are blunt tipped, relatively stout pegged. They are often straight and sometimes slightly curved towards the antennal shaft (Fig.3F).

#### 3.5.3 Sensilla trichodea

They show no special arrangement except that they are more frequent on the apical and basal areas of the segmented funnily on the flagellum.

##### 3.5.3.1 Sensilla trichodea 1

It is formed of straight hair, tapering to a fine end, growing out from a socket in the cuticle. The surface of the sensilla is smooth. It is measured about 240 μm in length and about 12.5 μm in diameter at its base (Fig.2D).

##### 3.5.3.2 Sensilla trichodea 2

It is a straight hair and most of them are blunt-tipped while few of them are bifurcate at the terminal end (Fig.2D). The hair bases inserted tightly into a large cuticular socket and the surface of the sensilla is similar to trichodea 1. They have a mean length about 105 μm and mean diameter about 12.5 μm.

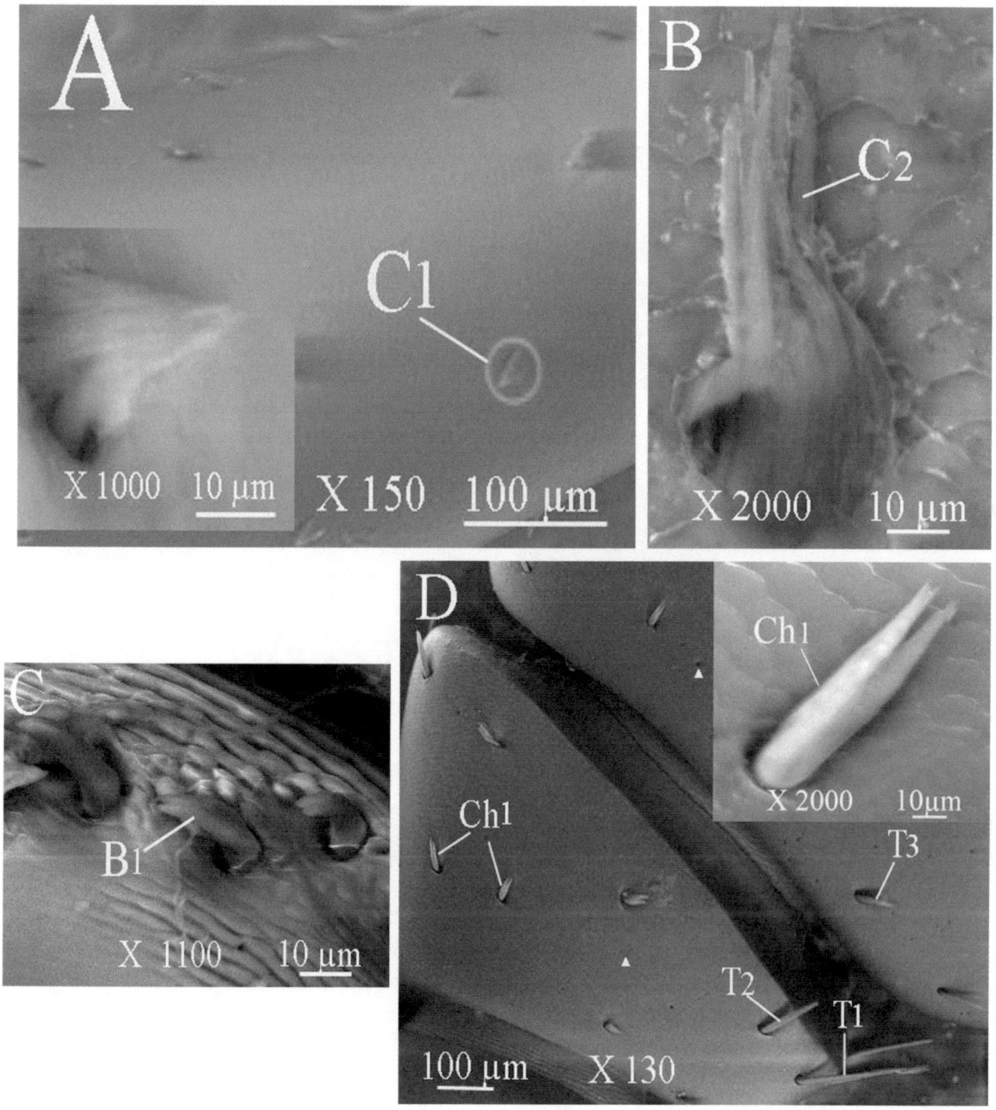


Fig. 2. SEM photomicrographs of non-irradiated adult antenna of *Rhynchophorus Ferrugineus.* (A): Sensilla coeloconica 1 (C1) on the scape, (B): Sensilla coeloconica 2 (C2) on the pedicel, (C): Sensilla basiconica 1(B1) on the ventral side between scape and pedicel, (D): Sensilla trichodea 1, 2, 3 (T1, T2, T3), Sensilla chaetica1 (Ch1) and Cuticular pores as arrowheads on the funicle. [Scale bars indicate 100 μm (A, D) and 10 μm (B, C), respectively]

### 3.5.3.3 Sensilla trichodea 3

The sensilla of this type are straight, and tapering to a fine point. In some sections, a socket of cuticle is found close to the base of the sensilla. The length measures about 70 μm and the diameter about 12.5 μm (Fig. 2D). Sensilla trichodea 1, 2, 3 observed on the lateral side of the club without any arrangement and have been described previously (Fig.2D).

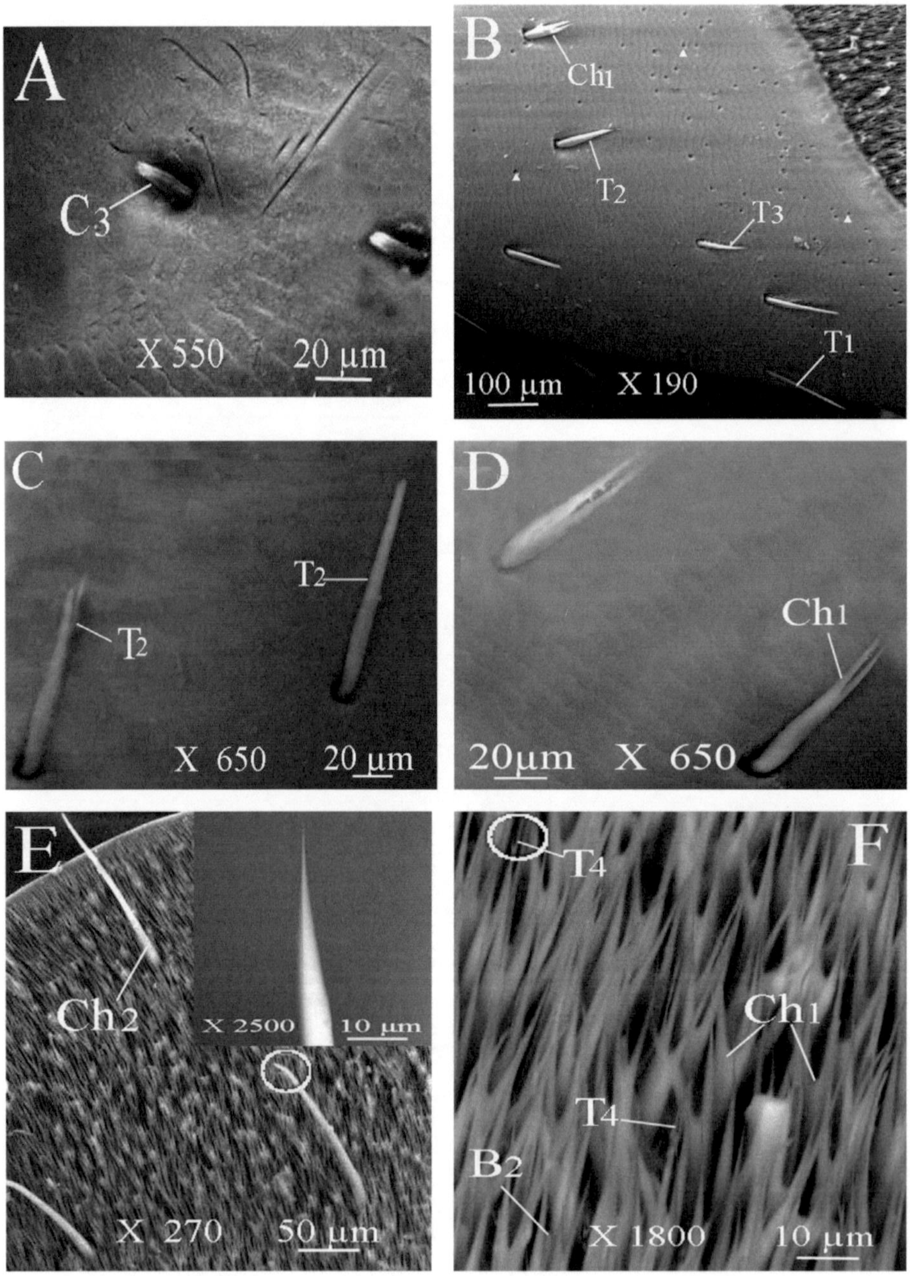


Fig. 3. SEM photomicrographs of non-irradiated adult antenna of *Rhynchophorus Ferrugineus.* (A): Sensilla coeloconica 3 (C3) on the funicle, (B): Sensilla trichodea 1,2,3 (T1,T2,T3) and Cuticular pores as arrowheads on the lateral side of the club, (C): Bifurcat trichodea 2 (T2) on the lateral side of the club, (D): Sensilla chaetica1 (Ch1) on the lateral side of the club, (E): Sensilla chaetica 2 (Ch2) on the surface of the club, (F): Sensilla Basiconica 2 (B2), Sensilla Trichodea 4 (T4) and Sensilla chaetica1(Ch1) on the surface of the club. [Scale bars indicate 20 µm (A, C, D), 100 µm (B), 50 µm (E) and 10 µm (F), respectively]

#### 3.5.3.4 Sensilla trichodea 4

These S. trichodea are present only in a few number on the surface of the club and are smooth -walled. Most of them have about the same length of S. chaetica 1 on the surface of club. The apical part of these sensilla is remarkable split into two branches. These branches are different in length and diameter and they are straight or slightly curved (Fig. 3F).

### 3.5.4 Sensilla chaetica

#### 3.5.4.1 Sensilla chaetica 1

These sensilla are thick and bifurcate at the half of length, and appear very sharply pointed at their apical margin of equal branches on the flagellum. They are smooth walled and distributed on all the surface of funnel .Some of them is extended in a straight way while the others are gently curved. The length ranged from 32.5 to 50.0 µm and the diameter about 12.5 µm (Fig.2D). These sensilla also, are most abundant on the surface of club; some of them are extended straight while the others gently curved (Fig.3F). S. chaetica 1 resemble that found on funicle (thick, bifurcate from about the middle part of sensilla, and appeared very sharply pointed at their apical margin of equal branches).Also, they could be observed on the lateral sides of the club, the two branches of these sensilla arisen from about middle or third total length and each branch are also divided into two parts (Fig.3D). The S. chaetica 1 on the lateral sides of the club not curved on the antennal shaft and measure about 33.5 µm in length and about 4 µm in diameter.

#### 3.5.4.2 Sensilla chaetica 2

They are long hair-like structures arranged in a line on the surface of the club. They are similar in shape to S. trichodea but much longer in length (approximately three or four times), and found fewer in number compared with other sensilla types. Some of their long axis forms a right angle with antennal shaft while others form an angle of around 60 °C with the antennal surface to protrude well above all the other sensilla. Distally, it tapers uniformly into simple extremities with pointed tips (Fig. 3E).

### 3.5.5 Cuticular pores

Abundant cuticular pores vary in their size and location and can be found in-between the sensilla on the whole surface of the lateral side of club and the first segment of the funnel (Figs.2D, 3D). Few numbers of medium-sized pores are noticed on the cuticular depression at the 2$^{nd}$ segment of the funnel (Fig.2D).

## 3.6 The effects of gamma irradiation on the antennal sensilla

### 3.6.1 The effects of the dose rate 15 Gray

The following sensilla showed certain alterations while the remaining ones did not show specific changes.

#### 3.6.1.1 Sensilla coeloconica

#### 3.6.1.1.1 Sensilla coeloconica 1

Changes in general appearance, enlarged in size and irregularity in shape were noticed. The base of sensilla became less dense especially in central part. The finger like processes at the apex also became flattened, diffuse, less dense, and not pointing .The sensilla measuring about 20.8 μm in length and about 14.8 μm in diameter (Fig.4A).

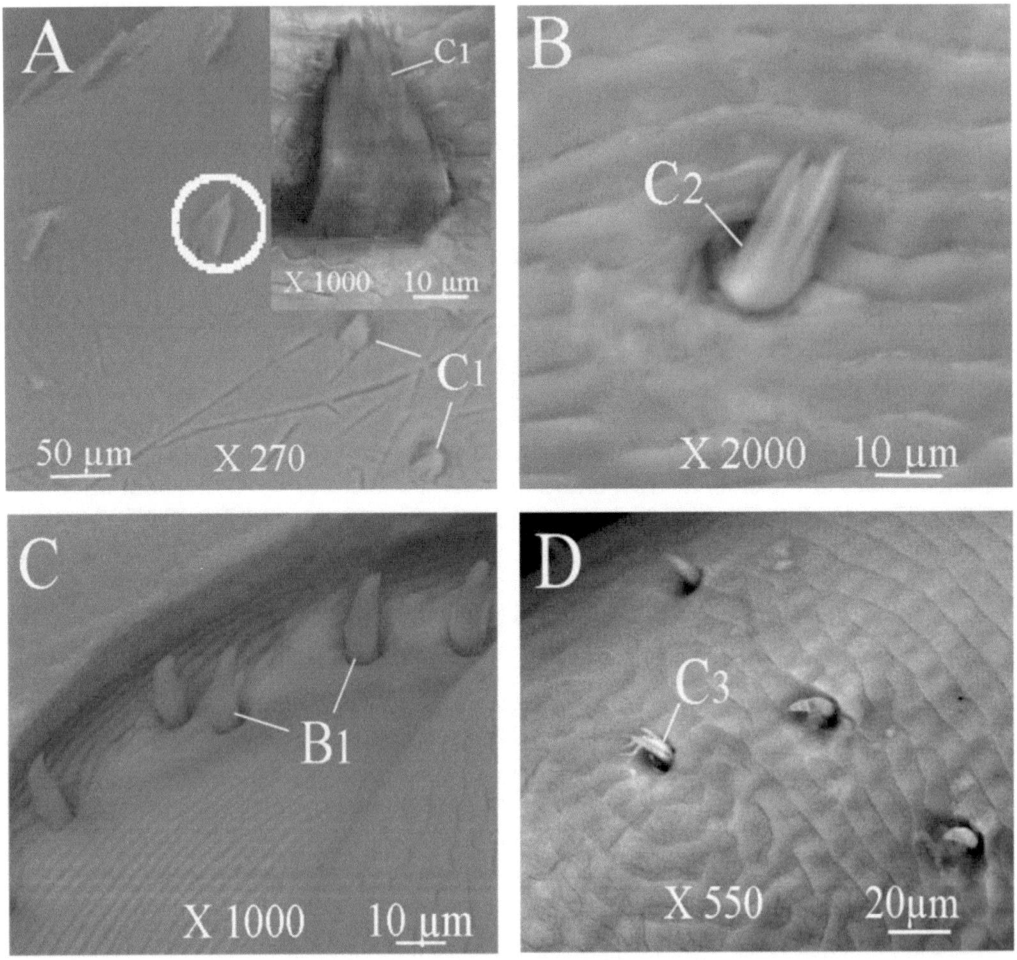


Fig. 4. SEM photomicrographs of irradiated adult antenna of *Rhynchophorus Ferrugineus* with 15 Gy of gamma rays. (A): Enlarged coeloconic1(C1) on the scape, (B): Shrunken sensilla coeloconic 2 (C2) on the pedicel, (C): Shrunken sensilla basiconica 1(B1)on the ventral side between scape and pedicel, (D): Shrunken sensilla coeloconic 3 (C3) on the funicel. [Scale bars indicate 50 μm (A), 10 μm (B, C) and 20 μm (D), respectively]

**3.6.1.1.2 Sensilla coeloconica 2**

The sensilla are markedly changed in appearance arising deeply from the socket and partly separated from the edges of the socket. The sensilla appear shrunken and diminished in size. Near to the base of sensilla the processes appear fused with each other, fewer in number and thickened out. Near to the apex the processes are fanned out. The processes arise from the socket then, curve upward giving L- shaped appearance. The sensilla varied in length from about 23.8 to 26.3 μm and about 7.5 μm in diameter (Fig.4B).

#### 3.6.1.1.3 Sensilla coeloconica 3

The sensilla observed very shrink. Most the finger -like processes of the S. coeloconica 3 were fused into blunt tip at apical margin, while in a few number were separated from each other and the separated processes were inflect apically (Fig.4D).The sensilla measures about 22.5 μm in length and about 7.3 μm in diameter.

### 3.6.1.2 Sensilla chaetica

#### 3.6.1.2.1 Sensilla chaetica 1

The S. chaetica 1 on the lateral sides of the club more curved on the antennal shaft and increased in size. The two branches of sensilla arise from the base of sensilla and are equal in length. The sensilla measure about 37.5 μm in length and about 8 μm in diameter (Fig.5A). The S. chaetica1 on the surface of the club become curved as if it needs support from nearby ones (Fig.5D).

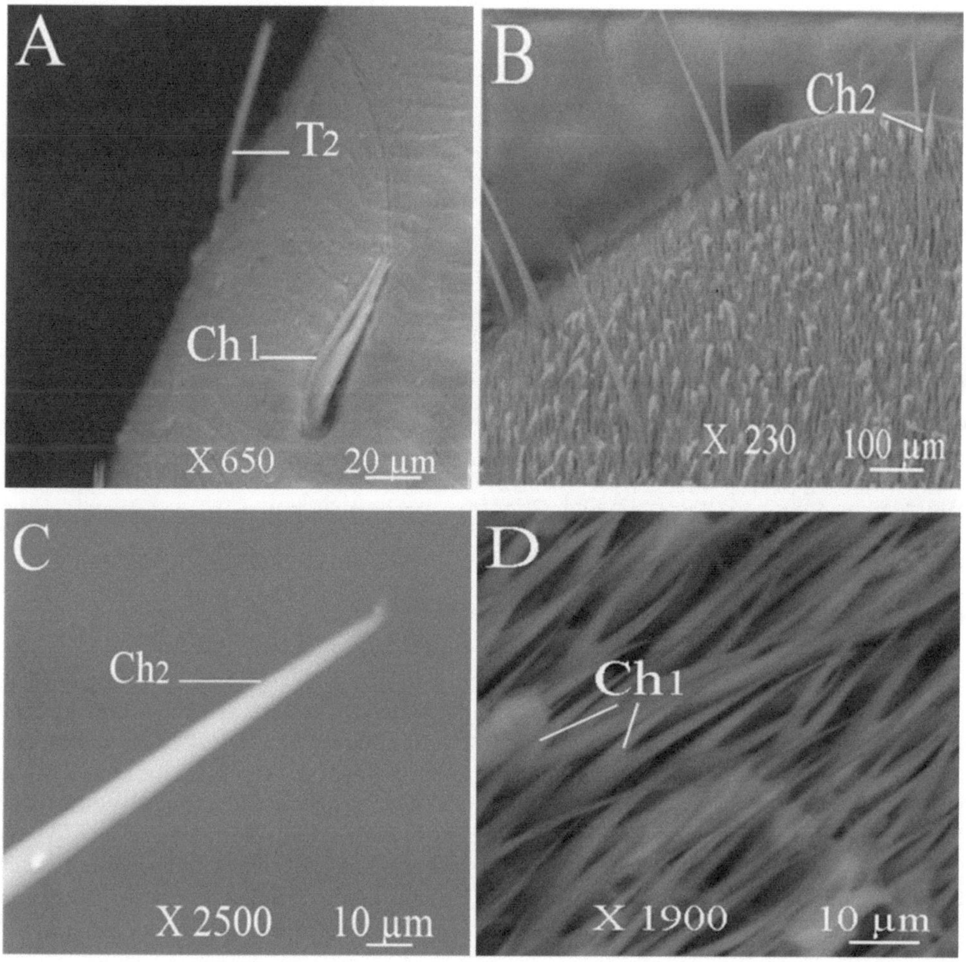


Fig. 5. SEM photomicrographs of irradiated adult antenna of *Rhynchophorus Ferrugineus* with 15 Gy of gamma rays. (A): Enflected sensilla chaetica1 (Ch1) on the lateral side of club,

(B): Swollen sensilla chaetica 2 (Ch2) on the surface of the club, (C): Blunt swollen tip of sensilla chaetica 2 (Ch2), (D): Irregularity of sensilla chaetica1(Ch1) on the surface of the club. [Scale bars indicate 20 μm (A), 100 μm (B) and 10 μm (C, D), respectively]

**3.6.1.2.2 Sensilla chaetica 2**

The sensilla became shorter, swollen near to the base. The apical part of the sensilla is suddenly interrupted into a blunt swollen tip (Fig. 5C).

**3.6.1.3 Sensilla basiconica 1**

The sensilla fewer in number and thickened out, shrunken and thinner near to the tip; the bifurcated end appears located more towards the apex. The length measures from about 6.5 to10 μm and the diameter about 2 μm (Fig.4C).

### 3.6.2 The effects of the dose rate 20 Gray

**3.6.2.1 Sensilla coeloconica**

**3.6.2.1.1 Sensilla coeloconica 1**

The sensilla appeared less in number but very tall. The fingers- like processes of the sensilla became very attachment and inflected around each other. The length measures about 35 μm and the diameter about 9.5 μm (Fig.6A).

**3.6.2.1.2 Sensilla coeloconica 2**

The fingers-like processes of sensilla fused into two groups. In each group these processes are less in number and broad basally which gradually decrease in diameter until reach to fine tipe. The sensilla arise straight and make 45$^{0}$C with the antennal shaft. It's length measures from about 28.8 to 30 μm and the diameter about 7 μm (Fig.6B).

**3.6.2.1.3 Sensilla coeloconica 3**

The fusion of finger- like processes of the sensilla disappeared from about the half length of most sensilla. Some of the separated processes were diverged, extend vertically on the antennal shaft, while the others were converged and form an angle of around 40 $^{0}$C with the antennal shaft. The sensilla measures about 14 μm in length and about 2 μm in diameter (Fig.6D).

**3.6.2.2 Sensilla chaetica**

**3.6.2.2.1 Sensilla chaetica 1**

The S. chaetica 1 on the lateral side of the club arises from a wide sachet and appeared increased in diameter in the half length reached to about 7 μm. The two branches of the sensilla are not equal in length. The sensilla ranged in length from about 31.5 to 35 μm and about 3.5 μm in diameter (Fig.7A).While the S. chaetica 1 on the surface of club gathered into dense collections and mostly destroyed. Some of the collected sensilla were mostly fused with loss of parts of each sensillum (Fig.7C).In other collection, the sensilla were markedly shrunken and curved into different directions.

**3.6.2.2.2 Sensilla chaetica 2**

The sensilla were few in number, markedly shrunken. The part near to the socket was swollen, while the apical part disappeared (Fig.7B).

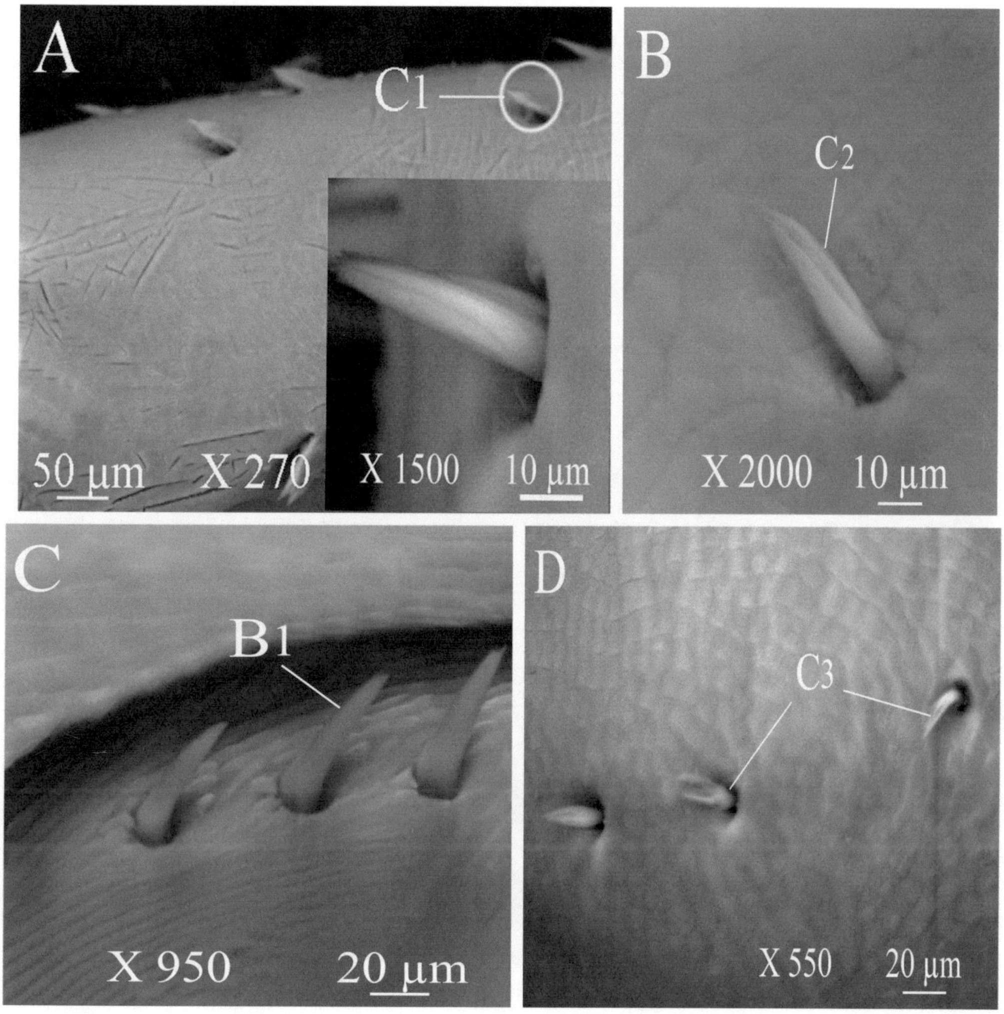


Fig. 6. SEM photomicrographs of irradiated adult antenna of *Rhynchophorus Ferrugineus* with 20 Gy of gamma rays. (A): Tall sensilla coeloconic1 (C1) on the scape, (B): Sensilla coeloconic 2 (C2) on the pedicel, (C): Sensilla basiconic 1 (B1) on the ventral side between scape and pedicel and (D): Sensilla coeloconic 3 (C3) on the funicel. [Scale bars indicate 50 µm (A), 10 µm (B) and 20 µm (C, D), respectively]

### 3.6.2.3 Sensilla basiconica 1

The sensilla thinned out, some of them lost the bifurcation at the apical part others fade into a blunt tip (Fig.6C). The length of sensilla ranged from about 18.5 to 27 µm in length and about 6 µm in diameter.

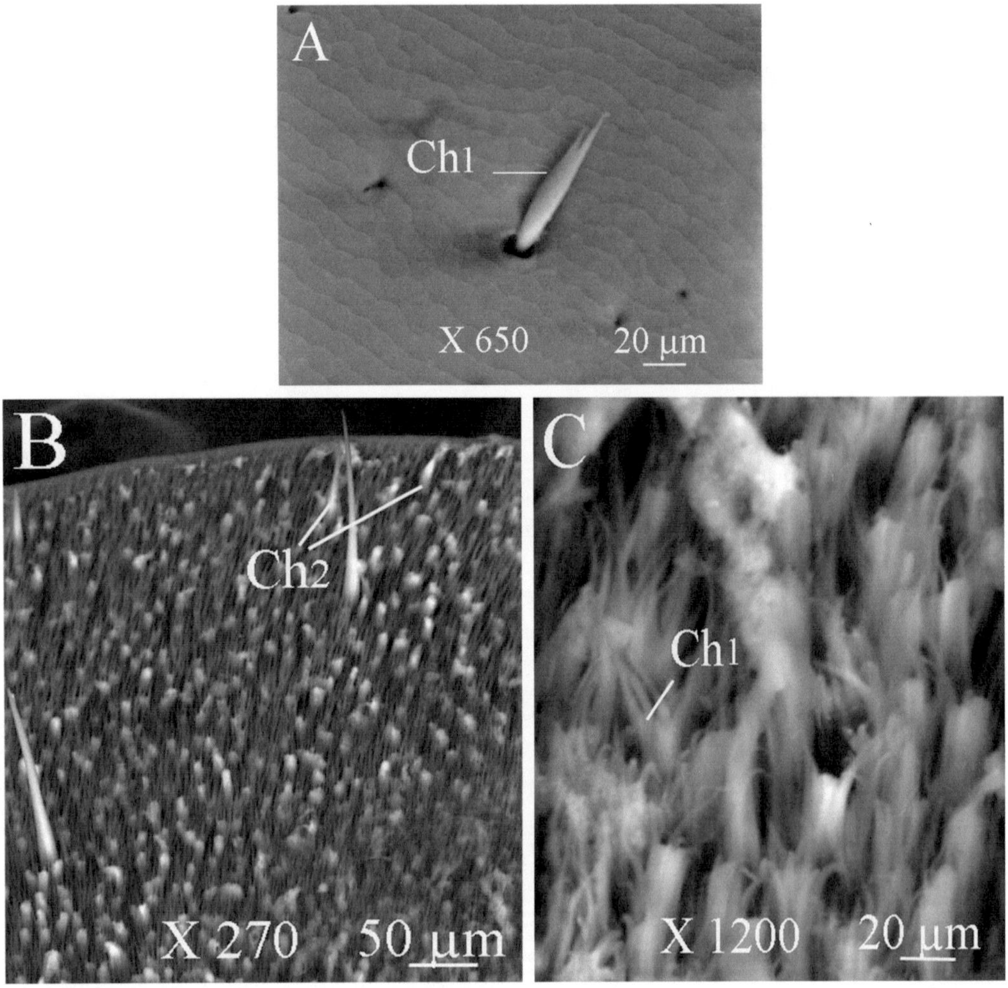


Fig. 7. SEM photomicrographs of irradiated adult antenna of *Rhynchophorus Ferrugineus* with 20 Gy of gamma rays. (A): Abnormal sensilla chaetica1 (Ch1) on the lateral side of club, (B): Shrunken and swollen sensilla chaetica 2 (Ch2) on the surface of club, (C): Sensilla chaetica 1 (Ch1) on the surface of club. [Scale bars indicate 20 µm (A, C) and 50 µm (B), respectively]

## 4. Discussion

The microscopic observations of the *R. Ferrugineus* antennae revealed four types (eleven subtypes) of hair- like structures and have not shown any sexual dimorphism. These observations are in agreement with behavioral responses of *R. palmarum* to odors in the laboratory and in the field, males and females respond with the same sensitivity to hynchophorol and sugarcane juice in a four-armed olfactometer (Said *et al.*, 2003).

The first type of sensilla is coeloconic which large abundance on the scape and pedicel. The morphology of S. C 1 is distinct from the other olfactory sense organs (Jhaveri *et al.*, 2000).

Coeloconic sensilla often consist of a peg whose wall is composed of numerous parallel-runing fingers (Keil, 1997). Also, Ameismeier (1985) illustrated that some of the fingerlike projections remained interstitial while some ones combined with each other and remained longitudinal grooves on the surface of the sensilla wall. Hunger and Steinbrecht (1998) observed the coeloconic sensella to be double walled, multiporous, rich in neurons and excessive contact interfaces. Other authors even considered that this type of sensilla may participate in receptivity to heat and humidity (Cuperus, 1983). Olson & Andow (1993) suggested an olfactory function of this sensillar type in *Trichogramma nubilale,* but as revealed in several insect species (Altner *et al.*, 1983), this sensillar type may be involved in thermo- or hygroreception. Coeloconica sensilla are found in many Hymenoptera belonging to several families and have been described by different terminologies, such as, small sub terminal sensilla, (Weselow, 1972), multiporous grooved sensilla (Barlin *et al.*, 1981), bulb sensilla (Cave & Gaylor, 1987) and smooth basiconica sensilla (Norton & Vinson, 1974).

Sensilla trichodea 1 has been described in different insects as having putative mechanoreceptive functions, such as in the perception of mechanosensory stimuli (Onagbola & Fadamiro, 2008). Keil (1999) cited that trichoid sensilla may be olfactory, but sensilla found on the pedicel are usually mainly mechanoreceptive. Schneider (1964) suggests that trichoid sensilla may be dye-permeable and so may posses chemoreceptivity. The great occurrence of the sensilla trichodea on the antennae of male *O. elegans* relative to the females may indicate a probable role in mate location, possibly for the detection of females sex pheromones (Onagbola, *et al.*, 2008).

Sensilla trichodea1 was found on the antennae of *Bembidion* species. These bristles are innervated by a single sensory neuron, ending with the typical tubular body, attached to the base of the hair shaft. This indicates a mechanosensory function (Zacharuk, 1985).

Sensilla similar to sensilla trichodea 2 have been described on several species of curculionid beetles (Alm & Hall 1986 and Isidoro & Solinas, 1992). In electrophysiological experiments with *H. abietis,* Mustaparta (1975) found that this sensillar type was electrophysiologically responsive to odours. Merivee *et al.,* (1999) suggested that they probably function as sex pheromone receptors. In the ground beetle, *Platynus dorsalis,* it might indicate that these sensilla probably respond to aggregation pheromone (Merivee *et al.,* 2001).

Sensilla similar in external morphology to the sensilla trichodea 3 on *R. Ferrugineus* are found on other alticinid species (Ritcey & McIver, 1990). The short sensilla trichodea, recorded from *Synempora* by Davis & Nielsen (1980) are in fact microtrichia, which are very frequently found on flagellomeres. They superficially resemble multiporous small sensilla basiconica which so far have been described only from Agathiphagidae among Lepidoptera (Faucheux, 1990).

The fourth type of trichodea sensilla are found only on the club of *R. ferrugineus* in few number are similar to type V sensilla found on the club of *R. palmarum* ,which characteristics of olfactory trichodea sensilla (Said *et al.,* 2003).Such sensilla were found to house neurons tuned to the pheromone component in *Ips typographus* L. and *Hylobius. abietis* (Mustaparta,1973,1975).

Shields & Hildebrand (2001) showed that every type -trichoid sensilla -of the female sphinx moth *M.sexta* was innervated by two olfactory receptor cells and could respond to aromatic or terpenoid odorants. Similarly, the sensory neuron membrane protein of the wild silk moth *Antheraea polyphemus* was most prevalent in neurons and was localized to receptor membrane of the dendrite cilia presumed to perform the role of olfactive conduction (Rogers *et al.*, 2001).

Sensilla chaetica 1 of *R. ferrugineus* are ascribed to mechano-chemoreception in coccinellids,*Semiadalia undecimnotata, Coccinella transverguttata* and *P.tsugae* (Jourdon *et al.*, 1995; Wipperfurth *et al.*,1987; Broeckling & Salom 2003). In neopseustids, it cannot be excluded that the hair is connected to the considerably elevated socket by a flexible joint membrane and, in this case, the sensilla would be contact mechanoreceptors (Faucheux *et al.*, 2006).

They are resemble in their external sensilla chaetica in *Coccinella transversalis* Fabricius (Coccinellidae) (James, 2001),"chetiform sensilla type I" in *Semiadalia undecimnotata* (Coccinellidae) (Jourdon *et al.*, 1995,"sensilla chaetica type I" in *Agriotes obscurus* Elateridae) (Merivee, 1992) .Some authors treat these also as sensilla trichodea. They are classified, for example, as sensilla trichodea type I in *Carabus fiduciaries saishutoicus* (Carabidae) (Kim & Yamasaki, 1996), and flea beetles (Alticidae) (Ritcey & McIver, 1990).

Sensilla chaetica II could be observed in the antenna of *Cawjeekelia pyongana* (Polydesmida: Paradoxmatidae) (Chung & Moon,2007,2009) and on the labrum of *Synempora andesae* (Neopseustidae),the aporous bifurcate sensilla chaetica could constitute an autapomorphy but would need to be described further in other species of that family (Faucheux 2008).

In the present study the sensilla basiconica 1 and 2 are found on the pedicel and the surface of club. Okada *et al.*, 1992 on the cigarette beetles; *Lasioderma serricorne* and Daly & Ryan, 1979 on the ground beetle, *Nebria brevicollis*, they demonstrated that the wall of these sensilla is perforated by numerous tiny pores. The numerous pores and branched dendrites are considered to be evidence that these basiconica sensilla function as olfactory receptors (Altner & Prillinger, 1980 and Zacharuk,1985).The sensilla basiconica of *Hylobius abietis* were responsive to odours in electrophysiological experiments (Mustaparta, 1975). Moreover, a small groove or depression, not characteristic for mechanoreceptive pegs, at the tip of tiny sensilla basiconica 3 indicate that they propaply function as chemoreceptors (Ploomi *et al.*, 2003).

Curculionid, scolytid and coccinellid beetles have been reported to bear antennal sensilla similar to the sensilla basiconica described here (Alm & Hall, 1986; Bland, 1981; Isidoro & Solinas, 1992, Jourdan *et al.*, 1995). Non-articulated blunt-tipped basiconica sensilla, which resemble sensilla basiconica 1, 2 of *Bembidion lampros, B. properans* and *Platynus dorsalis* are common on the antennal flagellum of most insects (Ploomi *et al.*, 2003).

Besides the sensory organs abundant cuticular pores, obviously openings of the antennal glands, penetrate the surface of the antennae of *R. Ferrugineus*. Differences in the size and placement of these pores may suggest differences in the function of respective cuticular glands. In some other insects, antennal glands may have enzymatic activity, degrading molecules of pheromones (Taylor *et al.*, 1981). In Chrysomelidae, antennal glands may produce pheromone (Bartlet *et al.*, 1994).

A few ultrastructurally obvious effects of irradiation in the features of normal antennal sensilla of *R. Ferrugineus* could be observed. Typically seen as shrunken, curved of sensilla into different directions, irregularity in shape, swollen in some parts and some sensilla gathered into dense collections. Similar works on the effects of gamma radiation have been recorded by many authors in other insects ; Oland *et al.*,(1988) illustrated the comparison between normal and irradiated *Manduca sexta* with 2.64 Gy revealed that features of normal antennal sensilla were present in irradiated ones with presence some cuticular disruption. Also, El-kholy & Mikhaiel (2008) revealed that gamma irradiation of full grown male pupae of *Galleria mellonella* with the sterile dose 400 Gy and the two substerile doses 100 and 150 Gy showed malformations in the $F_1$ male antennal sensilla less than that in the parents irradiated with the dose 400 Gy . Besides, the irradiated pupae with the sterilizing dose in *Bactrocera zonata* produced adults with different malformation on the antennal sensilla (El-Akhdar & Afia, 2009).

The present study showed that the malformations in the antennal sensilla induced from exposed of adult with the dose rate 15 Gy were very few, while the dose rate 20 Gy affected on high number of antennal sensilla. And so, the used of high dose (20 Gy) must be turned off because this dose will affect on the behavior of insects. These previous results are compatible with our study in using gamma radiations (15Gy) as part of an Integrated Pest Management program for controlling this pest.

## 5. Conclusion

The antennal sensilla of unirradiated and irradiated red palm weevile, *Rhynchophorus Ferrugineus* (Oliv.) (Coleoptera; Curculionidae) were investigated by using a scanning electrone microscope. The antenna was composed of three segments; scape, pedicel and flagellum (funicle, club). Four different sensillar types were distinguished. Eleven subtypes, these were; three subtypes of sensilla coeloconica, four subtypes of sensilla trichodea, two subtypes of sensilla basiconica, and two subtypes of sensilla chaetica. The position of these sensilla on the antenna was discussed. These types are used by insects as mechanoreceptor, sex pheromone, aggregation pheromone, olfactory, mate location, thermo-hygroreceptor, and receptivite to heat, and humidity.

There are differences in lengths and diameters of some types of sensilla were recorded as a result of irradiated adult with two doses of gamma rays (15, 20Gy). In the higher dose (20 Gy) more effects of sensilla were recorded for the sensilla chaetica followed by sensilla coeloconica.

## 6. References

Abraham, V.A. & Pillai, G.B.(1998). Red palm weevil - a dreaded enemy of coconut palm. *Indian Farmers` Digest* 7(1):15-20.

Albert, P. J. (1980). Morphology and innervation of the mouth-part sensillae in the larvae of the spruce budworm, *Choristoneura fumiferana* (Clem.) (Lepidoptera: Tortricidae). *Can.J. Zool.*, 58: 842-851.

Alm, S. R. & Hall, F. R. (1986). Antennal sensory structures of *Conotrachelus nenuphur* (Coleoptera: Curculionidae). *Ann. Entomol. Soc.Am.*, 79: 324-333.

Altner, H. & Prillinger, L.(1980).Ultrastructure of invertebrate chemo-, thermo-, and hygroreceptors and its functional significance. *International Revue of Cytology*, 67: 69-39.

Altner, H., Schaller-Selzer, L., Sletter, H. & Wohlrab, I. (1983). Poreless sensilla with inflexible sockets. *Cell. Tissue Res.*, 234: 279-307.

Ameismeier, F. (1985). Embryonic development and molting of the antennal coeloconic no pore-and double-walled wall pore sensilla in *Locusta migratoria* (Insecta, Orthopteroidea). *Zoo. Morphol.*, 105: 356-366.

Barlin, M.R., Vinson, S.B. & Piper, G.L. (1981). Ultrastructure of the antennal Sensilla of the cockroach-egg parasitoid. *Tetrasticus hagenowii* (Hymenoptera: Eulophidae). *Journal of Morphology*, 168: 97-108.

Bartlet, E., Isidoro, N. & Williams, I. H. (1994). Antennal glands in Psylliodes chrysocephala,and their possible role in reproductive behaviour. *Physiol. Entomol.*, 19: 241-250.

Bedford, G.O. (1980). Biology, ecology, and control of palm Rhinoceros beetle. *Annual Review of Entomology*, 25: 309-339.

Bland, R.G. (1981). Antennal sensilla of the adult alfalfa weevil (*Hypera postica* Gyllenhal) (Coleoptera: Curculionidae). *Int. J. of Insect Morphol. and Embryol.*, 10: 265-274.

Bland, R.G. (1984). Mouth part sensillae and mandibles of the adult alfalfa weevil, *Hypera postica* and the Egyptian alfalfa weevil, *H. brunneipennis* (Coleoptera: Curculionidae), *Ann. Entomol. Soc. Am.*, 77:720-724.

Bokhari, V.U.G. & Abuzuhira, R.A. (1992). Diagnostic tests for red palm weevil. *Rhynchophorous ferruginseus* infested date palm trees. *Arab Gulf Scientific Research* 10:93-104.

Broeckling, C. D. & Salom, S. M. (2003). Antennal morphology of two specialist predators of Hemlock woolly Adelgids, *Adelges tsugae* Annand (Homoptera: Adelgidae). *Ann. Entoml. Soc. Am.*, 96 (2): 153-160.

Cave, R.D. & Gaylor, M.J. (1987). Antennal Sensilla of male and female *Telenomus reynoldsi* Gordh and coker (Hymenoptera: Scelionidae). *Int. J. of Insect Morphol. and Embryol., 16*: 27-39.

Chadha, G.K. & Roome, R.E. (1980). Oviposition behaviour and the sensillae of the ovipositor of *Chilo partellus* and *Spodoptera littoralis* (Lepidoptera: Noctuidae). *J. Zool.*, 192: 169-178.

Chung, K.H. & Moon, M.J. (2007). Microstructure of the Antennal Sensory Organs in the Millipede *Cawjeekelia pyongana* (Polydesmida: Paradoxomatidae).*Korean J. Electron Microscopy*, 37 (2): 73-82.

Chung, K.H. & Moon, M.J. (2009). Microstructure of the Antennal Sensilla in the Millipede Anaulaciulus Koreanus Koreanus (Julida: Julidae).*Korean J. Electron Microscopy*, 39 (2): 141-147.

Cuperus, P. L. (1983). Distribution of antennal sense organs in male and female ermine moth *ypononeuta vigintipunctatus* (Retzius) (Lepidoptera: Ypononeutidae). *Int. J. Insect Morphol. Embryol.*, 12: 59-66.

Daly, P. J. & Ryan, M. F. (1979). Ultrastructure of antennal sensilla of *Nebria breuicollis* (Fab.) (Coleoptera: Carabidae). *Int. J. Insect Morphol Embryol.*, 8: 169-181.

Davis, D. R. & Nielsen, E. S.(1980). Description of a new genus and two new species of Neopseustidae from South America, with discussion of phylogeny and biological observations (Lepidoptera: Neopseustoidea). *Steenstrupia*, 6: 253-289.

El-Akhdar, E. A. H. & Afia. Y. E. (2009). Functional ultrastructure of antennae, wings and their associated sensory receptors of peach fruit fly, Bactrocera zonara (Saunders) as influenced by sterilizing dose of gamma irradiation. *J. R. Res. Appl. Sc.*, 2(4): 797-817.

El- Kholy, E. M. S. & Mikhaiel, A. A. (2008). Scaning electron microscopy on the male antennae of the greater wax moth, *Galleria mellonella* (L.), treated with gamma radiation. *Isotope and Rad.Res.*, 40(3): 603-613.

Faleiro, J.R., Al-Shuaibi, M.A. Abraham, V.A. & Prem Kumar, T. (1999). A technique to assess the longevity of the Pheromone (Ferrolure) used in trapping the date palm weevil *Rhynchophorous ferrugineus* Oliv. *Agricultural Sciences* 4(1): 5-9.

Faucheux, M. J.(1990). Antennal sensilla in adult *Agathiphaga vitiensis* Dumbl. and *A. queenslandensis* Dumbl. (Lepidoptera: Agathiphagidae). *Int. J. Insect Morphol. Embryol.*, 19: 257-268.

Faucheux, M. J., Kristensen, N.P. & Yen, S-H. (2006). The antennae of neopseustid moths: Morphology and phylogenetic implication, with special reference to the sensilla (Insecta,Lepidopetra ,Neopseustidae). *Zoologischer Anzeiger*, 245: 131-142.

Faucheux, M. J. (2008). Mouth parts and associated sensilla of a south american moth, *Synempora andesae* (Lepidoptera: Neopseutidae). *Rev.Soc. Entomol. Argent.*, 67(1-2): 21-33.

Fenemore, P.G. (1980). Oviposition of the potato tuber moth Phthorimaea operculella Zell. (Lepidoptera: Gelechiidae);identification of host-plant factors influencing oviposition response. *N. Z. J. Zool.*, 7: 435-439.

Gasser, M.S. , Mohsen, H.T. & Aly, H.F. (2008). Humic acid absorption onto Mg/Fe layered double hydroxide. Colloids and surfaces A. Physiochem. *Eng. Aspects*, 331,195.

Giblin-Davis, R. M., Oehlschlager, A. C., Perez, A.L., Gries, G., Gries, R., Weissling,T.J. Chinchilla,C.M., Peña, J.E. , Hallett,R.H., Pierce Jr, H.D. & Gonzalez, L.M. 1996. Chemical and behavioral ecology of palm weeevils (Curculionidae : Rhynchophorinae). *Florida Entomologist*, 79: 153-167.

Gush, H. (1997). Date with disaster. *The Gulf Today*, Sept. 29, p. 16.

Hunger,T. & Steinbrecht, R.A. (1998). Functional morphology of a double-walled multiporous olfactory sensillum: the sensillum coelocnicum of *Bombyx mori* (Insecta: Lepidoptera). *Tissue Cell*, 30: 14-29.

Isidoro, N. & Solinas, M. (1992). Functional morphology of the antennal chemosensilla of *Ceutorhynchus assimilis* Payk.(Coleoptera: Curculionidae).*Entomologica*, 27, 69-84.

James, B. E., (2001). Contribution on certain aspects of bioecology and behaviour of a ladybeetle, *Coccinella transversalis* Fabricius (Coccinellidae:Coleoptera).Ph.D. Thesis, University of Lucknow, pp .190.

Jhaveri, D., Sen, A., Reddy, G.V. & Rodrigues, V. (2000). Sense organ identity in the *Drosophila* antenna is specified by the expression of the proneural gene. *atonal. Mech. Dev.*, 99:101-111.

Jourdon, H., Barbier, R., Bernard, J. & Ferran, A. (1995). Antennal sensilla and sexual dimorphism of the adult ladybird beetle *Semwdulia undecimnotata* Schn. (Coleoptera: Coccinellidae ). *Int. J. Insect Morphol. Embryol., 24:* 307-322.

Keil, T. A. (1997). Comparitive morphogenesis of sensilla: a review. *Int. J. Insect Morphol. Embryol,* 26:151-160.

Keil, T. A. (1999). Morphology and development of the peripheral olfactory organs. In Insect olfaction, pp.5-47(ed. B. S. Hansson). Springer, New York.

Kim, J. L. & Yamasaki, T.(1996). Sensilla of *Carabus (Isiocarabus) Jiduciaries saishutoicus* Csiki (Coleoptera: Carabidae). *Int. J. Insect Morphol Embyol.* 25: 153-172.

Merivee, E. (1992). Antennal sensilla of the female and male elaterid beetle *Agriotes obscurus* L. ( Coleoptera: Elateridae) . *Proc. Estonian Acad .Sci. Biol.,* 41: 189-215.

Merivee, E., Rahi, M. & Luik, A.(1999). Antennal sensilla of the click beetle, *Melanotus villosus* (Geoffroy) (Coleoptera : Elateridae). *Int. J. Insect Morphol. Embryol.,* 28,41-51.

Merivee, E., Ploomi, A., Rahi, M., Luik, A. & Sammelselg, V. (2001). Antennal sensilla of the ground beetle, *Platynus dorsalis* (Pontoppidan, 1763) (Coleoptera: Carabidae). *Microsc. Res. Tech.,* 55: 339-349.

Mustaparta, H. (1973). Olfactory sensilla on the antennae of the pine weevil. Zeitsdchrift fÜr Zellforschung und Mikroskopische Anatomie,144: 559-571.

Mustaparta, H. (1975). Responses of single olfactory cells in the pine weevil *Hylobius abietis* (Coleoptera: Curculionidae). *Journal of Comparative Physiology,* 97: 271-290.

Nakamuta, K.(1985). Area-concentrated search in adult *Coccinella septempunctata* (Coleoptera: coccinellidae): releasing stimuli and decision of giving-up time. *Jpn. J. Appl. Entomol. Zool.,* 29: 55-60.

Norton, W.N. & Vinson, S.B.(1974). A comparative Ultrastructural and behavioral study of the antennal sensory Sensilla of the parasitoid *Cardiochiles nigriceps* (Hymenoptera: Braconidae). *Journal of Morphology,* 42: 329-349.

Obata, S.(1986). Mechanisms of prey finding in the aphidophagous ladybird beetle, *Harmonia axyridis. Appl. Entomol. Zool.,* 22: 434-442.

Okada, K, Mori, M.., Shimazaki, K. & Chuman, T. (1992). Morphological studies on the antennal sensilla of the cigarette Beetle, *Lasioderamu serricorne* (F.) (Coleoptera: Anobiidae). *Appl . Entoml . Zool.* 27: 269-276.

Olson, D.M. & Andow, D.A.(1993). Antennal Sensilla of female *Trichogramma nubilale* (Ertle and Davis) (Hymenoptera: Trichogrammatidae) and comparisons with other parasitic Hymenoptera. *Int. J. Insect Morphol. Embryol.,* 22: 507-520.

Oland, L., Tolbert, L., P. & Mossman, K., L. (1988). Radiation –induced reduction of the glial population during development disrupts the formation of olfactory glomeruli in an insect. *The Journal of Neuroscience,* 8(1): 353-367.

Onagbola, E.O. & Fadamiro, H.Y.(2008). Scanning electron microscopy studies of antennal sensilla of *Pteromalus cerealellae* (Hymenoptera:Pteromalidae). *Micron.,* 39:526-535.

Onagbola, E.O., Meyer, W.L., Raj Boina, D. & Stclinski, L.L. (2008). Morphological characterization of the antennal sensilla of the Asian citrus psyllid, *Diaphorina citri* Kuwayama (Hemiptera: Psyllidae), with reference to their probable functions. *Micron* 39 (8): 1184-1191.

Ploomi, A., Merivee, E., Rahi, M., Bresciani, J., Ravn, H.P., Luik, A. & Sammelselg,V. (2003). Antennal senailla in ground beetles (Coleoptera: Carabidae). *Agronomy Research,* 1(2): 221-228.

Rahalkar, G.W., Harwalkar, M. R., Rananavare, H.D., Tamhakar,A.J. & Shantharam, K. (1985). *Rhynchophorus ferrugineus.* In: Sing, P.Moore,R.F.(eds.),Handbook of Insect Rearing. Elsevier, Amsterdam, pp. 279-286.

Ritcey, G. M. & McIver, S.B. (1990). External morphology of antenna1 sensilla of four species of adult flea beetles (Coleoptera: Chrysomelidae: Alticinae). *Int. J. Insect Morphol. Embryol.,* 19: 141-53.

Rogers, M.E., Krieger, J. & Vogt, R.G., (2001). Antennal SNMPs (sensory neuron membrane proteins) of Lepidoptera define a unique family of invertebrate CD36-like proteins. *J. Neurocytol.,* 49: 47-61.

Salama, H.S.,Sharaby,A., Abd El-Aziz, S. E. , Shaarawy, F.& Azmy, N. (1987).Ultrastructure of chemoreceptors in the moth of the American bollworm, *Heliothis armigera* and their response to chemicals. *Bull. Ent. Soc. Egypt., Econ. Ser.,* 16: 237-263.

Salama, H. S. & Abd-El-Aziz, S.E. (2001). Ultra morphology structure of sensory sensillae on the legs and external genitalia of the red palm weevil *Rhynchophorus ferrugineus*(Oliv.). *Saudi J. Biol.Sci.,* 14(1): 29-36.

Said, I., Tauban, D., Renou, M., Mori, K. & Rochat, D.(2003). Structure and function of the antennal sensilla of the palm weevil, *Rhynchophorus palmarum* (Coleoptera: Curculionidae). *J. Insect Physiol.,* 49:857-872.

Schneider, D. (1964). Insect antennae. *Annu. Rev. Entomol.,* 116 (2): 178-186.

Shields, V.D.C. & Hildebr, J.G. (2001). Responses of a population of antennal olfactory receptor cells in the female moth *Manduca sexta* to plant - associated volatile organic compounds. *J. Comp. Physiol.,* A186: 1135-1151.

Srivastava, S., Omkar. (2003). Scanning electron microscopy of antennae of coccinella septempunctata ( Coleoptera: Coccinellidae). *Entomological* Sinica, 10(4): 271-279.

Stubbs, M. (1980). Another look at prey detection by coccinellids. *Ecol. Entomol.,* 5: 179-182.

Taylor, T., R., Ferkovich, S. M. & Von Essen, F. (1981). Increased pheromone catabolism by antennal esterases after adult eclosion of the cabbage looper moth. *Experientia,* 37: 729-731.

Vidyasagar, P.S.P.V., M. Hagi, R.A. Abozuhairah, O.E. Al-Mohanna, & A. Al-Saihati. (2000). Impact of mass pheromone trapping on the red palm weevil: Adult population and infested level in date palm Gardens of Saudi Arabia. *The Planter* 76: 347-355.

Weselow, R.M.(1972). Sense organs of hyperparasite *Cheiloneurus noxius* (Hymenoptera: Encyrtidae) important in host selection process. *Annals of the entomological Society of America,* 65: 41-46.

Wigglesworth, V.B.(1972). The principles of insect physiology. Chapman and Hall Publications. 827.

Wipperfurth, T. K., Hagen, K. S. & Mittler, T. E. (1987). Egg production by the coccinellid *Hippodamia convergens* fed on two morphs of the green peach aphid, *Myms persicae. Entono1. Exp. Appl.,* 44: 191-198.

Zacharuk, R.Y.(1985). Antennae and sensilla. Comprehensive Insect Physiology, Chemistry and Pharmacology (Kerkut, G .A. and Gilbert, L.I., eds.), Vol. 6, P. 1-69. Pergamon Press, Oxford.

# Part 5

## Agriculture and Forestry

# 12

# Gamma Irradiation for Fresh Produce

Agnes K. Kilonzo-Nthenge
*Department of Family and Consumer Sciences, Tennessee State University, Nashville, TN USA*

## 1. Introduction

Food irradiation is a promising food safety technology that has a significant potential to control spoilage and eliminate food-borne pathogens. The Food and Agricultural Organization (FAO) has estimated that approximately 25% of all worldwide food production is lost after harvesting due to insects, microbes, and spoilage. As the market for food becomes increasingly global, food products must meet high standards of quality and quarantine in order to move across international borders. The FAO has recommended that member states need to implement irradiation technology for national phytosanitary programs. There is a trend to use food irradiation mainly due to three main factors: the increase of foodborne diseases; high food losses from contamination and spoilage; and increasing global trade in food products. The ever increasing foodborne illness outbreaks associated with fresh produce continue to prove that traditional measures are not sufficient to eliminate food borne pathogens. More effective countermeasures are clearly needed to better manage the foodborne pathogen risks posed by contaminated produce. Fresh produce industries, government regulatory agencies, and consumers all are advocating for new technologies that will eliminate or significantly reduce foodborne pathogens on fresh produce. With increasing awareness of the foodborne idleness linked to fresh produce, gamma irradiation could be applied to mitigate human pathogens in fresh fruits and vegetables. Food irradiation is a safe and effective tool and could be used with other technology to control pathogenic bacteria in fresh produce. Irradiated foods are generally nutritious, better or the same as food treated by convectional methods such as cooking, drying, and freezing. Food irradiation also has other benefits such as delay in repining and sprouting. Further more food irradiation has a significant potential to enhance produce safety and if combined with other anti-microbial treatments; this technology is promising to solve some of the current produce pathogen problems. Although irradiation is safe and has been approved in 40 countries, food irradiation continues to be a debate and slows extensive acceptance and use in the food industries. Several foodborne pathogens have been linked to fresh produce and gamma irradiation could be applied to eliminate microbes before reaching the consumer. There is an urgent need to educate consumer on the principles and benefits of this promising technology.

## 2. Fresh produce and foodborne pathogens

Consumption of fresh produce in many countries has increased substantially in recent years, in part due to an increased awareness of the health benefits that fresh produce provides.

However, foodborne illness outbreaks linked to fresh produce are becoming more frequent and widespread (Warriner et al, 2009; Harris et al., 2003; Sivapalasingam et al, 2004). Foodborne illness outbreaks associated to leafy vegetables is an indication that increased consumption of fresh produce could present new challenges with regard to fresh produce safety (FDA, 2008). Recent outbreaks of foodborne illness linked with produce have raised concerns and underline the challenges to the public health as well as to fresh produce industry. The increased concern of fresh produce and its relation to food borne illness has been indicated by several surveillance studies (Ilic, Odomeru, & LeJeune, 2008; little & Gillespie, 2008). Other countries together with the United States of America have targeted foodborne pathogens in fresh produce as emerging issue in food safety and one of the most pressing public health needs. Fresh fruits and vegetables are frequently contaminated since they are grown in open fields with potential exposure to enteric pathogens from animals, soil, irrigation water, and manure. Cross contamination of fresh produce with foodborne pathogens may occur during the production cycle and can originate from soil, insects, equipment, animals or humans (Tracy and Harris, 2003; Liao and Fett, 2001; Ukuku and Sapers, 2001). It is indicated that pathogen contaminated water or surface run-off waters can lead to cross-contamination of fruits and vegetables in the field (Beuchat and Ryu, 1997). Similarly, the application of raw animal manure for fertilizer increases the threat of contamination on fruits and vegetables ( Brackett, 1992). Direct or indirect pathogen contamination of fresh produce can occur at many points in the production chain during growth and processing (Fenlon, et al, 1996; Beuchat and Ryu. 1997), thus presenting a food safety challenge to consumers.

Pathogenic microorganisms, such as Cyclospora *cayetanensis, Escherichia coli* O157:H7, Hepatitis A, *Listeria monocytogenes,* Norovirus, *Salmonella* spp., and *Shigella* spp. are the major foodborne microbial pathogens associated with fresh produce. *Salmonella, Escherichia coli* O157:H7, and *Listeria monocytogenes* have been associated with fresh produce over the past two decades. *E. coli* O157:H7 illnesses have also been linked to the consumption of fresh fruits and vegetables (Tauxe, et al 2000). *E. coli* O157:H7 is capable of causing hemorrhagic colitis and hemolytic uremic syndrome (HUS) and thus has gained attention from public health agencies and institutions. Proximity of domestic or wild animals to irrigation water systems may result *E. coli* O157:H7 (Wachtel, et al 2002) and other pathogenic bacteria being washed from manure to production fields. *E. coli* O157:H7 contaminated manure may get into the water system, and once present, can be applied to growing crops (Institute of Food Technologists, 2002). Foodborne pathogens in the fresh produce indicate a weakness in the fresh produce industry has was demonstrated by recent multi-state (Unites States) outbreaks in produce, including *E. coli* OH7:H7 outbreak from spinach that lead to 183 cases of illness, 29 cases of Hemolytic Uremic Syndrome (HUS), 95 hospitalizations, and one death (http://www.cdc. gov/foodborne/ecolispinach) ; *Salmonella* Typhimurium outbreak from tomatoes that involved 183 cases of illnesses http://www.fda.gov/bbs/topics/ NEWS/ 2006/NEW01504. html); and in December 2006, Taco Bell restaurants in the Northeast were also associated with *E. coli* O157:H7 and iceberg lettuce was considered to be the single most likely source of the outbreak, 8 cases of Hemolytic Uremic Syndrome (HUS), and 53 hospitalizations were reported to Center of Disease Control (http://www.cfsan.fda.gov/~news/whatsnew.html. *Salmonella enteritidis, S. infantis,* and *S. typhimurium* have also been reported to be capable of growth in chopped cherry tomatoes (Asplund, K., and E. Nurmi.1991). *Listeria monocytogenes* is a common contaminant and has also been associated with several produce recalls (Faber and Peterkin. 1991; Leverentz, 2004), including red bell peppers, romaine lettuce, sprouts, and apple slices. Presence of this pathogen has also been reported in potatoes, radishes, cabbage, cucumbers, and mushrooms obtained from the market (Heisick, et al 1989).

Many outbreaks have been traced to produce, and this will continue to occur until fresh produce growers, processing plants, retail stores, and consumers increase their knowledge and awareness of the risks and consequences of foodborne pathogens. Given that fresh produce is ready to eat, and is not subjected to further microbial killing steps, there is a call for produce industry to use effective methods to eliminate or reduce the risk of foodborne pathogens. Gamma irradiation could be applied as a control measures to help minimize food safety risks associated with foodborne pathogens.

## 3. Gamma irradiation

Gamma rays use irradiation given off by Cobolt-60, a radioisotope of cobalt (Steele and Engel, 1992). It is reported that all radiation facilities in the world use Cobalt-60 rather than Cesium-137 (WHO, 1987). Cobalt-60 is derived from cobalt-59 which is placed in a nuclear reactor and bombarded with neutrons until an extra neutron is absorbed forming the unstable radioisotope cobalt-60. Over 80% of the cobalt-60 available on the world market is being produced in Canada (Diehl, 1995). For use as a radiation source the activated cobalt pellets are encapsulated in a stainless steel linear in form of pin or pencil to minimize absorption of the cobalt and to minimize heat build up. The stainless rods are placed on racks which are stored in approximately 25 feet of water and raised into concrete irradiation chamber to dose the food (Jones, 1992). As the food go through the chamber, the stainless steel linears are raised above the water so that it is exposed to gamma rays. The irradiation dose applied to food is measured in terms of kiloGyray (kGy) and is usually measured in a unit called the Gray, abbreviated Gy. The newer unit,1 Gy = 100 rads; 1 kGy = 1,000 Grays). The practical working range of food irradiation is generally from 50 Gy to as high as 10,000 Gy, depending upon the food in question and the effect desired (Satin, 1993). There are three general application and dose categories that are referred to when foods are treated with ionizing radiation (Urbain, 1986): (1) Low dose (radurization) up to approximately 1 kGy for sprout inhibition, delay of ripening, and insect disinfection, (2) Medium dose (radicidation)-1 to 10 kGy for reduction of non-spore forming pathogens, delay of ripening, and reduction of spoilage microorganisms, and (3) High dose (radappertization)-10 to 50 kGy for reduction of microorganisms to the point of sterility. In the United States, the amount of irradiation dose applied to food is controlled by plant quality personnel and United States Department of Agriculture (USDA) and inspectors (Giddings and Marcotte, 1991).

## 4. Benefits of gamma irradiation

### 4.1 Penetrating sterilization

There has been mounting interest all over the world to utilize gamma irradiation to improve the shelf life of perishable foods as well as to ensure the microbiological safety of the products (Kamat A et al, 2003). According to Chervin and Boisseau, 1994 and Buchanan et al., 1998, ionizing irradiation is a fitting method to control the microorganisms on fruits, fresh fruit juices, fresh-cut vegetables, salads, sprouts, seeds and other, minimally processed foods. The efficacy of irradiation is not only limited to the surface, but it can penetrate the product and eliminate microorganisms that are present in crevices and creases of vegetables such lettuce (Prakash et al., 2000). According to Takeuchi and Frank 2000, Solomon et al. 2002, bacteria gets inside tissues of leafy vegetables through natural openings or through breakage caused by insect and mechanical in harvesting. Internalization of bacterial pathogens into the edible portions of plants is of particular concern as these microorganisms are unlikely to be detached by washing or surface sanitization methods (USFDA, 1999; Jablasone et al., 2005). Chlorinated water is widely used for disinfection of foods; but it does not completely inactivate bacteria in fresh produce (Seo and Frank, 1998) due to its limited penetrating power into plant tissues. Ionizing radiation is an effective non-thermal means of eliminating pathogenic bacteria in surface, subsurface, and interior regions of fresh produce. Unlike chlorine treatment, low dose irradiation may be effective method of reducing pathogen such as internalized *E. coli O157:H7* in and on the surface of fruits and vegetables. Irradiation technology, due to its ability to penetrate through the food, can be used to effectively control foodborne pathogens in fresh produce. The International Commission on Microbiological Specifications for Foods (ICMSF) in 1980 established that cobalt-60 rays penetrate approximately 20 cm of food (Frazier and Westhoff, 1988). Food irradiation using cobalt-60 is the mostly used method by most processors, because the deeper penetration enables administering treatment to entire industrial pallets, reducing the need for material handling Maurer K.F, 1958). Low dose (0.15-0.5 kGy) irradiation has been reported to be workable dosage range for fresh cut lettuce (Hagenmaier and Baker, 1997). Irradiation thus is potentially more effective than washing or other surface treatments against spoilage organisms and pathogens (Niemira and Fan 2006).

### 4.2 Gamma irradiation and survival of foodborne pathogens

In recent years, leafy vegetables and salads are gaining great importance in the human diet, in part due to due to the health concerns. The intake of fruit and vegetable juices are suggested for diverse health effects (Williams, 1995) and are considered a part of a healthy diet and assist in the protection against various diseases. However, fresh produce may harbor potential of foodborne pathogens (Beuchat, 1996; Sumner and Peters, 1997). The food safety regulations on fresh produce set by the Food and Drug Administration are expected for the producers and handlers to reduce the risk of future outbreaks caused by fruits and vegetables (Warner, 1997). Studies have also shown that irradiation as a means of controlling human pathogens such as *E. coli O157:H7* (Thayer and Boyd., 1993) and *Salmonella* (Thayer et al., 1991). Fortunately, the majority of food poisoning pathogens are sensitive to radiation and water is the principal target of ionizing radiation. Water radiolysis generates free radicals, which in turn damage microorganisms deoxyribonucleic acid DNA (Scott J. S, and P. Pillai, 2004). This "ionizing" effect splits water molecules into hydrogen (H+), hydroxyl (OH-) and oxygen (O-2) radicals and deactivate bacterial DNA, proteins, and cell membranes (Niemira and Sommers 2006). The gamma rays hit the double helix of the DNA and cause it to split which results in breaks. The severity and the number of the breaks determine the bacterial cells ability to repair and recover (Jay, 1992). The killing effect of irradiation on microbes is measured in D values. One D value is the amount of irradiation needed to kill 90% of that organism for example, it takes 0.3 kiloGrays to kill 90% of *E. coli* O157: H,7 so the D value of *E. coli* is 0.3 kGy (CAST, 1996). The D-values are different for each organism and change by temperature and the type of food. Irradiation is a treatment that is highly effective against microbial pathogens found in fresh produce. Anu Kamat et al. (2003) reported that a low-dose irradiation (1 kGy) was efficient enough for decontamination and elimination of potential pathogens. Irradiation has been successful in eliminating or greatly reducing the heavy load of spoilage microorganisms in vegetables, herbs, and spices (Satin, 1993). Kim et al. (2005) found that doses of 1.0 to 1.5 kGy reduced total aerobic count on fresh-cut green onions by about 3 logs. According to Lambert and Maxcy,(1984) *Camylobacter jejuni, Aeromonas hydrophila* (Palumbo et al., 1986), and *Yersinia enreocolitica* (El Zawahry and Rowley, 1979) have been found to have a low tolerance for irradiation.

The potential of gamma irradiation to inactivate foodborne pathogns on fresh produce has been investigated by various scientists. A dose of 5 kGy is reported to reduce a population of *Salmonella serotypes, Staphylococcus aureus, Shigella, E. coli,* and Vibrio species by at least 6 log cycles (Diehl, 1995). This also applies to *E. coli O157:H7* which is reported to be readily inactivated by irradiation (Clavero et al., 1994). Hagenmaier and Baker (1997) have also indicated that normal microflora on lettuce was reduced with a irradiation dose of 0.19 kGy. Irradiation at 0.3 and 0.6 kGy combined with Modified Atmosphere Packaging system MAP is reported to reduce *L. monocytogenes* on endive by 2.5 to 3 logs (Niemira et al. 2005). A 5 log reduction in *E.coli* O157:H7 and lack of adverse effects on sensory attributes was reported (Foley et al, 2002) when lettuce was subjected to 0.55 kGy. Farkas, et al, 1997 also observed a 4-log reduction of *L. monocytogenes* on the surface of sliced bell peppers irradiated at 1.0 kGy. Gamma irradiation is highly effective in inactivating micro-organisms in fresh produce and offers a safe alternative as a food decontamination method.

## 5. Effects of gamma irradiation on quality of fresh produce

Safety of irradiated foods involves four aspects: radiological safety, toxicological safety, microbiological safety, and nutritional adequacy. The Bureau of Foods Irradiated Food Committee (BFIFC) of Food and Drug Administration FDA established that more than 90% of all radiolytic compounds in irradiated foods were similar to those found in heating, drying, and freezing of food (Diehl, 1995 and FDA, 1988). Basing its recommendations on radiation chemistry, FDA has concluded that foods irradiated at dose levels up to 1 kGy and foods comprising no more than 0.01 % of daily diet irradiated at 50 kGy or less can be considered safe for human consumption without any toxicological testing (Diehl, 1995). Free radicals are formed when food is irradiated, but they are also formed by exposure to sunlight, frying, baking, grinding, and drying. In wet foods , free radicals disappear within a fraction of a second; in dry foods, the free radicals are much more stable and do not disappear as quickly (ACSH, 1988).

### 5.1 Nutrition quality of irradiated fresh produce

Irradiated foods are wholesome, nutritious, and nutrient losses are not significantly different from other alternative treatments. The extent of nutritional losses as a result of irradiation is comparable to or less than that of most other processing methods (Josephson and Peterson, 1983; Nawar, 1986). Generally, there is no effect of γ-radiation (up to 10kGy) on the nutrients of irradiated foods. In a previous study, papayas, rambutans, and Kau oranges were acceptable after subjecting to a quarantine level of 0.75kGy (Follet and Sanxter,

2002). According to Scott J. S, and P. Pillai, 2004, vitamins have been shown to keep considerable levels of activity post irradiation. In general, proteins, lipids, and carbohydrates quality is not get affected as a result of irradiation (Thayer, 1990; Thayer et al., 1987; WHO, 1999). The nature and extent of the effects of ionizing radiation on nutrients depends on the composition of food, radiation dose, and modifying factors such as temperature and presence or absence of oxygen. It has been documented that minerals are also stable to irradiation ((Diehl, 1995). According to a report by Fan and Sokorai, 2002, irradiation can reduce vitamin C in some vegetables, but the decrease is usually inconsequential and does not exceed the decline seen during storage. Research on vitamin $B_6$ has shown less destruction of this vitamin in products sterilized by ionizing energy than by heat (CAST, 1986). Follet and Sanxter (2002) studied the tropical fruits and found papayas, rambutans, and Kau oranges were acceptable when treated with a quarantine level of 0.75kGy (minimum dose required is 0.25 kGy). Irradiation is also reported to increase phenol compounds of certain vegetables consequently increasing their antioxidant ability (Fan, 2005).

Fresh produce may loss firmness after irradiation (Gunes et al., 2000; Palekar et al., 2004). However, the softening of fresh produce can be lessened by combining different treatments. According to Gunes et al., 2000; Prakash et al., 2007, dipping diced tomatoes and fresh-cut apples in a calcium solution prior to irradiation prevents the softening of the tissue.

### 5.2 Sensory quality of fresh produce

Generally, fresh produce indicate little change in appearance, flavor, color, and texture, after low doses (1 kGy or less). It has been reported that irradiation does not increase the temperature significantly and therefore, there is retention of color, flavor and textural properties (Willis, 1982). In a previous study, celery irradiated at 1kGy, was suggested to be better-quality in sensory qualities as compared to celery subjected to blanching, chlorination, and acidification (Prakash, 2000). Follet and Sanxter (2002) found that Chompoo and Biew Kiew fruit to be more satisfactory when treated with 0.40 kGy than with the currently used hot-water immersion.

## 6. Challenges of food irradiation

### 6.1 Consumer acceptance

Currently, there is not yet a large market demand for irradiated foods in the US and the rest of the world. In spite of additional of safety benefits offered by irradiation, marketing of irradiated foods has not been successful; in part due to consumers' believe that irradiated foods form harmful compounds in food (Oliveira & Sabato, 2002). The terms "radiation" and "radioactivity" have negative connotations to many consumers. Occasionally, consumers believe that food become radioactive after irradiation. Food does not become radioactive as the energy passes through; it only destroys bacteria and does not leave behind any residual radioactivity. There are anti-food radiation activists campaing against the public acceptance of irradiated food. It is crucial to educate the consumers and highlight the benefits of irradiation, particularly since the public has indicated to be more receptive to the negative argument (Fox, 2002; Hayes et al., 2002). It is indicated that given the preference and the access to irradiated products, consumers are willing to purchase them in noticeably great numbers (Bruhn, 1995).

### 6.2 Gamma Irradiation and cost of food

The principal economic challenge of food irradiation is the projecting of market demand for irradiated fresh produce. A strong market demand, will attract investors absorb the large up-front costs needed to support food irradiation. Definitely, economic considerations are some of the factors that slow the widespread use of food irradiation. As any other food process, food irradiation adds a few cents per pound to the cost of production (http://www.fipa.us/q%26a.pdf Food Irradiation Processing Alliance).

Contributing to the limited marketing of irradiated food in USA, is the insufficient food irradiation facilities. As of August 2000, there were only two facilities in the United States used primarily for gamma irradiation of food. (GAO-10-309R. 2010. Federal Oversight of Food Irradiation (http://www.gao.gov/new.items/d10309r.pdf). It is costly to build an irradiation facility. A commercial food irradiation plant is in the range of $3 million to $5 million, depending on its size and processing capacity. Consumer reception of novel food technologies depends on risks and benefits associated to the new technology and reachable alternatives. Some consumer are attracted to purchase irradiated produce by the discernment that irradiated it is safer. Irradiated produce tend to have longer shelf life hence less storage losses. The cost of irradiated food could be offset by benefits such as keeping a product fresher longer and enhancing its safety (http://ag.arizona.edu/pubs/health/az1060.pdf.

### 6.3 Regulatory approval and labeling

Labeling of irradiated food has been considered indispensable in order to inform the consumers. Labeling laws of irradiated foods differ from country to country. In the U.S., as in many other countries, irradiated food are labeled as "Treated with irradiation" or "Treated by radiation" and require the use of the radura symbol at the point of sale in (Xuetong et al, 2007). Fresh produce should have the radura symbol displayed at the point of sell. For fruits and vegetables, radura symbol can be on each piece, on the shipping container, or on a sign near the merchandise. Analytical methods are used by government and regulatory agencies to determine irradiated foods in the market place (CODEX STAN 231, 2003). Using these methods, existing labeling principles are imposed and consumers' confidence is strengthened

### 6.4 Consumer education

Consumers' knowledge about food irradiation is insufficient and therefore, education is desired to improve the acceptance of irradiated food by the public. Consumers are confused over what food irradiation and studies time and again display that when provided with science-based information, a high percentage of consumers favor irradiated foods. Food irradiation, pasteurization, canning, freezing, and drying are means of treating food in order to make it safer to eat and longer lasting (Satin, 1993). Despite its advantages, consumer knowledge about it is very limited. Many consumers' fears or misunderstanding of food irradiation are from reports of nuclear incidents at Hiroshima (Japan), Chernobyl (USSR), and the Three Mile Island (United States) and from nuclear waste disposal. For many, food irradiation is a process that creates the same fear as the word, radiation. Some consumers mistake the association of food irradiation with nuclear radiation, and food irradiation opponents use this as their most effective tool of negative influence. During food irradiation, food is not in contact with radioactive source and therefore, food can not be radioactive (FIFA, 2006).

Conley (1992) advocates that in the US, the Food Safety and Inspection Service FSIS and the National Agricultural Library in cooperation with other food-related agencies such as the FDA should provide education materials to consumers regarding food irradiation. Consumers favor food irradiation after they are given science-based information including product benefits, safety and wholesomeness of irradiated products (Bruhn, 1998; Bruhn, C.M., and Schutz, H.G. 1989). Previous studies indicate that educational activities and science-based information increase consumer acceptance of irradiated foods (Resurreccion et al., 1995; Bruhn, 1998; Fox & Olson, 1998). Nayga et al. (2004) reported that consumers are "willing to pay" premiums for irradiated food depending on the awareness and background information of food irradiation.

## 7. Conclusion

Food irradiation involves many intricate issues, but if consumers are educated on the benefits, it can be an effective process of reducing microorganisms which cause food spoilage and human illness. It is time to educate the consumers and increase their knowledge and awareness of gamma irradiation as a technology intended to increase the quality and the safety of food, especially fresh produce. Consumer acceptance of food irradiation will definitely influence the intensity to which irradiation gets accepted as an alternative food processing technology.

## 8. References

[1] ACSH. 1988. Irradiated foods. American Council on Science and Health, New York.

[2] Anu Kamat, A., Pingulkar, K., Bhushan, B., Gholap, A., Thomas, P. 2003. Potential application of low dose gamma irradiation to improve the microbiological safety of fresh coriander leaves. *Food Control* 14:529–537.

[3] Asplund, K., and E. Nurmi. 1 991. The growth of *Salmonellae* in tomatoes. Int. J. *Food Microbiol.*13:177-182.

[4] Beuchat, L. R., and J. H. Ryu. 1997. Produce handling and processing practices. *Emerg. Infect. Dis.* 3:65–67.

[5] Beuchat, L.R., 1996. Pathogenic microorganisms associated with fresh produce. J. *Food Prot.* 59, 204–216.

[6] Brackett, R.E., 1992. Shelf stability and safety of fresh produce as influenced by sanitation and disinfection. *J. Food Prot.* 55:808–814.

[7] Bruhn, C. M. (1998). Consumer acceptance of irradiated food: Theory and reality. *Radiation Physics and Chemistry*, 52:129–133.

[8] Bruhn, C.M. 2007. Effect of an educational program on attitudes of California consumers toward food irradiation. *Food Protection Trends* 27: 744-748.

[9] Bruhn, C.M., and H. G. Schutz. 1989. Consumer awareness and outlook for acceptance of food irradiation. *Food Technol.* 43: 93-94.

[10] Buchanan, R.L., Edelson, S.G., Snpes, K., Boyd, G., 1998. Inactivation of *E. coli* O157:H7 in apple juice by irradiation. *Appl. Environ.Microbiol.* 64: 4533–4535

[11] CDC. 1993. Multistate outbreak of *Salmonella* serotype *montevidio* infections. *EPI-AID* 93:89

[12] Chervin, C., P. Boisseau. 1994. Quality maintenance of "ready-to-eat" shredded carrots by gamma irradiation. *J. Food Sci.* 59:359–361.

[13] Clavero, M. R. S., J. D. Monk, L. R. Beuchat, M. P Doyle, and R. E. Brackett. 1994. Inactivation of *Escherichia coli* O157: H7, *Salmonella,* and *Campylobacter jejuni* in raw ground beef by gamma irradiation. Appl. *Environm. Microbiol.*60:2069-2075.

[14] Conley, S.T. 1992. What do consumers think about irradiated foods? *FSIS Food Safety Review* (Fall): 11-15.

[15] Diehl, J. F. 1995. Safety of Irradiated Foods. Second edition. Marcel Decker, Inc. New York.

[16] El Zawahry, Y. A., and D. B. Rowley. 1979. Radiation resistance and injuring of *Yersinia enreocolitica.* Appl. Environ. Microbial. 37: 50-54.

[17] Faber, J. M, and P. I. Peterkin. 1991. *Listeria monocytogenes,* a foodborne pathogen. *Microbiol. Rev.* 55:476–511.

[18] Fan, X. 2005. Antioxidant capacity of fresh-cut vegetables exposed to ionizing radiation. *J. Sci. Food Agric.* 85: 995-1000.

[19] Fan, X. and Sokorai, K. J.B. 2002. Sensorial and chemical quality of gamma irradiated fresh-cut iceberg lettuce in modified atmosphere packages. *J. Food Protect.* 65: 1760-1765.

[20] Farkas, J., Saray, T., Mohacsi-Farkas, C., Horti, C., & Andrassy, E. 1997. Effects of low-dose gamma irradiation on shelf life and microbiological safety of precut/prepared vegetables. *Advances in Food Science,* 19, 111–119.

[21] FDA News. (2007). FDA finalizes report on 2006 spinach outbreak. http://www.fda.gov/bbs/topics/NEWS/2007/NEW01593.html Accessed 22.02.08.

[22] FDA. 1988. Irradiation in the production, processing, and handling of food; final rule; denial of request for hearing and response to objection. Food and Drug Admin., *Fed. Reg.* 53:251: 53176-53209.

[23] FDA News. 2006. http://www.fda.gov/bbs/topics/ NEWS/ 2006/NEW01504. html

[24] FDA. 2007b. Irradiation in the production, processing and handling of food. Proposed rules. *Fed. Reg.* 72:16291-16306. www.cfsan.fda.gov/~lrd/fr070404.html. A accessed October 17, 2011.

[25] Fenlon, D. R., J. Wilson and W. Donachie. 1996. The incidence and levels of *Listeria monocytogenes* contamination of food sources at primary production and initial processing. *J. Appl. Microbiol.* 81:641–650.

[26] FIFA. 2006. Food irradiation Questions and Answers. Available. http//www/fifa.us/q%26a.pdf. A accessed October 16, 2011.

[27] Foley D.M., A. Dufour, L. Rodriguez, F. Caporaso and A. Prakash. 2002. Reduction of *Escherichia coli* 0157:H7 in shredded iceberg lettuce by chlorination and gamma irradiation. *Radiation Physics and Chemistry.* 63:391–396.

[28] Follett, P.A and S. S. Sanxter. 2002. Longan quality after hot-water immersion and X-ray irradiation quarantine treatments. *Hort. Sci.* 37: 571-574.

[29] Food and Drug Administration. 2006. Guidance for industry: guide to minimize microbial food safety hazards for fresh fruits and vegetables Available at http://www.cfsan.fda.gov/_dms/prodgui2.html#prime. March 2006. Last viewed

*July 2006 US July* US Department of Health and Human Services. Center for Food Safety and Applied Nutrition (CFSAN).

[30] Food and Drug Administration's (FDA's). FDA Investigating *E. coli* O157 Infections Associated with Taco Bell Restaurants in Northeast http://www.cfsan.fda.gov/~news/whatsnew.html

[31] Fox, J. A. and D. G. Olson. 1998. Market trials of irradiated chicken. *Radiation Physics and Chemistry*, 52, 63–66.

[32] Fox, J.A. 2002. Influence on purchase of irradiated foods. *Food Technol.* 56(11): 34-37.

[33] Fox, J.B., L. Lakritz, J. Hampson, R. Richardson, K. Ward and D. W. Thayer. 1995. Gamma irradiation effects on thiamin and riboflavin in beef, lamb, pork, and turkey. *J. Food Sci.* 60:596-598, 603.

[34] Frazier, C. W., and C. D. Westhoff. Food Microbiology 1988, 4th ed. McGraw- Hill, Inc. New York, NY.

[35] Giddings, G. G,. and M. Marcotte. 1991. Poultry irradiation: for hygiene/safety and market-life enhancement. *Food Rev. Intl.* 7 (3): 259-283.

[36] Hagenmaier, D. R. and R. A. Baker. 1997. Low-dose irradiation of cut iceberg lettuce in modified atmosphere packaging. *J. Agric. Food Chem.* 45: 2864-2868.8 (5).

[37] Han, Y., J. D. Floros, R. H. Linton, S. S. Nielsen and P. E. Nelson. 2001. Response surface modeling for the inactivation of *Escherichia coli* O157:H7 on green peppers (*Capsicum annuum* L.) by chlorine dioxide gas treatments. *J. Food Prot.* 64, pp. 1128–1133.

[38] Hancock, D. D., T. E. Besser, M. L. Kinsel, P. I. Tarr, D. H. Rice, and M.G. Paros. 1994. The prevalence of *Escherichia coli* O157: H7 in dairy and beef cattle in Washington State. *J. Epidemiol. Infect.* 113:199-207.

[39] Harris, L.J. and J. N. Farber. 2003. Outbreaks associated with fresh produce: incidence, growth, and survival of pathogens in fresh and fresh-cut produce. *Compreh Rev. Food Sci. Food Safety.*2:78-141.

[40] Hayes, D.J., J.A., Fox, and J.F. Shogren. 2002. Experts and advocates: How information affects the demand for food irradiation. *Food Policy* 27:185-193.

[41] Heisick, J. E., D. E. Wagner, M. L. Nierman, and J. T. Peeler. 1989. *Listeria* spp. found on fresh market produce. *Appl. Environ. Microbiol.*55:19–25.

[42] Ho, J. L., Shandas, K. N., Friedland, G., Ecklind, P., & Graser, D. W. 1986. An outbreak of type 4 *L. monocytogenes* infection involving patients from eight Boston hospitals. *Archives of International Medicine*, 146:520–524.

[43] Ilic, S., J. Odomeru, and J. LeJeune. 2008. Coli forms and prevalence of *Escherichia coli* and foodborne pathogens on minimally processed spinach in two packing plants. *J. Food Prot.* 7: 2398–2403.

[44] Institute of Food Technologists. 2002. IFT expert report on emerging microbiological food safety issues: implication for control in the 21St century. Institute of Food Technologists, Chicago, ILL.

[45] Jablasone, J., K. Warriner, M. Griffiths. 2005. Interactions of *Escherichia coli* O157:H7, *Salmonella typhimurium* and *Listeria monocytogenes* plants cultivated in a gnotobiotic system. *International Journal of Food Microbiology* 99, 7–18.

[46] Jay, J. M. Modern microbiology. 2000. 4th ed. Chapman and Hall, New York, NY.

[47] Jones, J. M. 1992. Chapter 12. Food Irradiation in Food Safety. Eagan Press. St. Paul, MN

[48] Kamat A., P. Kiran, B. Brij, G. Achyut, T. Paul. 2003. Potential application of low dose gamma irradiation to improve the microbiological safety of fresh coriander leaves. Food Control 14:529–537

[49] Kudva, I. T., P. G. Hatfield, and C. J. Hovde. 1997. *Escherichia coli* O157:H7 in microbial flora of sheep. *J. Clin. Microbiol. 34:431-433.*
[50] Lambert, J. D., and R. B. Maxcy. 1984. Effect of gamma radiation on *Campylobacter jejuni. J. Food Sci.* 49:665.
[51] LeJeune J.T., and N.P. Christie. 2004. Microbiological quality of ground beef from conventionally-reared cattle and 'Raised without antibiotics' label claims. *J. Food Prot.* 67: 1433-1437.
[52] Leverentz, B., W. S. Conway, W. Janisiewicz, and M. Camp. 2004. Optimizing concentration and timing of phage spray application to reduce *Listeria monocytogenes* on honeydew melon tissue. *J. Food.Prot.* 67:1682–1686.
[53] Liao, C.-H. and W. F. Fett. 2001. Analysis of native microflora and selection of strains antagonistic to human pathogens on fresh produce. *J. Food Prot.* 64:1110–1115.
[54] Little, C. L and I. A. Gillespie. 2008. Prepared salads and public health. *J. of Appl Microbiology*, 105(6), 1729–1743.
[55] Maurer K.F., Zur Keimfreimachung von Gewürzen, Ernährungswirtschaft 5(1958) nr.1, 45-47.
[56] Nayga Jr., R.M., A. Poghosyan, and J. Nichols. 2004. Will consumers accept irradiated food products? Intl. J. Consumer Studies. 28:178-185.
[57] Odumeru, J. A., S. J. Mithecu, D. M. Alves, J. A. Lynch, A. J. Yee, S. L. Wang, S. Styliadis and J. M. Farber. 1997. Assessment of the microbiological quality of ready to use vegetables for health care food services. *Journal of Food Protection*, 60:954–960.
[58] Oliveira, I. B. and S. F. Sabato. 2002. Brazilian consumer acceptance of irradiated food: initial trials. The Americas Nuclear Energy Symposiums (ANES 2002), http://www.anes.fiu.edu/Pro/s7oli.pdf, (reached on June 18, 2004).
[59] Palekar, M.P., E. Cabrera-Diaz, A. Kalbasi-Ashtari, J. E. Maxim, R. K. Miller, L. Cisneros-Zevallos and A. Castill. 2004. Effect of electron beam irradiation on the bacterial load and sensorial quality of sliced cantaloupe. *J. Food Sci.* 69: M267-M273.
[60] Palumbo, S. A., R. K. Jenkins, R. L. Buchanan, and D. W. Thayer. 1986. Determination of irradiation D values for *Aeromonas hydrophila. J. Food Prot.* 49:189-191.
[61] Prakash, A., A. R. Guner, F. Caporaso and D. M. Foley. 2000. Effect of low-dose gamma irradiation on the shelf life and quality characteristics of cut romaine lettuce packaged under modified atmosphere. *J. Food Sci.* 65: 549-553.
[62] Prakash, A., Chen, P.C., Pilling, R., Johnson,N. and Foley, D. 2007. 1% Calcium chloride treatment in combination with gamma irradiation improves microbial and physicochemical properties of diced tomatoes. *Foodborne Pathogens Dis.* 4: 89-97.
[63] Resurreccion, A. V. A., F.C. F. Galvez, S. M. Fletcher and S. K. Misra. 1995. Consumer attitudes toward irradiated food: results of a new study. *Journal of Food Protection*, 58, 193–196.
[64] Roberts, Tim.1998. Cold Pasteurization of Food. Virginia Tech; Publication Number 458-300. http://www.ext.vt.edu/pubs/foods/458-300/458-300.html
[65] Satin, M. 1993. Food irradiation. Lancaster, PA
[66] Schlech, W. F., Lavigne, P. M., Bostolussi, R. A., Allien, A. C.,Haldane, E., Wort, A. J., Hightower, A. W., Johnmson, S. E.,Kingh, S. H., Nicholls, E. S., & Broome, C. V. 1983. Epidemic listeriosis--evidence for transmission by food. *New Engineering Journal of Medicine*, 308:203–206.
[67] Scott J. S, and P. Pillai. 2004. Irradiation and Food Safety. *Food Tech.* 58:48-55

[68] Seo, K. H., and J. F. Frank. 1999. Attachment of *Escherichia coli* 0157:H7 to lettuce leaf surface and bacterial viability in response to chlorine treatment as demonstrated by using confocal scanning laser microscopy. *J. Food Prot.* 6:3-9.

[69] Sivapalasingam S., C. R. Friedman, L. Cohen, R. V. Tauxe. 2004. Fresh produce: a growing cause of outbreaks of foodborne illness in the United States. 1973 through 1997. *J Food Prot.* 67:2342-2353.

[70] Solomon, E.B., S. Yaron and K. R. Matthews. 2002. Transmission of *Escherichia coli* O157:H7 from contaminated manure and irrigation water to lettuce plant tissue and its subsequent internalization. *Applied and Environmental Microbiology* 68:397-400.

[71] Steele, J. H., and R. E. Engel. 1992. Radiation processing of food. J. Am. Vet. Med: Assoc. 201:522-1529.

[72] Sumner, S., D.L., Peters. 1997. Microbiology of vegetables. In: Smith,D.S. (Ed.), Processing Vegetable Science and Technology. Technomic Publishing Co. Inc., Lancaster, PA, USA, pp. 87–106

[73] T. Setsukoi., B. Latiful, K. Kazumi, O. Mika, I. YasuhiroI, T. Yuka, K. Norihito, Y. Erika, M. Tatsuya, K. Yukio, and K. Shinichi. 2009. Effect of gamma-irradiation on the survival of Listeria monocytogenes and allergenicity of cherry tomatoes. Radiation Physics and Chemistry 78:619–621

[74] Tauxe, R. V. 1997. Emerging food borne diseases: an evolving public health challenge. *Emergc. Infect. Dis. 3:425 434.*

[75] Ukuku, D.O. and G. M. Sapers. 2001. Effect of sanitizer treatments on *Salmonella* Stanley attached to the surface of cantaloupe and cell transfer to fresh-cut tissues during cutting practices. *J. Food Prot.* 64:1286–1291.

[76] Urbain, W. M. 1986. Food irradiation. Orlando: Academic Press, Inc.

[77] USFDA. U. S. Food and Drug Administration.1999. Potential for infiltration, survival and growth of human pathogens within fruits and vegetables. http://www.cfsan.fda.gov/~comm/juicback.html.

[78] Wachtel, M. R., L. C. Whitehand and R. E. Mandrell. 2002. Association of *Escherichia coli* O157:H7 with pre-harvest leaf lettuce upon exposure to contaminated irrigation water. *J. Food. Prot. 65:18-25.*

[79] Warner, G. 1997. Fresh packers await new food safety guidelines. Good fruit Grower 48:7-8.

[80] WHO. 1981. Wholesomeness of irradiated foods. Technical Report Series 659. World Health Organization, Geneva, Switzerland.

[81] WHO. 1987. Food Irradiation. In Point of Fact. No. 40. Geneva, Switzerland.

[82] WHO. 1999. High-dose irradiation: Wholesomeness of food irradiated with doses above 10kGy.

[83] Report of a joint FAO/IAEA/WHO study group. WHO technical report series 890. World Health

[84] Organization, Geneva, Switzerland

[85] Williams, C., 1995. Healthy eating: clarifying advice about fruit and vegetables. Brit. Med. J. 310, 1453–1455.

[86] Wimberly R.C., B. J. Vander, B. L Wells. 2003. The globalization of food and how Americans feel about it. North Carolina State University 1–17.

[87] Xuetong, Fan, A. Brendan A. Niemira, and A. Prakash. Irradiation of fresh fruit and vegetables.http://www.wga.com/DocumentLibrary/scienceandtech/Irradiation%20of%20Fresh%20Fruits%20and%20Vegetables-IFT%20report. Food Tech.

# 13

# Current Importance and Potential Use of Low Doses of Gamma Radiation in Forest Species

L. G. Iglesias-Andreu, P. Octavio-Aguilar and J. Bello-Bello
*Instituto de Biotecnología y Ecología Aplicada, Universidad Veracruzana, Xalapa, Veracruz, México*

## 1. Introduction

It is well known that ionizing radiation is currently a very important way to create genetic variability that is not exists in nature or that is not available to the breeder (Ahloowalia & Maluszynski, 2001; Lemus *et al.*, 2002). Therefore, there are many papers aimed to determine the best radiation dose to applied in plant breeding work. As a result it has been defined intervals gamma radiation useful for many cultivated species, though the determination of the radiosensitivity of tissues by exposure to different intensities of radiation (De la Fe *et al.*, 1996; Castillo *et al.*, 1997; Fuchs *et al.*, 2002; Lemus *et al.*, 2002; Fuentes *et al.*, 2004; Ramírez *et al.*, 2006). However, most studies have been conducted have been designed to evaluate the biological response to high doses of radiation, while in relatively few studies have used low doses to stimulate physiological processes (radiostimulation) although the ionizing radiation hormesis has been widely supported (Luckey, 1980). Hormesis is the excitation, or stimulation, by small doses of any agent in any sistem (Luckey, 2003). The beneficial effect of hormesis has been well documented in species of agricultural importance (Zaka *et al.*, 2004; Kim *et al.*, 2005). However, there is not enough information about its use in forestry. Although little is known about the basic nature of this phenomenon, Vaiserman (2010) had indicated the possible relationship between the hormesis and epigenetic effects. The application of low-dose ionizing radiation could produce in coniferous species hormetics radiostimulants effects through genetic and epigenetic changes that manifest as adaptive responses.

In Mexico and especially in many natural populations of conifers from Veracruz such as *Pinus hartwegii* Lindl., and *Abies religiosa* Kunth (Schltdl.) *et.* Cham., both located in Cofre de Perote, Ver., are seriously affected mainly by the high load of lethal alleles which are causing a serious reduction in reproductive rate and a significant decrease in the production and quality of its seed (Iglesias *et al.*, 2006).

Despite the usefulness of using ionizing radiation to increase the germination potential and generating useful mutations in forestry, there are not many references in the literature on the use of nuclear techniques in these species (Iglesias *et al.*, 2010). Therefore, in this chapter will be give a review on the use of low doses of ionizing radiation on forest species, and it will perform a particular consideration to the effect of low doses of gamma radiation on germination and growth of some variables forest species such as *P. hartwegii* and *A. religiosa*.

## 2. Uses of low doses of ionizing radiation on plant species

Ionizing radiation is defined as the energy that propagates in the form of photons (X-rays and γ) or in the form of subatomic particles (α, β, neutrons and protons). Among them, gamma rays have been reported to be the most efficient ionizing radiation of creating mutants in plants. Gamma rays belonging to ionizing radiation group are the most energetic form of electromagnetic radiation (Ikram *et al.*, 2010). This kinds of rays possesses the energy level from 10 keV to several hundred kiloelectron volts, and they are considered as the most penetrating physical mutagenic agent in comparison to other radiation source such as alpha and beta rays (Kovács & Keresztes, 2002). Like other ionining radiation gamma rays interacts with atoms or molecules to produce free radicals in cells. These radicals can induce high mutation in plants because it could produce serious cell damage or afectations in important plant cells components (Kovács & Keresztes, 2002).

| Species | Gamma rays doses | Effects | Reference |
|---|---|---|---|
| *Vigna radiata* (L.) Wilczek. | 40 - 80 kR | Increase polygenic variability. | Sangwan & Singh (1977) |
| *Triticum aestivum* L. | 0.5–7 kR | Stimulatory effect on height, tillering and grain yield. | Iqbal (1980) |
| *Sorghum vulgare* L. | 1–10 kR | Large reduction in mean seedling height and tillering. | Iqbal (1980) |
| *Salix nigra* Marsh. | 0.1 - 100 kR | Low doses increasing the growth rate. | Gehring (1985) |
| *Tectona grandis* L. f. | 10, 20, 30, 40 and 50 kR | Improve the germination rate of the seeds. | Bhargava & Khalatkar (1987) |
| *Allium cepa* L. | 10, 20, 40, 80, and 100 kR | Percentage of abnormal seedlings increased with increase in radiation dose. | Amjad & Akbar (2003) |
| Basmati rice varieties | 150 - 300 Gy | Increase of total spikelets above the non-irradiated rice. | Ali & Manzor (2003) |
| *Chrysanthemum morifolium* cv. | 15, 30 and 60 Gy | The regeneration rate decrease with increase in the total dose of radiation. | Yamaguchi *et al.* (2008) |
| *Triticum aestivum* L. | 10, 20, 30 and 40 kR | Irradiated seeds showd superiority over control population for several traits. | Singh & Balyan (2009) |
| *Sesamum indicum* L. | 200, 400, 600 and 800 Gy | Mutagenic effects by intergenomic chromosomal rearrangements. | Kumar & Singh (2010) |

Table 1. Application of high-doses of ionizing radiation in plant breeding.

Ionizing radiation can be quantified in terms of absorbed dose (D) which is the amount of ionizing radiation energy deposited per unit mass of irradiated material. Initially, the magnitude was measured in roentgens, called "radiation dose". A Roentgen is a unit of measure used to quantify the radiometric exposure, in other words, the total charge of ions released per unit mass of dry air at standard conditions of pressure and temperature, but now the most often unit used to quantify the biological effects of ionizing radiation is the gray (Gy), which is a unit derived from the international system of units to measure the absorbed dose of ionizing radiation for a certain material. One gray is equivalent to the absorption of one joule of radiation energy per kilogram of irradiated material. Radiation doses are divided into three broad categories:high (> 10 kGy), medium (1 to 10 kGy), and low (<1 kGy).

Gamma rays are widely used for mutation induction in plants. Therefore many studies have been done on the dose-response effects of ionizing radiation, specifically gamma radiation on several growth and yield traits in plants with high-doses of ionizing radiation (Table 1).

| Species | Gamma rays doses | Effect | Reference |
|---|---|---|---|
| *Daucus carota* L. | 0.5 and 1 kR | Irradiation accelerated germination of carrot seeds. | Bassam & Simon (1996) |
| *Capsicum annuum* L. | 2, 4, 8 and 16 Gy | Low doses stimulated the growth and stress resistance. | Kim *et al.* (2005) |
| *Solanum tuberosum* L. | 2.5, 5, 10 and 15 Gy. | Low doses increase the number of microtubes *in vitro*. | Bassam *et al.* (2000) |
| *Beta vulgaris* L. | 20 kR | Induced abnormal floral structures by mutation. | Chauhan *et al.* (2009) |
| *Secale montanum* Guss. | 0,2,4,6,8,10,12,14,16,20,25 and 30 kR | Increase the frequency of cells in anaphase and metaphase. | Akgün & Tosun (2004) |

Table 2. Examples of hormetic effects in plants by low doses of ionizing radiation.

Some reports (Gunckel & Sparrow, 1961; Ikram *et al.*, 2010) have been shown higher exposures of gamma rays produce generally negative effects on plant growth and development although the effect of dose rate on mutation frequency might differs among plant species. These effects include cytological, anatomy, genetical, biochemical, physiological and morphogenetic changes in cells and tissues. Many changes in the plant cellular structure and metabolism e.g., dilation of thylakoid membranes, alteration in photosynthesis, modulation of the antioxidative system and accumulation of phenolic compounds had been documented in different plant species (Kim *et al.*, 2004; Wi *et al.*, 2005). Higher exposures of gamma rays usually produce inhibitor effects on Gymnosperm and Angiosperm seed germination (Kumari & Singh, 1996) whereas lower exposures produce sometimes a stimulatory effect (Raghava & Raghava, 1989; Thapa, 1999).

It is important define the threshold between high-doses, in several cases dangerous, and low-doses with stimulatory effects. A radiostimulant low-dose is defined as any doses from environmental radiation levels and the threshold that marks the boundary between positive and negative biological effect (Luckey, 2003). This radiostimulatory effects which has been observed in different plant species (Table 2), through the use of low-doses of ionizing gamma radiation could be considered an interesting alternative somewhat unexplored in agriculture and forestry practice.

It has been recognized low-doses of radiation promote increased of cell respiration, enzyme activation, increase in the threshold of lethal doses of radiation, increasing the production of reproductive structures, higher growth, early maturation, accelerated development and disease resistance (Luckey, 1980; 1998). However, most of the works done in this way have been addressed to find the boundaries between hormesis and tissue damage (Castillo *et al.*, 1997; Fuchs *et al.*, 2002; Lemus *et al.*, 2002; Fuentes *et al.*, 2004; Ramírez *et al.*, 2006). There is few researches have been conducted to evaluate the effect of the radiation on the full cycle of the organisms (Cepero *et al.*, 2002; Ramírez *et al.*, 2006). The results obtained in the works of radiostimulation or radiohormesis revealed increases (10-40%) in agricultural yields, seed germination, contents of carotenes and vitamin C in some vegetables and protein and fat in cereals, finally resistance to diseases and abiotic factors (González *et al.*, 2002; Vasilevski, 2003). On other hand, chronic radiation is another kind of irradiation treatment used to increase variation in different plant species (Sparow & Woodwell, 1962). This type of irradiation could produce at the plant population level, most severe effects on sexual reproduction because during and after meiosis: (1) nuclear volume is high; (2) chromosome number is reduced after meiosis; (3) the rate of nuclear division may be low, some species requiring two years between meiosis and full maturation of seeds; (4) meiotic pairing and reduction tend to enhance the damage wrought by aberrations which may survive in diploid somatic cells. In forestry was evaluated too the effects of chronic ionizing radiation of low intensity (3–15 r/20 hr day) over a period of several years on the reproductive capacity of the trees, floral abnormalities, as well as growth of their progenies (Mergen & Stairs, 1962). These authors were found for *Pinus rigida* a decrease in cone length and in seed germination and seedling height for plants grown from irradiated cones that was associated with an increase in the chronic gamma radiation accumulated by the trees. For *Quercus alba* was observed visual aberrations in flower morphology in trees receiving from 6 to 12 r/day and a decrease on survival percentage and height growth of seedlings with radiation level (Mergen & Stairs, 1962). In this case like others mutagenic treatments the relative dosage levels necessary to produce specified responses in growth rate, reproductive capacity or in degree of mortality vary greatly within a species.

Irradiation treatments performed at *in vitro* culture has been also employed to increase genetic variability and mutants as a potential source of new commercial cultivars (Rasheed *et al.*, 2003; Orbovic´ *et al.*, 2008). So, tissue culture techniques offer a wide choice of explants (initial plant material) for gamma radiation treatment (cells, tissues, somatic embryos and organs). These explants will give origin to complete plants composed of a few or even of one cell with a higher probability for find mutated cells. *In vitro* culture also allows for the handling of unlimited vegetative material for radiation treatment, aseptic and controlled *in vitro* selection, and micropropagation of selected variants (so called somaclones). In addition, according to Predieri (2001), tissue culture increases the efficiency of mutagenic treatments for variation induction, handling of large populations, use of ready selection methods, and rapid cloning of selected variants.

Recent developments in biotechnology—especially in understanding the structure and function of plant genomes—confirms *in vitro* mutation induction as one of the most efficient and cost-effective tools for functional genomics projects dealing with both forward and reverse genetics strategies (Jain, 2001; Shu & Lagoda, 2007). The high number of research reports suggests also that mutagenesis in combination with tissue culture has high potential in plant breeding programs. It has been indicated (Maluszynski *et al.*, 1995), that the use of tissue culture techniques can overcome some of the limitations in the application of mutation techniques; these are the lack of effective mutant screening techniques and the unrealistically large but necessary size of the mutated population, calculated on the basis of an expected mutation frequency for a desired trait. The determination of radio-sensitivity tests, irradiation with optimal doses and multiplication of irradiated material through in vitro mutation techniques has assumed a new dimension (Ahloowalia & Maluszynski, 2001). An example of setting the boundary between and tissue damage from ionizing radiation was shown by Fuchs *et al.* (2002). These authors found in callus culture of *Saccharum sp.* (sugarcane), dose of greater than 4 kR of gamma radiation eliminated any possibility to induce an organogenetic process in this tissue. In this case, the hormetic threshold (2 to 4 kR) was much lower that applied to seeds of other species (4 to 20 kR) (Lemus *et al.*, 2002; Ramírez-Calderón *et al.*, 2003; González *et al.*, 2004).

## 3. Hormesis and molecular mechanisms of the adaptive response

The hormesis term comes from greek meaning "to excite". Exposure of sublethal doses of ionizing radiation can induce protective mechanisms against a subsequent higher dose irradiation. So, the hormesis is the excitement and stimulation by small doses of any agent on any system. Luckey (1980), in his book entitled "Hormesis", documented thousand experiments where fungi and other lower life forms were seen to prosper with doses of radiation exceeding their normal background exposures with ionizing radiation. In a second book entitled "Radiation Hormesis" (Luckey, 1991) examined hundreds of studies on animals and humans, showing that low levels of radiation were beneficial to health, longevity, and reproduction.

Many studies have been also indicated that pre-expose to low dose radiation (or some other genotoxic agent) can change radiosensitivity, reducing score of chromosomal aberrations, micronuclei and mutations. This phenomenon is called adaptation and could be related with defense mechanisms some of them have evolved to minimize genotoxic damage. One of these is induced radioresistance or adaptive response (AR). The term "adaptive response" usually means that a relatively small "conditioning" radiation dose induces increased radioresistance when the cells are irradiated with higher doses several hours later (Hillova & Drasil, 1967). Thus, radioadaptive response induction expresses the ability of low dose radiation to induce cellular changes that alter the level of subsequent radiation-induced or spontaneous damage (Amundson *et al.*, 1999). The exposure to minimal stress inducing a very low level of damage can trigger an AR resulting in increased resistance to higher levels of the same or of other types of stress (Patra *et al.*, 2003; Asad *et al.*, 2004; Girigoswami & Ghosh, 2005; Yan *et al.*, 2006). The AR could be considered a nonspecific phenomenon and have been confirmed but not explained by many studies. Adaptation after preexposure to chronic or prolonged exposure to low-level radiation doses was not often described.

Several types of cellular responses to ionizing radiation, such as the adaptive response, suggest that low-dose radiation may possess characteristics that distinguish it from its high-dose counterpart. Accumulated evidence also implies that the biological effects of low-doses and high-doses of ionizing radiation are not linearly distributed. This is an important physiological effect of exposure to low doses of radiation. The radioadaptive was first documented in a convincing way to protect against chromosomal aberrations (Scott *et al.*, 2009).

The capability of forest tree species to adapt to the new environments not only will depend on their genetic background, but also rely on their phenotypic plasticity. Several reports have shown the involvement of epigenetic modifiers as the basis of the phenotypic plasticity, and in particular to the adaptation to abiotic stresses. DNA methylation is one the most important epigenetic modification in eukaryotes. It is involved in specific biological processes such as gene transcription regulation, gene silencing, mobile element control or genome imprinting. Therefore, there is a great interest in analyzing methylation levels and distribution within the genome. Epigenetic regulation of gene activity is widespread in the genome of eukaryotic cells and can lead to silencing or activation of gene expression. According to Scott *et al.* (2009), high doses of radiation can promote epigenetically silencing of adaptive response genes, for example via promoter associated DAN and /or histone methylation or deacetylation.

Adaptive-response genes can be stabilized and activated in response to cellular stress (e.g., low dose radiation) through post translational modifications that include acetylation (Ito *et al.*, 2002). This radiation, above a stochastic threshold stimulate intracellular and intercellular signaling that leads to activated natural protection (ANP) against cancer and other genomic-instability-associated diseases (Scott, 2005; Scott & Di Palma, 2006).

The AR has been observed in bacteria (Assis *et al.*, 2002; Sedgwick & Lindahl, 2002; Rohankhedkar *et al.*, 2006), yeast (Boreham & Mitchel, 1991; Gajendiran & Jeevanram, 2002), algae (Chankova & Bryant, 2001; Rubinelli *et al.*, 2002; Chankova *et al.*, 2007), insect cells (Savina *et al.*, 2003), mammalian cells (Wang & Cai, 2000; Tiku & Kale, 2001; Ulsh *et al.*, 2004; Zhou *et al.*, 2004), human cells (Schlade-Bartusiak *et al.*, 2002; Atanasova *et al.*, 2005; Coleman *et al.*, 2005; Friesner *et al.*, 2005; Lanza *et al.*, 2005; Seo *et al.*, 2006) and higher plants models (Rieger *et al.*, 1993; Panda *et al.*, 1997; Jovtchev & Stergios, 2003; Patra *et al.*, 2003). A study of the conditions essential for the induction of an adaptive response is of critical importance to understanding the novel biological defense mechanisms against the hazardous effects of radiation. The results statistically significant with microorganisms, plants, non vertebrates and other animals of experimentation, showed the existence of a radiogenic metabolism, in other words, a metabolism promoted by ionizing radiation.

Little is currently known about the precise mechanisms of AR. There is evidence that different stress conditions can activate similar defense mechanisms in biological systems (Joiner *et al.*, 1996; 1999; Babu *et al.*, 2003). The AR probably involves the transcription of many genes and the activation of numerous signaling pathways that trigger cell defenses more efficient detoxification of free radicals, DNA repair systems, induction of new proteins in irradiated cells with a conditioning dose, and enhanced antioxidant production (Wolff, 1998; Mendez-Alvarez *et al.*, 1999; Pajovic *et al.*, 2001; Assis *et al.*, 2002; Chankova & Bryant, 2001; Coleman *et al.*, 2005; Lanza *et al.*, 2005). There is evidence that

DNA repair underlies the AR induced by low radiation doses in human and plant cells (Lambin *et al*, 1994; Patra *et al.*, 2003) by increasing the amount and rate of DNA repair (Joiner *et al.*, 1996; 1999). It has been proposed that these effects could be related to the induction of an AR. A clue as to the nature of the underlying process was provided by results showing a dependence on de novo protein synthesis. The synthesis of DNA-binding proteins (MWs 50, 74 and 130 kdal) was found in radiation-conditioned cells of *C. reinhardtii* (Bryant, 1979). The induction of new protein synthesis by low doses could be caused by an effect of low doses on chromatin conformation near genes coding for DNA repair proteins (Belyaev *et al.*, 1996). For example, there are earlier observations that hydrogen peroxide induced a cross-adaptive response to cumene hydroperoxide in *E. coli* which did not require novel gene products but involved modification of the small subunit of Ahp, a protein involved in the protection against alkyl hydroperoxides (Asad *et al.*, 1998). On the other hand, Reactive Oxygen Species (ROS) could serve as signal transducers in plant and animal cells (Babu *et al.*, 2003; Matsumoto *et al.*, 2004). As signaling molecules, ROS might affect the development of AR through participation in the damage-sensing process after conditioning dose exposure.

## 4. Current and potential use of low ionizing radiation in forestry: A case study

At present there are virtually no studies of hormesis by ionizing radiation in forest species. Most of the work focused radiation treatment of species of agronomic interest since they have shorter lifetime and germination time, the tissues of this type of seeds have a greater amount of water, which maximizes the effect of radiation; and the generation of seedlings is much easier in the herbaceous form.

One of the few jobs that exist in tree species was conducted with *Araucaria angustifollia* (Bert) O. Kuntze (Ferreira *et al.*, 1980). The study showed a hormetic effect on seed germination and seedling growth at low doses of gamma radiation (0.1 to 0.4 kR). This first study showed the effectiveness of ionizing radiation to improve seed germination in species of trees, one of the main agronomic traits for forest management.

*Abies religiosa* (fir) and *Pinus hartwegii* are two conifer species that develop on the National Park Cofre de Perote, Veracruz, Mexico. Both have great ecological (*P. harwegii* is the conifer species taking place at higher altitudes) and economical importance (in particular, the fir is valuable for its timber, trementine production and as an ornamental Christmas trees). These forests have protective functions to other resources to cushion the effects of environmental pollution and contribute as a regulation of the hydrological cycle (Solís, 1994).

In Veracruz, these populations develop principally in the National Park Cofre de Perote and Pico de Orizaba between 2 800 y 3 500 m.a.s.l. in 17°30′ to 20°00′ N and 97°104′ W (Manzanilla 1974; Sánchez-Velásquez & Pineda-López 1993). Both species have been seriously affected by fire and logging clandestine, has resulted in a reduction of the effective size of the same low viability and high percentage of abortive seeds, and a significant decrease in reproductive rate, apparently due to manifestation of the phenomenon of inbreeding depression, common in coniferous species (Williams & Savolaienen, 1996). But, in both species, is common a low reproductive rate (Franklin, 1974).

Despite the potential utility of the low doses of ionizing radiation for the induction of hormetic effects in these species, to date have not implemented these techniques to increase the germination potential and generating useful mutation in forestry. Therefore, we study the mutagenic effect of gamma radiation and know whether low doses of radiation can have a stimulatory effect on germination and development of *P. hartwegii* and *A. religiosa*. Twenty-five to thirty cones of *A. religiosa* and *P. hartwegii* were collected in populations that develop on 3510 m.a.s.l., in the locality "El Conejo", both located in the National Park "Cofre de Perote" Veracruz, Mexico. These populations are fragmented and have been affected by significant changes in land use to agricultural crops (Sánchez-Velásquez *et al.*, 1991).

To apply the mutagenic treatments, seeds were extracted from the cones collected, and kept under controlled conditions at a relative humidity of 8%. Two replicates of 100 seeds of each species at low doses (2, 5, 10, 15 and 20 Gy) of gamma radiation ($Co^{60}$) were made. The seeds for irradiation were placed at a distance of 80 cm (for a field of 30 x 30 cm), with the help of a head of the Cobalt 60 Unit (Theraton 780e) in the "Centro de Cancerologia de Xalapa, Veracruz, Mexico; and plastic tray, which was secured in a cage at 50 cm from the radiation source (Figure 1). Subsequently, the irradiated seeds of each species and their corresponding controls were planted separately under greenhouse conditions in trays containing a mixture of forest soil and sand in a 1:1 ratio. In both cases we used a randomized complete block design with three replications. Seedlings were transferred to plastic bags for study.

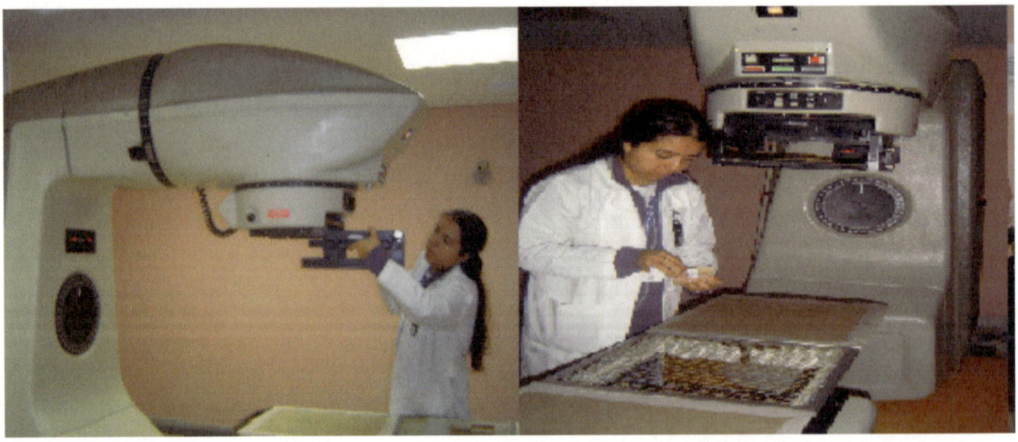

Fig. 1. Cobalt 60 Unit (Theraton 780e), used for irradiation of seeds of *Abies religiosa* and *Pinus hartwegii*, located in Centro Estatal de Cancerología de Xalapa, Veracruz, Mex.

To evaluate the effect of the applied radiation dose was counted the number of seeds germinated at 90 days for each dose studied; the percentage of germination was calculated. Was considered germinated seeds showed a greater than 5 mm radical length. At 45 days, was evaluated the height and number of needles of each seedling. Height (cm) was measured with a millimeter rule from the base of the stem of the root to the terminal bud. From the measurements we calculated the percentages of germination, plant height and number of needles as a relative value with respect to control.

The results showed a significant radiostimulating effect on the germination of the seeds of *Abies religiosa* and *Pinus hartwegii* treated with low doses of gamma radiation (Figure 2).

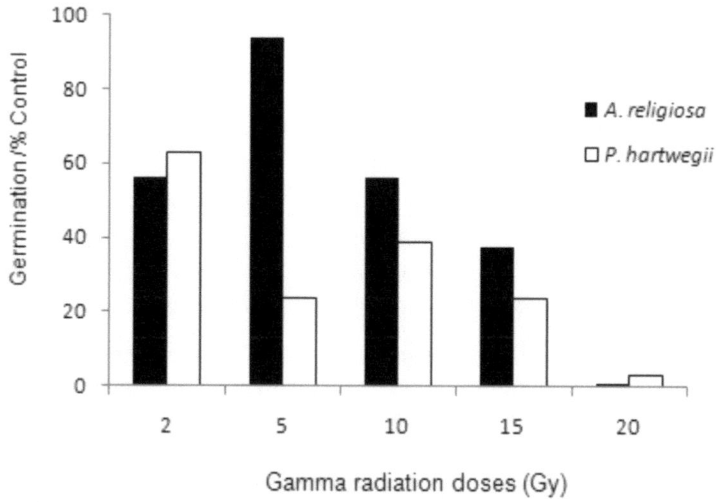


Fig. 2. Effect of the gamma radiation on seed germination in *Abies religiosa* and *Pinus hartwegii*.

According to these results the radiostimulatory effect was more pronounced in *Pinus hartwegii* since treatment of 2 Gy resulted in a high percentage of germination. 5 Gy dose was most effective to induce a similar effect on the fir.

These results are consistent with the radiostimulatory effect observed by Rudolph (1979) and Sokolov *et al.* (1998) to evaluate the germination of seeds of *Pinus bankasiana* and *P. sylvestris*, respectively. On the other hand, showed that, like as detected by Nwachukwu *et al.* (1994) and Lemus *et al.* (2002); the frequency of mutation increased with the percentages of germination. The few seedlings that managed to germinate at the highest doses of gamma radiation showed thickened short roots and therefore were less vigorous and did not survive.

Low doses of gamma radiation used in both species showed similar radiostimulatory effects on the characters of the seedling height and number of needles (Figure 3 a,b).

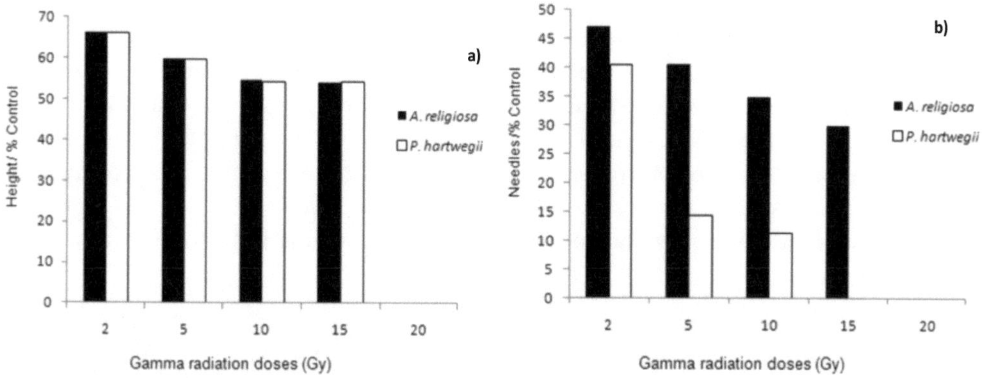


Fig. 3. Effect of the low-doses of gamma radiation on height (a) and number of needles in seedlings (b), *in Pinus hartwegii* and *Abies religiosa.*

As for the variable germination, was found at 2 Gy a radiostimulatory effect on height and number of needles of the seedlings of *P. hartwegii*. In fir, this effect is slightly higher.

In addition, there was a trend towards reduction in height and number of needles in both species as they increased the dose of radiation. This effect was more pronounced for the number of needles of *P. hartwegii* (Figure 3 b).

According to Olvera & West(1985); reducing the growth of seedlings generated from seeds treated with high doses of radiation is mainly due to the destruction of auxin and its precursors. It should be noted that the height variable is used in this type of study as the most sensitive indicator of radiation.

Based on our results consider the possibility of using the aforementioned dose to induce mutations that may be of interest in these species. However, all the applied doses produced a negative effect on the number of needles per plant, which are fundamental in the production of Christmas trees. Doses of 2, 5 and 10 Gy gave average values ranging from 3.3 to 6.8 needles per seedlings, well below the average control value, which was 27.7 in *P. hartwegii*. Contravention in the production because mutations are required to encourage a more fodder for a demanding market of this product.

It has been suggested in this regard that high doses of radiation cause damage that affects physiological character related to growth, especially with the number of needles. High doses of radiation can alter in a direct or indirect the DNA, causing damage of bases, single strand breaks and chromosomal alterations, serious and irreversible destruction of the membrane system of mitochondria and chloroplasts (Ladanova, 1993). However, it will take more repetitions to achieve seedlings with large needles and branches, or select for traditional breeding seedlings with large needles for future generations.

In summary, prolonged exposure to radiation by gamma rays produced a severe effect on almost all variables; this effect was greater in the seeds irradiated with 15 and 20 Gy. In the range of medium dose (5 Gy), radiation induced lesions can eventually lead to an important radiobiological response, which at the cellular level can alter the viability and even cause cell death (Ward, 1988).

This response was manifested by affectations in traits related to germination and seedling growth, as the germination percentage and number of needles that were most affected. On the other hand, not all variables were impacted in the same direction, since a dose of 5 Gy showed the presence of a radiopositive effect in the percentage of germination for fir, and at doses of 5 - 10 Gy there was a negative effect at the height of the plant and number of needles, with respect to control. However, when analyzing these results it is recommended to fully explore dose of 5 Gy for *A. religiosa*, and other below 2 Gy for *P. hartwegii*, in accordance with the sensitivity shown by this species to gamma radiation for improve the germination rate.

## 5. Conclusions

With all these examples we can say that low doses of ionizing radiation could improve the crop by increasing production, reducing the time of germination, accelerate growth of seedlings and generate interest new varieties of some plant species, including trees.

However, to achieve these results it is important to set the threshold hormetic species specific depending upon the type of tissue that is irradiated, the quantity of humidity inside the tissue and establish an appropriate model depending on the type of production that is required.

Radiation hormesis provides the basis for appropriate utilization of ionizing radiation as a useful tool in our society. It can provide more efficient use of resources, maximum production of grain, vegetables, and meat, and increased health and longevity. Efficient utilization of nature's resources demands support to explore the practical application of radiation hormesis.

## 6. References

Ahloowalia, B. & Maluszynski, M. (2001). Induced mutations- A new paradigm in plant breeding. *Euphytica*, 118(2):167-173.

Akgün, I. & Tosun, M. (2004). Agricultural and Cytological Characteristics of M1 Perennial Rye (*Secale montanum* Guss.) as effected by the application of different doses of Gamma Rays. *Pakistan Journal of Biological Science*, 7(5):827-833.

Ali, C. & Manzoor, A. (2003). Radiosensitivity studies in basmati rice. *Pakistan Journal of Botany*, 35(2):197-207.

Amjad, M. & Akbar, A. (2003). Effect of post-irradiation storage on the radiation-induced damage in onion seeds. *Asian Journal of Plant Science*, 2(9):702-707.

Amundson, S.; Do, K. & Fornace, A. (1999). Induction of stress genes by low doses of gamma rays. *Radiation Research*, 152:225-231.

Asad, N.; Asad, L.; Silva, A.; Felzenszwalb, I. & Leitão, A. (1998). Hydrogen peroxide induces protection against lethal effects of cumene hydroperoxide in *Escherichia coli* cells: An *Ahp* dependent and *OxyR* independent system? *Mutants Research* 407:253-259.

Asad, N.; Asad, L.; De Almeida, C.; Felzenszwalb, I.; Cabral-Neto, J. & Leitão, A. (2004). Several pathways of hydrogen peroxide action that damage the *E. coli* genome. *Genetic and Molecular Biology*, 27:291-303.

Assis, M.; De Mattos, J.; Caceres, M.; Dantas, F.; Asad, L.; Asad, N.; Bezerra, R.; Caldeira-de-Araujo, A. & Bernardo-Filho, M. (2002). Adaptive response to $H_2O_2$ protects against $SnCl_2$ damage: The *OxyR* system involvement. *Biochemistry*, 84:291-294.

Atanasova, P.; Hadjidekova, V. & Darroudi, F. (2005). Influence of conditioning on cell survival and initial chromosome damage in X-irradiated human cells. *Trakia Journal of Science*, 3:37-42.

Babu, S.; Akhtar, T.; Lampi, M.; Tripuranthakam, S.; Dixon, G. & Greenberg, B. (2003). Similar stress responses are elicited by copper and ultraviolet radiation in the aquatic plant *Lemna gibba*: Implication of reactive oxygen species as common signals. *Plant Cell Physiology*, 44:1320-1329.

Bassam, A. & Simon, P. (1996). Gamma irradiation-induced variation in carrots (*Daucus carota* L.). *American Society for Horticultural Science*, 121(4): 599-603.

Bassam, A.; Ayyoubi, Z. & Jawdat, D. (2000) The effect of gamma irradiation on potato microtuber production *in vitro. Plant Cell, Tissue and Organ Culture*, 10.1023/A:1006477224536.

Belyaev, I.; Spivak, I.; Kolman, A. & Harms-Ringdahl, M. (1996). Relationship between radiation induced adaptive response in human fibroblasts and changes in chromatin conformation. *Mutant Research*, 358:223-230.

Bhargava, Y. & Khalatkar, A. (2004). Improve performance of *Tectona grandis* seeds with gamma irradiation. *Acta Horticulturae*, 215:51-54.

Boreham, D. & Mitchel, R. (1991). DNA lesions that signal the induction of radioresistance and DNA repair in yeast. *Radiation Research*, 128:19-28.

Bryant, P. (1979). Evidence for inducible DNA-associated proteins formed during the development of increased resistance to radiation in *Chlamydomonas. Progress Physic and Theoretical Chemistry*, 6:305-313.

Castillo, J.; Estévez, A.; González, M.; Castillo, E. & Romero, M. (1997) Radiosensibilidad de dos variedades de papa a los rayos gamma de 60Co. *Cultivos Tropicales*, 18(1): 62-65.

Cepero, L.; Mesa, A.; García, M. & Suárez, J. (2002). Efecto de la radiación láser en semillas de *Albizia lebbeck*. I. fase de vivero. *Pastos y Forrajes*. 25 (3):181.

Chankova, S. & Bryant, P. (2001). Acceleration of DNA-double strand rejoining during the adaptive response of *Chlamydomonas reinhardtii. Radiation Biology and Radioecology*, 42(6):600-603.

Chankova, S.; Dimova, E.; Dimitrova, M. & Bryant, P. (2007). Induction of DNA double-strand breaks by zeocin in *Chlamydomonas reinhardtii* and the role of increased DNA double-strand breaks rejoining in the formation of an adaptive response. *Radiation and Environmental Biophysics*, 46:409-416.

Chauhan, S.; Nakashima, H. & Kinoshita, T. (2009). Gamma-ray induced abnormal floral mutants in sugar beet (*Beta vulgaris* L.) *The International Journal of Plant Reproductive Biology*, 1(2):137-140.

Coleman, M.; Yin, E.; Peterson, L.; Nelson, D.; Sorensen, K.; Tuckera, J. & Wyrobeka, A. (2005). Low-dose irradiation alters the transcript profiles of human lymphoblastoid cells including genes associated with cytogenetic radioadaptive response. *Radiation Research*, 164:369-382.

De la Fe, C.; Romero, M. & Castillo, E. (1996). Radiosensibilidad de semillas de papa a los rayos gamma 60Co. *Cultivos Tropicales*, 17(3): 77-80.

Ferreira, C.; Do Nascimento, V.; Ferreira, M. & Vencovscky, R. (1980). Efeito de baixas doses de radiacao fama na conservacao do poder germinativo de sementes de *Araucaria angustifolia* (Bert) O. Kuntze. *IPEF*, 21:67-82.

Franklin, F. (1974). *Abies Mill. (Fir) Gen. Tech. rep W/N. USDA*. Forest Service Pacific Northwest Forest and Range Experiment Station. USA.

Friesner, J.; Liu, B.; Culligan, K. & Britt, A. (2005). Ionizing radiation- dependent *γ-H2AX* focus formation requires ataxia telangiectasia mutated and ataxia telangiectasia mutated and *Rad3*-related. *Molecular Cell Biology*, 16:2566-2576.

Fuchs, M.; González, V.; Castroni, S.; Díaz, E. & Castro, L. (2002). Efecto de la radiación gamma sobre la diferenciación de plantas de caña de azúcar a partir de callos. *Agronomía Tropical*, 52:311-323.

Fuentes, J.; Santiago, L.; Valdés, Y.; Guerra, M.; Ramírez, I.; Prieto, E.; Rodríguez, N. & Velázquez, B. (2004). Mutation induction in zygotic embryos of avocado (*Persea americana* Mill). *Biotecnología Aplicada,* 21:82-84.

Gajendiran, N. & Jeevanram, R. (2002). Environmental radiation as the conditioning factor for the survival of yeast *Saccharomyces cerevisiae. Indian Journal of Experimental Biology,* 40:95-100.

Gehring, R. (1985). The effect of gamma radiation on *Salix nigra* Marsh. Cuttings. *Arkansas Academy of Science Proceedings,* 39:40-43.

Girigoswami, B. & Ghosh, R. (2005). Response to gammairradiation in V79 cells conditioned by repeated treatment with low doses of hydrogen peroxide. *Radiation Environmental Biophysics,* 44:131-137.

González, L.; Ramírez, R. & Camejo, Y. (2002). Estimulación del crecimiento y desarrollo de plántulas de tomate del cultivar Santa Clara a los rayos gamma del 60Co. *Alimentaria,* 331: 67-70.

González, G.; Alemán, S.; Barredo, F.; Keb, M.; Ortiz, R.; Abreu, E. & Robert, M. (2004). Una alternativa de la recuperación henequenera de Cuba, mediante el uso de técnicas biotecnológicas y moleculares. *Biotecnología Aplicada,* 21 (1): 44-49.

Gunckel, J. & Sparrow, A. (1961). Ionizing radiation: Biochemical, Physiological and Morphological aspects of their effects on plants. In: *Encyclopedia of Plant Physiology,* Ruhland,W. pp. 555-611, Springer-Verlag, Berlin.

Hillova, J. & Drasil, V. (1967). The inhibitory effect of iodoacetamide on recovery from sublethal damage in *Chlemydomonas reinbardi. International Journal of Radical Biology,* 12:201-208.

Iglesias, L.; Mora, I.; Casas, J. (2006) Morfometría, viabilidad y variabilidad de las semillas de la población de *Pinus hartwegii* del Cofre de Perote, Veracruz, México. *Cuadernos de Biodiversidad,* 19:14-22.

Iglesias, L.; Sánchez-Velásquez, L.; Tivo-Fernández, Y.; Luna-Rodríguez.; Flores-Estévez, N.; Noa-Carrazana, J.; Ruiz-Bello, C. & Moreno-Martínez, J. (2010). Efecto de radiaciones gamma en *Abies religiosa* (Kunth) Schltd. (et Cham). *Revista Chapingo. Serie Ciencias Forestales y del Ambiente,* 16(1): 5-12.

Ikram, N.; Dawar, S.; Abbas, Z. & Javed, Z. (2010). Effect of (60cobalt) gamma rays on growth and root rot diseases in mungbean (Vigna radiata l.). *Pakistan Journal of Botany,* 42(3):2165-2170.

Iqbal, J. (1980). Effects of acute gamma irradiation, developmental stages and cultivar differences on growth and yiel of wheat and sorghum plants. *Environmental and Experimental Botany,* 20(3):219-231.

Ito, D.; Walker, J.; Thompson, C.; Moroz, I.; Lin, W. & Veselits, M. (2004). Characterization of stanniocalcin 2, a novel target of the mammalian unfolded protein response with cytoprotective properties. *Molecular Cell Biology,* 24:9456–69.

Jain, S. (2001). Tissue culture-derived variation in crop improvement. *Euphytica,* 118:153-166.

Joiner, M.; Lambin, P.; Malaise, E.; Robson, T.; Arrand, J.; Skov, K. & Marples, B. (1996). Hypersensitivity to very-low single radiation doses: Its relationship to the adaptive response and induced radioresistance. *Mutation Research,* 358:171-183.

Joiner, M.; Lambin, P. & Marples, B. (1999). Adaptive response and induced resistance. *Critical Academic Science*, 322:167-75.

Jovtchev, G. & Stergios, M. (2003). Genotoxic and adaptive effect of cadmium chloride in *Hordeum vulgare* meristem cells. *Comptes Rendus Academic Bulgarian Science*, 56:75-80.

Khalatkar, A. & Bhargava Y. (1987) Effect of gamma radiations on the nuts (Seeds) of *Anacardium occidentale. ISHS Acta Horticulturae*, 215:45-50.

Kim, J.; Baek, M.; Chung, B.; Wi, S. & Kim, J. (2004). Alterations in the photosynthetic pigments and antioxidant machineries of red pepper (*Capsicum annuum* L.) seedlings from gamma-irradiated seeds. *Journal of Plant Biology*, 47: 314-321.

Kim, J.; Chung, B.; Kim, J. & Wi, S. (2005). Effects of *in planta* gamma-irradiation on growth, photosynthesis, and antioxidative capacity of red pepper (*Capsicum annuum* L.) plants. *Journal of Plant Biology*, 48(1): 47-56.

Kovács, E. & Keresztes, A. (2002). Effect of gamma and UV-B/C radiation on plant cells. *Micron*, 33: 199-210.

Kumar, G. & Singh, Y. (2010). Induced intergenomic chromosomal rearrangements in *Sesamum indicum* L. *CYTOLOGIA*, 75 (2):157-162.

Kumari, R. & Singh, Y. (1996). Effect of gamma rays and EMS on seed germination and plant survival of *Pisum sativum* L., and *Lens culinaris. Medical Neo Botanica*, 4(1): 25-29.

Ladanova, N. (1993). The ultrastructural organization of pine needles after radiation exposure. *Radiobiologia*, (33(1):25-30.

Lambin, P.; Fertil, B.; Malaise, E. & Joiner, M. (1994). Multiphasic survival curves for cells of human tumor cell lines: Induced repair or hypersensitive subpopulation? *Radiation Research*, 138:32-36.

Lanza, V.; Pretazzoli, V.; Olivieri, G.; Pascarella, G.; Panconesi, A. & Negri, R. (2005). Transcriptional response of human umbilical vein endothelial cells to low doses of ionizing radiation. *Journal of Radiation Research*, 46:265-276.

Lemus, Y.; Méndez-Natera, J.; Cedeño, J. & Otahola-Gómez, V. (2002). Radiosensibilidad de dos genotipos de frijol (*Vigna unguiculata* (L.) Walp. a radiaciones gamma. *Revista UDO Agrícola*, 2: 22-28.

Luckey, T. (1980). *Hormesis with ionizing radiations*. CRC press. Boca Raton, FLO, USA.

Luckey, T. (1991). *Radiation Hormesis*. CRC press. Boca Raton, FLO, USA.

Luckey, T. (1998). *Radiation Hormesis: Biopositive effect of Radiation. Radiation Science and Health*. CRC press. Boca Raton, FLO, USA.

Luckey, T. (2003). Radiation for health. *Radio Protection Management*, 20:13-21.

Maluszynski, M.; Ahloowalia, B. & Sigurbjörnsoon, B. (1995). Application of *in vitro* and *in vivo* mutation techniques for crop improvent. *Euphytica*, 85:303-3-15.

Manzanilla, H. (1974). *Investigaciones Epidométricas y Silvícolas en Bosques Mexicanos de Abies religiosa*. Dirección General de Información y Relaciones Públicas de la SAG. México, D. F.

Matsumoto, H.; Takahashi, A. & Ohinishi, T. (2004). Radiationinduced adaptive response and bystander effects. *Biological Science Space*, 18:247-254.

Mendez-Alvarez, S.; Leisinger, U. & Eggen, R. (1999). Adaptive responses in *Chlamydomonas reinhardtii*. *International Microbiology*, 2:15-22.

Mergen, F. & Stairs, G. (1962). Low level chronic gamma irradiation of a Pitch Pine-Oak forest-its physiological and genetical effects on sexual reproduction. *Radiation Botany*, 2(3-4):205-206.

Nwachukwu, E.; Ene, L. & Mbanaso, E. (1994). Radiation sensitivity of two ginger varieties (Zingiber officinale Rosc.) for gamma irradiation. In: *Der Tropenlandwirt, Zeltschrift für Die Lndwirtschaft In Den Tropen and Suptropen*. Jahrgang, S. 93-103. South Africa.

Olvera, E. & West, S. (1985). Aspects of germination of *Leucaena*. *Tropical Agricultural*, 62(1):68-72.

Orbovic´, V.; Cálovic´, M.; Viloria, Z.; Nielsen, B.; Gmitter, F.; Castle, W. & Grosser, J. (2008). Analysis of genetic variability in various tissue culture-derived lemon plant populations using RAPD and flow cytometry. *Euphytica*, 161:329–335.

Pajovic, S.; Joksic, G.; Pejic, S.; Kasapovic, J. & Cuttone, L. (2001). Antioxidant dose response in human blood cells exposed to different types of irradiation. *The Sciences*, 1:133-136.

Panda, K.; Patra, J. & Panda, B. (1997). Persistence of cadmium- induced adaptive response to genotixicity of maleic hydrazide and methyl mercuric chloride in root meristem cells of *Allium cepa* L.: Differential inhibition by cycloheximide and buthionine sulfoximine. *Mutation Research*, 389:129-139.

Patra, J.; Sahoo, M. & Panda, B. (2003). Persistence and prevention of aluminium- and paraquat-induced adaptive response to methyl mercuric chloride in plant cells *in vivo*. *Mutation Research*, 538:51-61.

Predieri, S. (2001). Mutation induction and tissue culture in improving fruits. *Plant Cell, Tissue and Organ Culture*, 64:185–210.

Raghava, R. & Raghava, N. (1989). Effect of gamma irradiation on fresh and dry weight of plant parts in *Physallis* L. *Geobios*, 16(6): 261-264.

Ramírez, R.; González, L.; Camejo, Y.; Zaldivar, N. & Fernández, Y. (2006) Estudio de radiosensibilidad y selección del rango de dosis estimulantes de rayos X en cuatro variedades de tomate (*Lycopersicon esculentum* Mill). *Cultivos Tropicales*, 27(1):63-67.

Ramírez-Calderón, J.; Cervantes-Santana, T.; Villaseñor-Mir, H. & López-Castañeda, C. (2003). Selección para componentes del rendimiento de grano en triticale irradidado. *Agrociencia*, 37(6): 595-603.

Rasheed, S.; Tahira, F.; Khurram, B.; Tayyab, H. & Shiekh, R. (2003). Agronomical and physiochemical characterization of somaclonal variants in Indica basmati rice. *Pakistan Journal of Biological Science*, 6:844–848.

Rieger, R.; Michaelis, A.; Jovtchev, G. & Nicolova, T. (1993). Copper sulphate and lead nitrate pretreatments trigger "adaptive responses" to the induction of chromatid aberrations by maleic hydrazide (MH) and / or JEM in *Vicia faba*, *Hordeum vulgare*, and human peripheral blood lymphocytes. *Biology Zentralbl*, 112:18-27.

Rohankhedkar, M.; Mulrooney, S.; Wedemeyer, W. & Hausinger, R. (2006). The AidB component of the *Escherichia coli* adaptive response to alkylating agents is a flavincontaining, DNA-binding protein. *Journal of Bacteriology*, 188:223-230.

Rubinelli, P.; Siripornadulsil, S.; Gao-Rubinelli, F. & Sayre, R. (2002). Cadmium- and iron-stress-inducible gene expression in the green alga *Chlamydomonas reinhardtii*: Evidence for *H43* protein function in iron assimilation. *Planta*, 215:1-13.

Rudolph, T. (1979). Effects of gamma irradiation of *Pinus banksiana* Lamb. seed as expressed by M 1 trees over 10-year period. *Environmental and Experimental Botany*, 19(2):85-92.

Sánchez-Velásquez, L. ; Pineda-López, M. & Martínez-Hernández, A. (1991). Distribución y estructura de la población de *Abies religiosa* Schl. et Cham., en el Cofre de Perote, Estado de Veracruz, México. *Acta Botánica Mexicana*, 16: 45-55.

Sánchez-Velásquez, L. & Pineda-López, M. (1993). Conservación y desarrollo rural en zonas de montaña: El manejo forestal como un elemento potencial en Veracruz. *BIOTAM*, 5:35-44.

Sangwan, H. & Singh, R. (1977). Pattern of gamma ray-induced polygenic variability in mung (*Vigna radiata* (L) Wilcrek). *Journal of Genetics*, 63(2):83-88.

Savina, N.; Dalivelya, O. & Kuzhir, T. (2003). Adaptive response to alkylating agents in the Drosophila sex-linked recessive lethal assay. *Mutation Research*, 535:195-204.

Schlade-Bartusiak, K.; Stembalska-Kozlowska, A.; Bernady, M.; Kudyba, M. & Sasiadek, M. (2002). Analysis of adaptive response to bleomycin and mitomycin C. *Mutation Research*, 513:75-81.

Scott, B. & Di Palma, J. (2006). Sparsely ionizing diagnostic and natural background radiations arelikely preventing cancer and other genomic-instability-associated diseases. *Dose-Response*, 5:230-255.

Singh, N. K. & Balyan H. S. (2009) Induced mutations in bread wheat (*Triticum aestivum* L.) CV. "Kharchia 65" for reduced plant height and improve grain quality traits. *Advances in Biological Research*, 3(5-6):215-221.

Shu, Q. & Lagoda, P. (2007) Mutation techniques for gene discovery and crop improvement. *Molecular Plant Breeding*, 5:193-195.

Scott, R. B., Belinsky, S. A., Leng, S., Lin, Y., Wilder, J. A., Damiani, L. A. (2009) Radiation-stimulated epigenetic reprogramming of adaptative-response genes in the lung: an evolutionary gift for mounting adaptative protection against lung cancer. *Dose-Response*, 7:104-131.

Scott, B. (2005). Stochastic thresholds: A novel explanation of nonlinear dose-response relationships. *Dose-Response*, 3:547-567.

Scott, B.; Belinsky, S.; Leng, S.; Lin, Y.; Wilder, J. & Damiani, L. (2009). Radiation-stimulated epigenetic reprogramming of adaptive-response genes in the lung: an evolutionary gift for mounting adaptive protection against lung cancer. *Dose-Response*, 7:104–131.

Sedgwick, B. & Lindahl, T. (2002). Recent progress on the *Ada* response for inducible repair of DNA alkylation damage. *Oncogene*, 21:8886-8894.

Seo, H.; Chung, H.; Lee, Y.; Bae, S.; Lee, S. & Lee, Y. (2006). *p27Cip/Kip* is involved in *Hsp25* or inducible *Hsp70* mediated adaptive response by low dose radiation. *Journal of Radiation Research*, 47:83-90.

Shu, Q. & Lagoda, P. (2007). Mutation techniques for gene discovery and crop improvement. *Molecular Plant Breeding*, 5:193-195.

Singh, N. & Balyan, H. (2009). Induced mutations in bread wheat (*Triticum aestivum* L.) CV. "Kharchia 65" for reduced plant height and improve grain quality traits. *Advances in Biological Research*, 3(5-6), 215-221.

Sokolov, M.; Isayenkov, S. & Sorochynskyi, B. (1998). Low-dose irradiation can modify viability characteritics of common pine (*Pinus sylvestris*) seeds. *Tsitologiya Genetika*, 32(4): 65- 71.

Solís P. (1994). *Monografía de Pinus hartwegii Lindl.* Tesis de licenciatura. División de Ciencias Forestales, Universidad Autónoma Chapingo. Chapingo, México.

Sparrow, A. & Woodwell, G. (1962). Prediction of the sensitivity of plants to chronic gamma irradiation. *Radiation Botany*, 2(1): 9-12.

Thapa, C. (1999). Effect of acute exposure of gamma rays on seed germination of *Pinus kesiya* Gord and *P. wallichiana* A.B. Jacks. *Botanica Orientalis Journal of Plant Science*, 120-121.

Tiku, A. & Kale, R. (2001). Radiomodification of glyoxalase I in the liver and spleen of mice: Adaptive response and split-dose effect. *Molecular Cell Biochemistry*, 216:79-83.

Ulsh, B.; Miller, S.; Mallory, F.; Mitchel, R.; Morrison, D. & Boreham, D. (2004). Cytogenetic dose-response and adaptive response in cells of ungulate species exposed to ionizing radiation. *Journal of Environment Radioactive*, 74:73-81.

Vaiserman, A. (2010). Hormesis, adaptive epigenetic reorganization, and implications for human health and longevity. *Dose Response*, 8(1):16–21.

Vasilevski, G. (2003). Perspectives of the application of biophysical methods in sustainable agriculture. *Bulgarian Journal of Plant Physiology*, Special Issue:179-186.

Wang, G. & Cai, L. (2000). Induction of cell proliferation hormesis and cell-survival adaptive response in mouse hematopoietic cells by whole-body low-dose radiation. *Toxicology Science*, 53:369-376.

Ward, J. (1988). DNA damage produced by ionizingradiation in mammalian cells: identities, mechanisms of formation, and repairability. *Programing Nucleic Acid Research and Molecular Biology*. 35: 96–128.

Wi, S.; Chung, B.; Kim, J.; Baek, M.; Yang, D.; Lee J. & Kim, J. (2005).Ultrastructural changes of cell organelles in Arabidopsis stem after gamma irradiation. *Journal of Plant Biology*, 48(2): 195-200.

Williams, C. & Savolaienen, O. (1996). Inbreeding Depression in Conifers: Implications for Breeding Strategy. *Foretry Science*, 42: 102–117.

Wolff, S. (1998). The adaptive response in radiobiology: evolving insights and implications. *Environ Health Perspective*, 106(1):277–283.

Yamaguchi, H.; Shimizu, A.; Degi, K. & Morishita T. (2008). Effect of dose and dose rate of gamma ray irradiation on mutation induction and nuclear DNA content in chrysanthemum. *Breeding Science*, 58:331-335.

Yan, G.; Hua, Z.; Du, G. & Chen, J. (2006). Adaptive response of Bacillus sp. F26 to hydrogen peroxide and menadione. *Current Microbiology*, 52:238-242.

Zaka, R.; Chenal, C. & Misset, M. (2004). Effects of low doses of short-term gamma irradiation on growth all development through two generations of *Pisum sativum*. *Science Total Environment*, 320:121-129.

Zhou, H.; Randers-Pehrson, G.; Waldren, C. & Hei, T. (2004). Radiation- induced bystander effect and adaptive response in mammalian cells. *Advance Space Research*, 34:1368-1372.

14

# Changes in Selected Properties of Wood Caused by Gamma Radiation

Radovan Despot[1], Marin Hasan[1], Andreas Otto Rapp[2], Christian Brischke[2], Miha Humar[3], Christian Robert Welzbacher[2] and Dušan Ražem[4]

[1]*Faculty of Forestry, University of Zagreb,*
[2]*Leibniz Universität Hannover, Faculty of Architecture and Landscape Sciences, Hannover,*
[3]*Biotechnical Faculty, University of Ljubljana,*
[4]*Ruđer Bošković Institute, Zagreb,*
[1,4]*Croatia*
[2]*Germany*
[3]*Slovenia*

## 1. Introduction

Wood as a natural organic material is susceptible to biodeterioration by insects, fungi, and bacteria. There is almost no wooden artefact or old wooden element, which is not infected and not damaged, at least partly, by wood-destroying organisms. As ancient wood and wooden artefacts are invaluable, their appropriate restoration is of particular importance. The first step in restoration is the detection and quantification of wood pests and decay and accordingly, under certain conditions, the disinfestation of ancient wood and old wooden artefacts could be needed.

Besides disinfestation prior to restoration, sterilization of wood is applied for testing the resistance of wood and wooden products against wood-destroying organisms. For both purposes, restoration and resistance testing, gamma radiation is considered as a suitable decontamination method.

## 2. Gamma radiation as a sterilisation method

Gamma radiation, as a high energy, ionising electromagnetic radiation, easily penetrates through wooden objects. It is known to be very effective in the context of disinfestation of wooden artefacts (Unger *et al.*, 2001; Katušin-Ražem *et al.*, 2009; Fairand and Ražem, 2010) but also for wood sterilisation (Sharman and Smith, 1970; Shuler, 1971; Freitag and Morrell, 1998; Pratt *et al.*, 1999; Severiano *et al.*, 2010). In contrary to alpha and beta rays, which penetrate only very thin layers, gamma radiation fully penetrates wooden objects (Fengel and Wegener, 1989; Tišler and Medved, 1997). The energy-rich gamma rays modify molecular structures and lead to unexpected function of living cells or to their death.

### 2.1 Doses of gamma radiation necessary for wood sterilisation

According to Kunstadt (1998), insects do not withstand doses between 0.7 and 1.3 kGy, while elimination of fungi requires significantly higher doses. Unger *et al.* (2001) mentioned doses between 0.25 and 3 kGy to be adequate for extermination of wood-destroying insects, depending on species and developmental stage. Extermination of wood decay fungi in wood usually requires higher doses, ranging from 2 to 18 kGy depending on the fungus species. Mycelium of the fungus *Serpula lacrymans* can be killed off with 2 - 3 kGy, but it can be reduced to 0.5 kGy if the temperature rises to 50 °C. Unger *et al.* (2001) also stated that the bacteria elimination requires doses of 3 - 15 kGy. Lester *et al.* (2000) sterilised New Zealand soft woods including radiata pine (*Pinus radiata*) against Huhu beetle larvae (*Prionoplus reticularis*) and concluded that doses between 2.5 and 3.7 kGy are enough to control wood destroying insects and that moisture content of wood during irradiation has an important role in radiation dose determination. Magaudda *et al.* (2001) stated that the dose of 10 kGy is sufficient for practical disinfestations of Whatman paper-destroying insects. Freitag and Morrell (1998) reported on gamma radiation doses around 15 kGy to be adequate to combat pests in wood (predominantly insects). They have also stated that the doses for extermination of *Xylophagous* microorganisms in wood are much higher than for sterilisation of other materials. Csupor *et al.* (2000) irradiated wood decayed by *Xylophagous* fungi in the range from 2 to 1400 kGy, and they concluded that 12 kGy is sufficient for safe sterilisation of wood against fungi. The European Standard EN 113 (CEN, 1996) requires doses between 25 and 50 kGy for wood sterilisation in lab testing procedures.

The treatment time depends on the power of the irradiation source, and there is no significant difference if the wood was irradiated with a weaker source for a longer time or with a stronger source for a shorter time (Unger *et al.*, 2001; Hasan, 2006 mentioned Ražem, 2005). The important quantity is the absorbed dose (the amount of absorbed energy per mass unit) which irradiated substance receives (Fairand and Ražem, 2010). On the contrary Curling and Winandy (2008) reported that dose rate and total dose of gamma radiation differently affect both, bending strength and some chemical components in tested wood.

## 3. Other applications of gamma radiation to wood

One of the interesting areas of application of gamma rays and X-rays is in non-destructive analysis of density and water content and their distribution in solid wood, and the wood-based materials, i.e. wood panels (Davis *et al.* 1993, Lu and Lam, 1999). Karsulovič *et al.* (2002) used gamma irradiation in their studies for non-destructive detection of decay and other defects in logs. In this way, the quality of the timber can be to some extent predicted prior to mechanical processing. Because of extremely high energy of gamma radiation, it penetrates through the entire cross section of wood. Consequently, gamma rays can be used as a catalyst for polymerization of monomers in monomer-impregnated wood as well as for chemical modification of wood to create wood-plastic composites - WPC (Chawla, 1985; Sheikh and Afshar Taromi, 1993; Bakraji *et al.* 2001; Bakraji *et al.*, 2002). Struszczyk *et al.* (2004) mentioned three possible uses of gamma radiation:

a. as pre-treatment in the chemical modification of cellulose;
b. as the initiator of the catalytic polymerization of monomers in the cellulose chain and
c. as pre-treatment for further chemical processing of cellulose in order to improve its solubility.

Klimentov and Bysotskaia (1979); Chawla (1985) and Šimkovic *et al.* (1991) reported that gamma irradiation as a pre-treatment for the "softening" of wood prior to chemical processing, can significantly reduce the use of chemicals. Oldham *et al.* (1990) used gamma radiographic technique for analysing wood destruction by fire. They determined the efficiency of fire retardants by measuring the difference in absorbed gamma radiation energy in wood protected by these retardants before and after burning the wood. Bogner (1993) and Bogner *et al.* (1997) used different surface activators in order to achieve greater adhesion of adhesives to wood. Among other used activators, gamma radiation at different doses was also used. The results show that after treatment of wood by gamma radiation in doses range of 0 and 100 kGy, glue improves adhesion to wood and thus increases the strength of glued bonds. The first investigations of the influence of gamma radiation on lignocellulose materials, in terms of increasing the solubility of insoluble high-polymerized sugars such as cellulose, were performed by Klimentov and Bysotskaia (1979) and Klimentov *et al.* (1981).

## 4. The influence of gamma radiation on wood

Sterilisation by gamma radiation is very easy, fast and effective, but at doses higher than disinfestation doses it changes the molecular structures not only in wood decaying organisms but also in wood cell walls. Although Severiano *et al.* (2010) reported no influence of gamma radiation on some wood physical, thermal and mechanical properties in the radiation dose range between 25 and 100 kGy, the majority of other studies reported significant influence of gamma radiation on wood properties.

### 4.1 Chemical properties of gamma irradiated wood

The random break-up of cellulose chains in gamma irradiated wood is a typical reaction (Seaman *et al.*, 1952; Kenaga and Cowling, 1959). Seifert (1964), Tabirih *et al.* (1977) and Cutter *et al.* (1980) found that the holocellulose portion of cell walls was degraded by gamma irradiation. After exposure of cypress wood (*Pseudotsuga mensziessi*) and tulipwood (*Liriodendron tulipifera*) to gamma rays, Lhoneaux *et al.* (1984) confirmed the occurrence of ultra structural changes in the cell walls of tested wood species. Following the above research results, Fengel and Wegener (1989) reported that gamma irradiation changes the anatomical and chemical structure of wood, but also the physical and mechanical properties. Seifert (1964) also found that small doses of gamma radiation lead to a destruction of hemicelluloses' pentose creating new compounds and new chemical bonds. Chawla (1985) reported that up to the dose of 500 kGy the increase of wood solubility occurs primarily because of hemicelluloses depolymerisation and destruction. The influence of gamma radiation was also investigated on compounds of cellulose-acetate and cellulose nitrate (Fadel and Kasim, 1977; Zamani *et al.*, 1981; Subrahmanyam *et al.*, 1998). They came to the conclusion that also modified cellulose chains broke up. Seifert (1964) came to the conclusion that, an average increase of 25 kGy of gamma radiation caused a loss of 1 % cellulose in the dose range of 0 - 1 MGy. This proportional loss in cellulose crystallinity with increasing radiation dose was confirmed by Zamani *et al.* (1981) in a narrower interval of doses from 0 to 0.5 MGy. Fedel and Kasim (1977) successfully used cellulose-acetate as an indicator of the amount of absorbed energy of gamma irradiated objects, which reaffirms the proportionality between the reduction of cellulose crystallinity and gamma radiation dose.

Summary of data collected from the literature on chemical changes of wood caused by gamma radiation are presented in Table 1 and in Figure 1.

| Characteristics of compounds | Change | Dose | Reference |
|---|---|---|---|
| Degree of crystallinity of cellulose | 50 % reduction | 0,95×$10^6$ kGy | Cutter & McGinnes (1980) |
| | unchanged | up to 300 kGy | Tsutomu *et al.* (1977) |
| | rapid reduction | above 1000 kGy | Tsutomu *et al.* (1977) |
| | 100 % reduction | 1,9×$10^6$ kGy | Cutter & McGinnes (1980) |
| Degree of polymerisation of cellulose | cellulose cross linking | up to 1×$10^{-3}$ kGy | Seifert (1964) |
| | strong decrease of DP | above 10 kGy | Fengel & Wegener (1989) |
| Cellulose | no change | up to 31,6 kGy | Seifert (1964) |
| | 6 to 12 % degradation | 31,6 kGy | Seifert (1964) |
| | 82 % degradation | 1778,3 kGy | Seifert (1964) |
| | complete degradation | 6,55×$10^3$ kGy | Fengel & Wegener (1989) |
| Lignin | no change | up to 31,6 kGy | Seifert (1964) |
| | 10 % degradation | 1778,3 kGy | Seifert (1964) |
| | 15 % degradation | 19×$10^3$ kGy | Cutter & McGinnes (1980)<br>Tabirih *et al.* (1977) |

Table 1. Summary of data on the influence of various doses of gamma radiation onto main wood compounds.

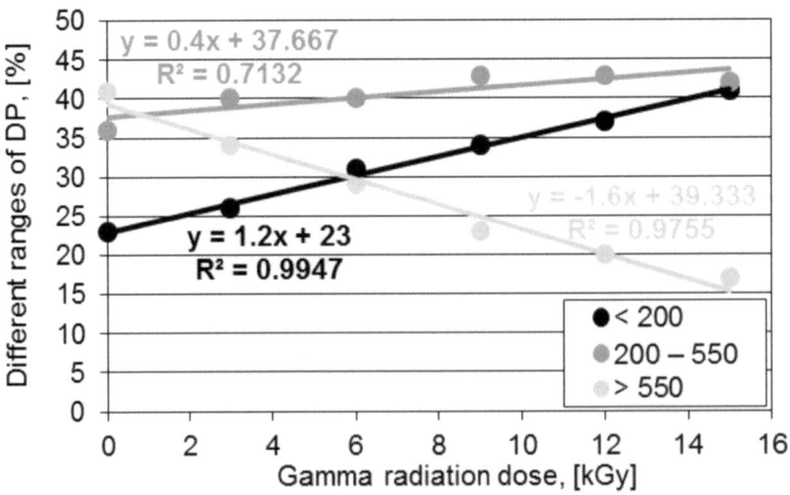


Fig. 1. Different ranges of polymerisation degree (DP) of cellulose depending on gamma radiation dose (according to Struszczyk *et al.* (2004)).

Seifert (1964) found an increasing amount of free radicals in wood after gamma radiation, which could be effective after irradiation for wood modification in terms of their repolymerisation. Fengel and Wegener (1989) and Tišler and Medved (1997) discussed that the cellulose chains continue breaking down during 100 days after finishing the gamma treatment. Curling and Winandy (2008) gamma irradiated southern pine sapwood specimens (*Pinus* spp.) at a range of nine total irradiation doses (from a $^{60}Co$ source) applied at two different dose rates: 8.5 kGy/h (total doses of 15.0, 25.0, and 50.0 kGy) and 16.9 kGy/h (total doses of 15.0, 20.0, 25.0, 37.5, 50.0 and 75.0 kGy). Gamma radiation at higher (16.9 kGy/h) dose rate clearly had more negative effect on Mason lignin levels than did either the steam sterilization or gamma radiation at lower (8.5 kGy/h) dose rate. Gamma radiation at the 16.9 kGy/h dose rate reduced Klason lignin content from 29.5 % to 28.1 and to 27.5 %, but seemed to have little on-going effect related to dose accumulation. They also found that galactans, which are side-chain constituents of the hemicelluloses, are affected sooner and to a greater extent than xylans or mannans that represent the primary backbone of the hemicellulose polymers. Furthermore arabinan was probably unaffected by various radiation regimes. The authors suspected that the β-(1→3) linkage of the arabinan-xylan bond was less affected by radiation than the β-(1→6) linkage of the galactan-mannan bond.

### 4.1.1 Total amount of water-soluble carbohydrates, (TSC)

Hasan (2006) and Despot *et al.* (2007; 2008) measured the total amount of water-soluble carbohydrates (TSC) according to Rapp *et al.* (2003). They used the following procedure: Five oven-dried specimens of each group were milled together for 60 s in a heavy vibratory disc mill (Herzog Maschinenfabrik, Osnabrück, Germany). 400 mg of wood powder were mixed up with one drop of detergent and 20 ml distilled water in 50 ml flasks. Afterwards, the flasks were shaken at 20 °C for 60 min with a vibration of 120 $min^{-1}$ and filtered through glass filter paper. The filtrates were diluted with distilled water in proportion 1:5.

The reagent for optical spectrophotometry was prepared by dissolving 2 g of dihydroxytoluol (Orcin) in 1l of 97 % sulphuric acid. For analysing the carbohydrate content 1 ml of dilution was mixed with 2 ml reagent in a test-tube and heated for 15 min at 100 °C in a Thermo-block (Merck TR205). After cooling to room temperature, the dilution was analysed at 540 nm with a Merck SQ 115 filter-photometer. The TSC was calculated after calibration with a glucose standard between 20 and 200 ppm leading to extinction between 0.12 and 1.32.

Hasan (2006) and Despot *et al.* (2007; 2008) found a strong influence of gamma irradiation on TSC. As gamma radiation causes destruction of cellulose chains and the resulting smaller cellulose fragments are easily soluble, increasing of TSC with increasing radiation dose was expected. They found no significant influence of time after gamma treatment on TSC (Figures 2 and 3).

The sensitiveness of the TSC content to reveal changes in the carbohydrate structure of wooden materials was shown in earlier studies with thermally modified timber (TMT) where TSC was significantly reduced with ascending heat treatment temperature and increasing heat-induced loss of mass (Welzbacher *et al.*, 2004; 2009). Thus, the extraction of soluble carbohydrates appeared to be a sensitive method to display degradation of carbohydrates (Hasan, 2006; Despot *et al.*, 2007; 2008) and is therefore also considered as a suitable tool to characterize the treatment intensity of gamma radiation.

According to Fengel and Wegener (1989) and Tišler and Medved (1997), the highest TSC should be observed 150 days after gamma radiation. In contrast to mass loss by leaching, TSC did not decrease over time. Accordingly, the radical recombination is probably limited to higher mol mass fragments but not to lower fragments, which appear as TSC.

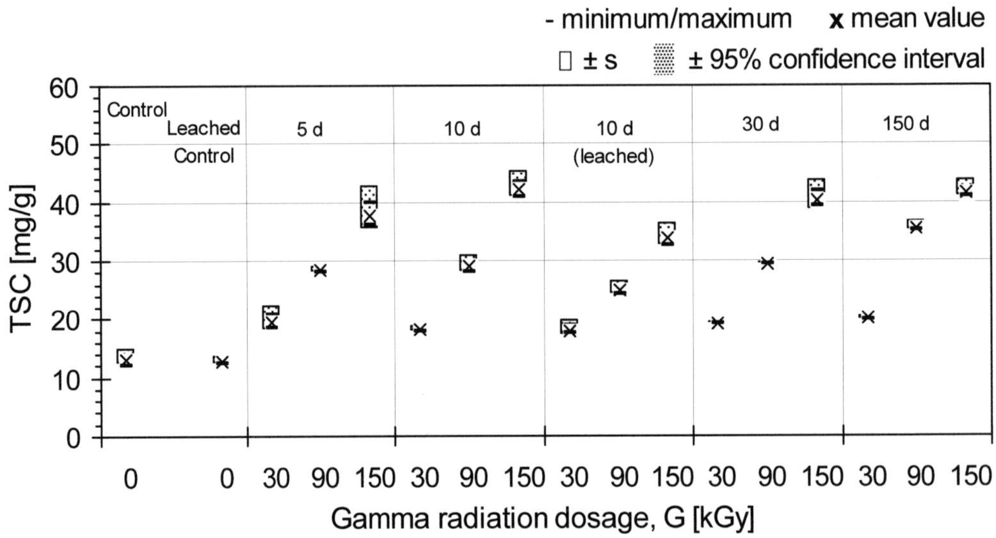


Fig. 2. Total amount of water-soluble carbohydrates (TSC) of leached and non-leached specimens irradiated with different gamma doses for different time intervals after gamma irradiation; n = 3 (Despot *et al.*, 2007).

After leaching, irradiated specimens still had significantly greater TSC than non-irradiated and non-leached controls. The TSC relative to the controls ($TSC_r$) was calculated for each radiation dose as a ratio of mean TSC of gamma irradiated specimens and the mean TSC of non-irradiated and non-leached controls. $TSC_r$ of non-leached and leached specimens showed linear correlation with the radiation dose (Figure 3).

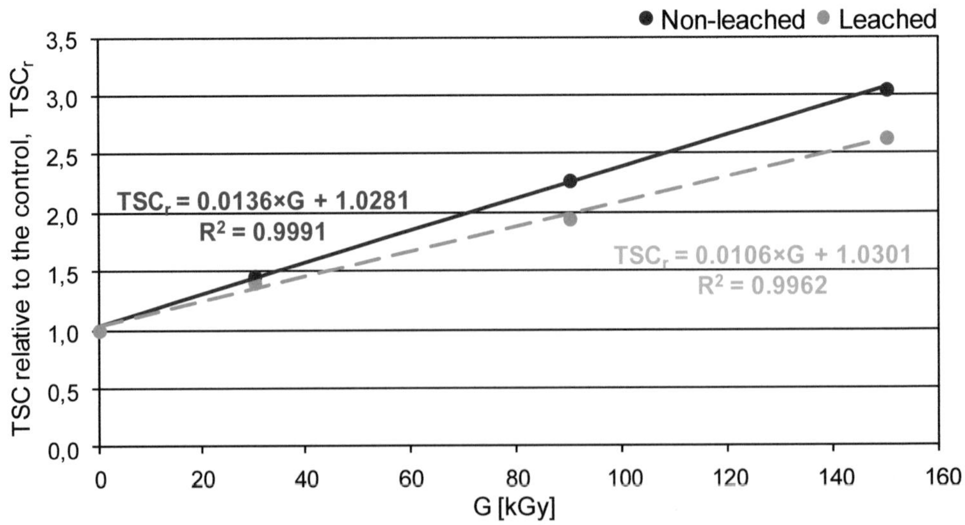


Fig. 3. Correlation between total amount of water-soluble carbohydrates relative to controls ($TSC_r$) and gamma radiation dose (G) for non-leached and leached specimens (Despot *et al.*, 2007).

All these studies lead to the conclusion that due to breaking the chains of cellulose in wood cell wall, there must be some changes in physical and mechanical properties of gamma-irradiated wood. These changes should mostly influence the hygroscopicity (Fengel and Wegener, 1989) and tension strength of gamma-irradiated wood, because it becomes more brittle (Divos and Bejo, 2005). Furthermore, gamma irradiated wood is more susceptible to chemical and enzymatic degradation (Seifert, 1964; Klimentov and Bysotskaia, 1979; Klimentov *et al.*, 1981; Ardica *et al.*, 1984; Šimkovic *et al.*, 1991; Magaudda *et al.*, 2001; Struszczyk *et al.*, 2004; Despot *et al.*, 2006; Hasan, 2006 and Hasan *et al.*, 2006a; 2008).

### 4.2 Physical properties of gamma irradiated wood

Gamma radiation lead to significant colour changes of wood. With increasing radiation dose the darkening of the specimens increased as can be seen from Figure 4.

Fig. 4. Colour change of specimens used for mass loss determination, **3AA** - control group, **3AB** - group irradiated with 30 kGy, **3AC** - group irradiated with 90 kGy and **3AD** - group of specimens irradiated with 150 kGy.

#### 4.2.1 Decrease in mass caused by gamma irradiation

On the question whether there is any loss of mass or density of the wood due to gamma radiation, Loos (1962) in his studies noted no significant changes in the density of wood. Seifert (1964) stated the possibility of negligible loss of $CO_2$ from wood due to radiation induced chemical reactions. However, wood irradiated in the presence of air might absorb atmospheric nitrogen in small quantities (Seifert, 1964). Tsutomu *et al.* (1977) reported on a a very small effect of gamma radiation on specific gravity of wood and cellulose. Curling and Winandy (2008) reported that density of southern pine sapwood stayed unchanged by any tested level of irradiation dose or dose rate of gamma radiation. Hasan (2006) and Despot *et al.* (2007) measured oven dry mass of Scots pine (*Pinus sylvestris*) specimens. They determined no statistically significant changes in mass before and after irradiation at a confidence interval of 95 % although specimens were gamma irradiated in plastic bags with the presence of air (Figure 5).

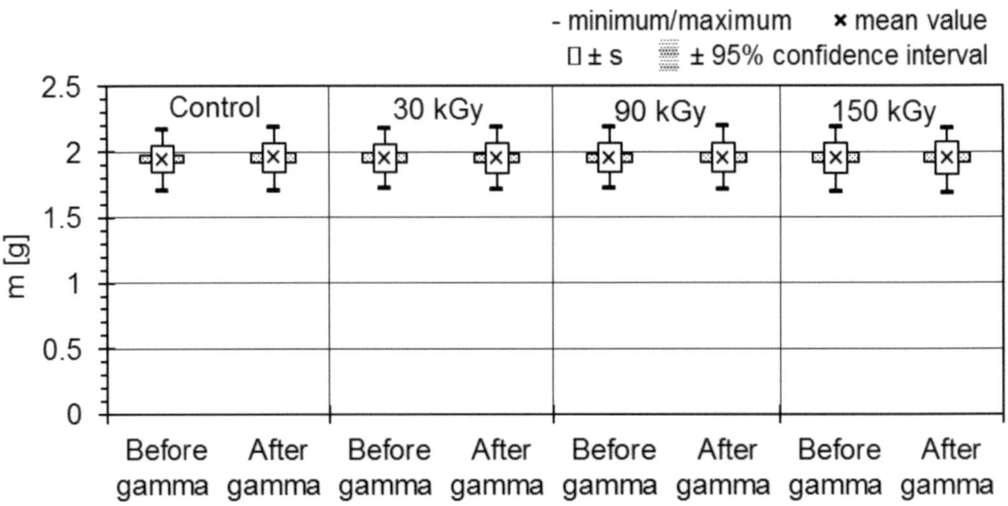


Fig. 5. Comparison of dry mass of specimens before and after gamma radiation.

#### 4.2.2 Decrease in mass by leaching, (dm)

Leaching in water of Scots pine sapwood specimens caused a significant decrease in mass (dm). A strong linear correlation was found between gamma radiation doses (G) and dm (Despot *et al.*, 2007; Figure 6). This might be explained by random break up of cellulose

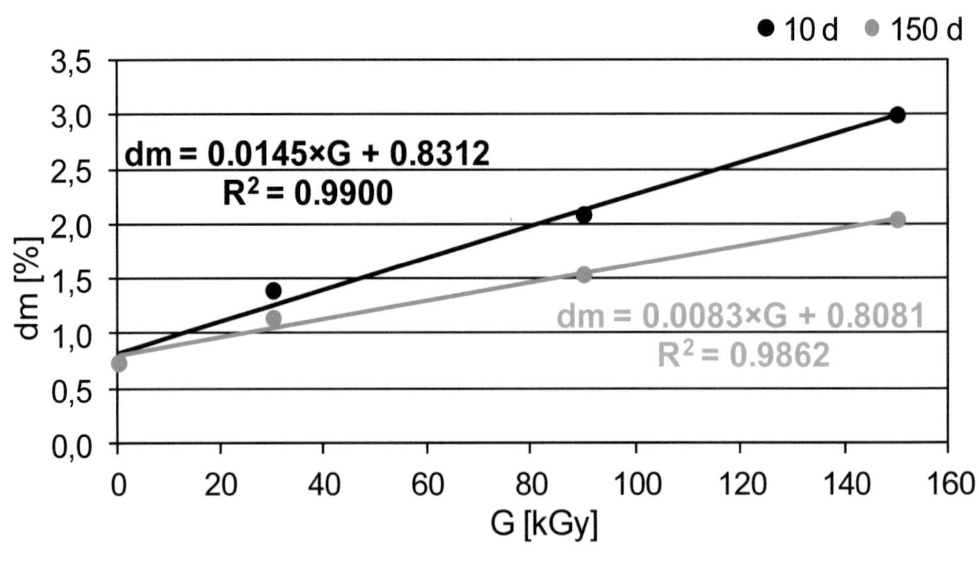


Fig. 6. Correlation between decrease in mass (dm) by leaching and gamma radiation dose (G) for 10 and 150 days after gamma treatment.

chains and the solubility of smaller cellulose fragments in water (Seifert, 1964; Tabirih *et al.*, 1977; Šimkovic *et al.*, 1991). Despot *et al.* (2007) reported that for 10 days after gamma treatment dm was significantly higher than after 150 days for each applied dose (Table 2). This finding is in contrast to the results of Fengel and Wegener (1989) and Tišler and Medved (1997), but it can easily be explained by repolymerisation reactions between the free radicals caused by irradiation (Seifert, 1964; Bogner *et al.*, 1997).

| **Time after gamma treatment [days]** | **10** | | | **150** | | |
|---|---|---|---|---|---|---|
| Gamma radiation dose, G [kGy] | 30 | 90 | 150 | 30 | 90 | 150 |
| Minimum dm [%] | 1.1773 | 1.6206 | 2.2598 | 0.9146 | 1.2781 | 1.8308 |
| Mean dm [%] | 1.3999 | 2.0900 | 3.0026 | 1.1471 | 1.5423 | 2.0438 |
| Maximum dm [%] | 1.9200 | 3.0700 | 4.4500 | 1.2916 | 1.8732 | 2.3433 |
| Standard deviation [%] | 0.2675 | 0.5449 | 0.9213 | 0.1126 | 0.1741 | 0.1805 |
| Coefficient of variation [%] | 19.11 | 26.07 | 30.68 | 9.82 | 11.29 | 8.83 |
| T - value | 6.2780 | 8.3589 | 9.0220 | 4.8216 | 8.8164 | 14.2067 |
| Probability of mistake [%] | 0.0020 | 0.0001 | 0.00 | 0.0271 | 0.00 | 0.00 |
| Signification degree | +++ | +++ | +++ | +++ | +++ | +++ |

Table 2. Results of statistical analysis on differences of decrease in mass (dm) by leaching between non-irradiated controls and with different doses irradiated specimens 10 and 150 days after the gamma treatment for 95 % confidence interval; n=15; (+ significant, ++ high significant, +++ highest significant).

#### 4.2.3 Maximum swelling, ($\alpha_{MAX}$)

Panshin and de Zeeuw (1980) stated that the arrangement of the cellulose crystals in microfibrils can be observed by the existence of amorphous zones along the microfibril length, in which the crystallinity is interrupted. These zones allow the penetration of chemicals into the microfibrils. Furthermore, the gamma radiation caused break-up of cellulose to shorter chains, which are water-soluble, and it most likely leads to an "opening of additional microcracks", in which water molecules can easily penetrate. Consequently it is to expect that gamma irradiated wood swells faster but also more than non-irradiated wood. Results by Hasan (2006), Hasan *et al.* (2006b) and by Despot *et al.* (2007; 2008) showed no significant influence of radiation dose on maximum swelling, neither in radial ($\alpha_{R\,MAX}$) nor in tangential ($\alpha_{T\,MAX}$) direction (Figure 7). Leached irradiated wood showed significantly higher $\alpha_{R\,MAX}$ compared to non-leached wood. With increasing radiation dose the difference in $\alpha_{R\,MAX}$ between leached and non-leached specimens became more significant (Despot *et al.*, 2007). Obviously, cellulose chains and lignin (Curling and Winandy, 2008) were affected by gamma radiation and the "plywood effect" of the wood rays has been reduced and $\alpha_{R\,MAX}$ increased. In contrast, $\alpha_{T\,MAX}$ of leached irradiated wood significantly decreased. It can be explained by the decreasing of tangential vessel wall thickness, ray cell double-wall thickness and latewood fibre double-wall thickness, as described by Tabirih *et al.* (1977) and Cutter *et al.* (1980). Despot *et al.* (2007; 2008) found no significant influence of time after gamma radiation on $\alpha_{MAX}$.

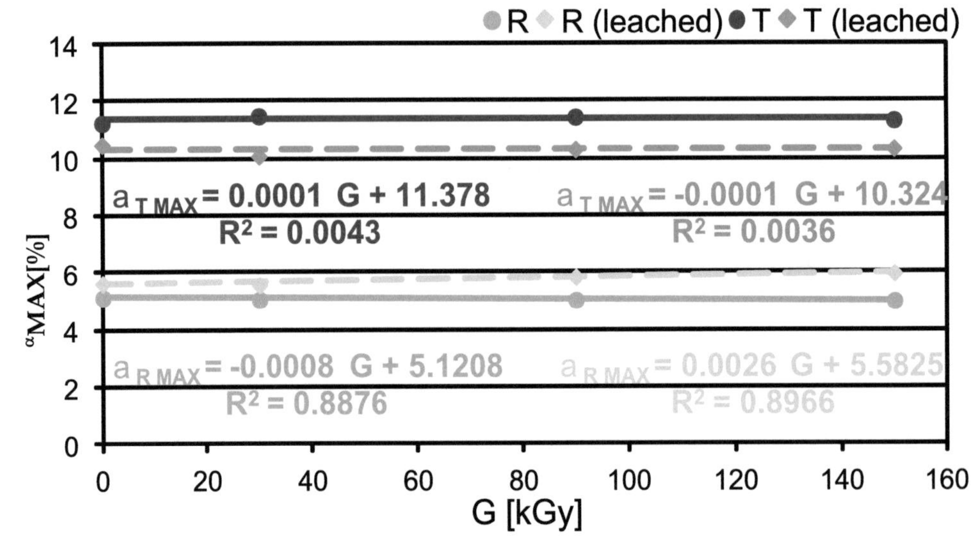


Fig. 7. Correlation between maximum swelling ($\alpha_{MAX}$) in radial (R) and tangential (T) direction of non-leached and leached specimens and gamma radiation dose (G).

#### 4.2.4 Maximum moisture content ($MC_{MAX}$)

Maximum moisture content, ($MC_{MAX}$) of each irradiated group of specimens was equal or higher than the MC of controls (Hasan, 2006; Hasan *et al.*, 2008). It is well known, that glucose and other simple sugars make wood considerably more hygroscopic (Fengel and Wegener, 1989). No significant influence of gamma radiation dose on $MC_{MAX}$ was found (Figure 8). Radiation dose did not have significant influence on maximum MC, the same as

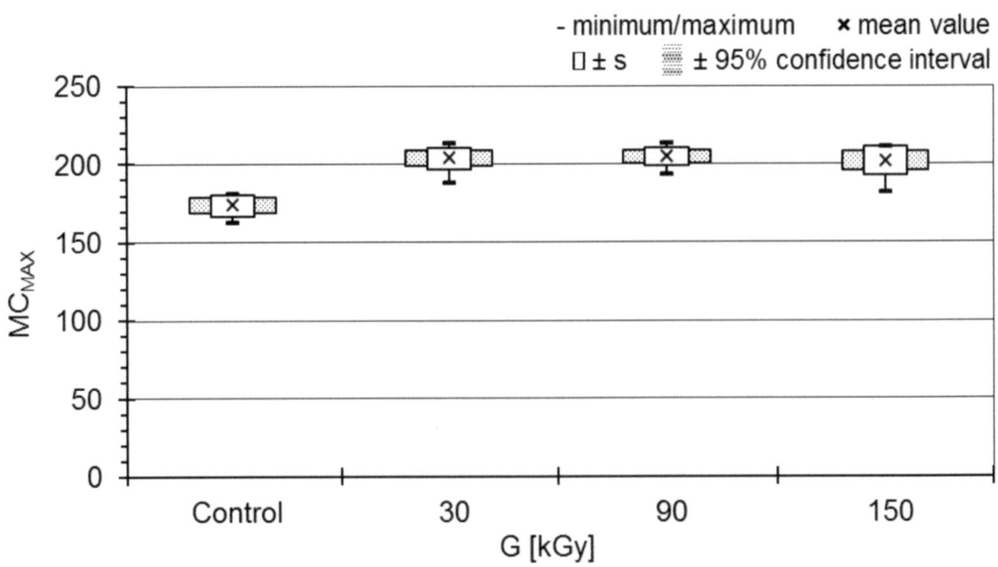


Fig. 8. Correlation between maximal moisture content ($MC_{MAX}$) and gamma radiation dose (G).

on maximum swelling (Despot *et al.*, 2007). It is probable that specimens irradiated with higher radiation dose reaches faster the $MC_{MAX}$.

### 4.3 Mechanical properties of gamma irradiated wood

Loos (1962), Ifju (1964), Shuler *et al.* (1975), El-Osta *et al.* (1985), Hasan (2006), Hasan *et al.* (2006b) and Despot *et al.* (2007; 2008) verified that gamma-radiation-induced depolymerisation causes a significant decrease in wood strength. Curling and Winandy (2008) reported that dose rate and total dose of gamma radiation differently affected bending strength of pine wood (*Pinus* sp.). They found that cumulative duration of radiation exposure period appeared to be more critical than total dose in determining overall strength loss. When total gamma radiation dose was directly compared at or near the critical doses required to achieve sterilization, shorter exposures using higher dose rates affected strength less than longer exposures using lower dose rates (Curling and Winandy, 2008). Loos (1962) comes to 1 kGy as a threshold - wood toughness had a tendency of linear increase, followed by a progressive decrease. Gamma radiation depolymerised wood had significantly reduced tensile strength, although early wood showed an initial increase in tensile strength (Ifju, 1964). Tsutomu *et al.* (1977) reported on considerable decrease in strength of wood with increasing irradiation dosage, depending remarkably on loading modes. Testing the tensile and compression strengths parallel to the grain of gamma-irradiated wood, El-Osta *et al.* (1985) have found a slight increase in tensile and compression strength till the dose of 1.4 kGy. Increasing the gamma radiation dose, tensile and compression strengths constantly decreased. Decrease in tensile strength was more pronounced than that in compression strength. Researching dynamic modulus of elasticity of spruce wood (*Picea* sp.) Shuler (1971) noted the increase of the modulus in the range of radiation doses from 0 to 1 kGy while further increase of radiation dose decreased dynamic modulus of elasticity.

Shuler *et al.* (1975) measured bending strength of American elm (*Ulmus americana*), which was exposed to gamma radiation during growth. The results showed that small doses of radiation up to 0.22 kGy increased bending strength on average by 30 %, and further increase in dose lead to a continuous and significant decrease of bending strength. Measuring the bending strength, Shuler *et al.* (1975) noted the increase of the modulus of elasticity in the range of radiation doses from 0 to 0.22 kGy while further increase of radiation dose lead to the decrease of the modulus of elasticity and decrease of the proportionality limit.

Csupor *et al.* (2000) irradiated wood in the range of 2 - 1400 kGy and concluded that a dose of 12 kGy was sufficient for wood sterilization while modulus of elasticity at this dose was not significantly reduced (0.2 %). Tests have also shown that an initially higher modulus of elasticity decreased more rapidly with increasing radiation dose. Divos and Bejo (2005) reported on a steady and a continuous decrease of modulus of elasticity (MOE) for all tested wood species in the range of gamma radiation dose from 130 to 770 kGy. Results also showed a difference in the intensity decrease of MOE between the wood species, and wood density had significant influence on decrease intensity of the MOE. Lower densities resulted in faster decrease in MOE. Interpolating the obtained data Divos and Bejo (2005) concluded that the dose sufficient for wood sterilization not significantly decreased MOE. In the graphs (Figure 9) an inversely proportional correlation between the gamma radiation dose and dynamic MOE was visible with high degrees of determination.

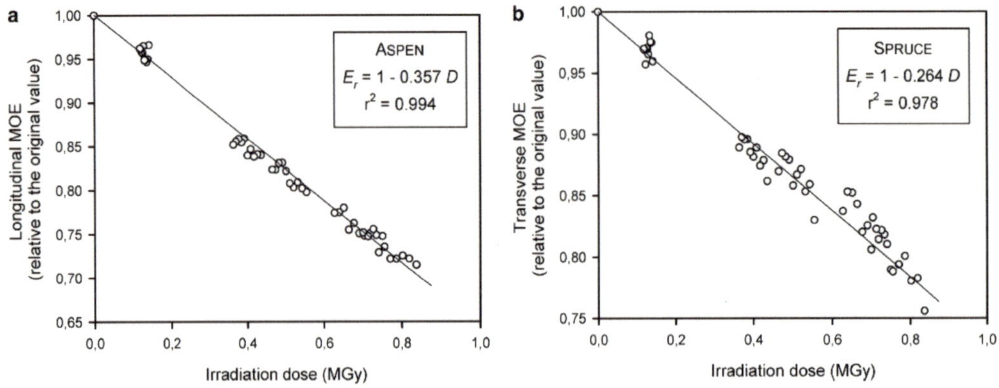


Fig. 9. Correlation between dynamic MOE and gamma radiation dose: a) longitudinal MOE of aspen; b) transverse MOE of spruce (taken from: Divos and Bejo, 2005).

### 4.3.1 Structural integrity of gamma irradiated wood

The High-energy multiple impact (HEMI) test has been developed to characterise the effect of thermal and chemical modification procedures as well as fungal and bacterial decay on the structural integrity of wood. The development and optimization of the HEMI-test have been described by Brischke *et al.* (2006a). In studies on the effect of gamma radiation onto structural integrity of gamma irradiated wood (Hasan, 2006; Hasan *et al.*, 2006b; Despot *et al.*, 2007; 2008) the following procedure was applied:

Five of ten specimens were placed in the bowl (140 mm inner diameter) of a heavy vibratory impact ball mill, together with one steel ball of 35 mm Ø, 3 of 12 mm Ø, and 3 of 6 mm Ø. The bowl was shaken for 60 s at a rotary frequency of 23.3 $s^{-1}$ and a stroke of 12 mm. This crushing procedure was repeated on another five specimens (Figure 10). The fragments of the ten specimens were fractionated with slit screens on an orbital movement shaker (amplitude: 25 mm, rotary frequency: 250 $min^{-1}$, duration: 10 min).

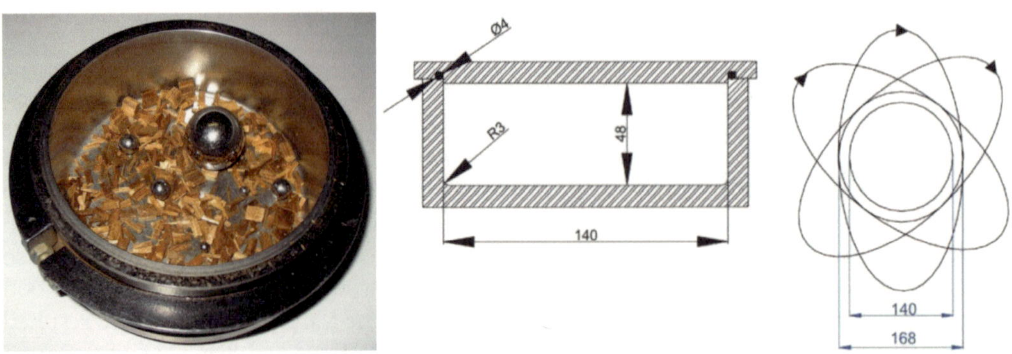


Fig. 10. Experimental set up of the High-energy multiple impact (HEMI) test. Left: Bowl with steel balls of different sizes and wood fragments. Right: Cutting scheme of the steel bowl and movement schedule.

The following five fractions were separated and weighed: Fraction 1 (F1, > 5 mm), Fraction 2 (F2, 3-5 mm), Fraction 3 (F3, 2-3 mm), Fraction 4 (F4, 1-2 mm) and Fraction 5 (F5, < 1 mm). The following values were calculated:

The degree of integrity *I*, which is the ratio of the mass of the 10 biggest fragments to the mass of all fractions $m_{all}$ after crushing:

$$I = \frac{m_{10\ biggest\ fragments}}{m_{all}} \times 100[\%] . \quad (1)$$

The fine fraction F5 is the ratio of the mass of fraction 5 (<1 mm) to the mass of all fractions $m_{all}$ multiplied by 100. Finally, the resistance to impact milling (RIM) was calculated from *I* and F5 as follows:

$$RIM = \frac{I - (3 * F5) + 300}{4}[\%] . \quad (2)$$

### 4.3.2 Resistance to impact milling (RIM)

The resistance to impact milling (RIM) decreased significantly with increasing radiation dose. Time after gamma treatment had no significant influence on RIM neither of non-leached nor of leached specimens (Despot *et al.* 2007; Figure 11).

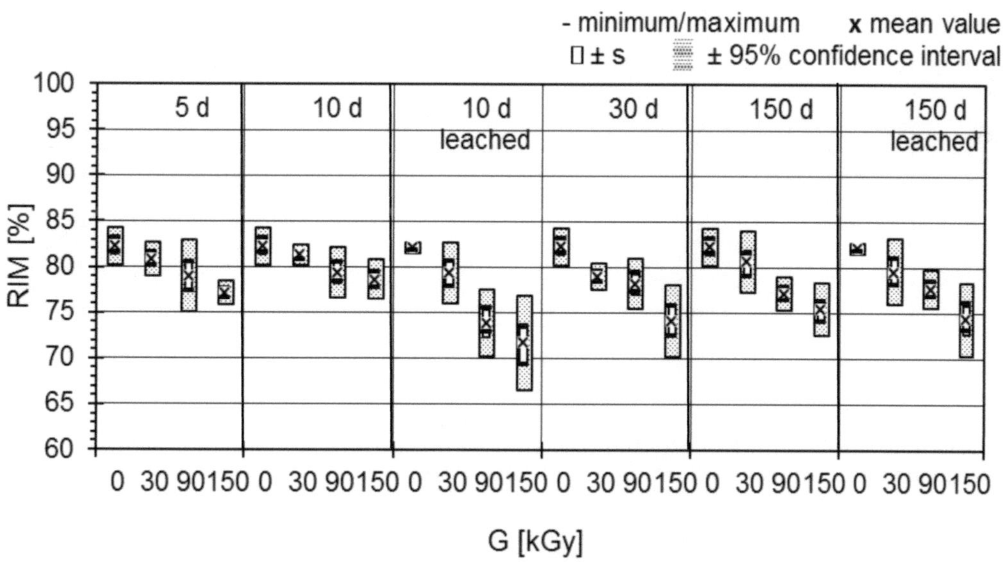


Fig. 11. Resistance to impact milling (RIM) of leached and non-leached specimens for different gamma doses (G) and different periods after gamma treatment (n = 3 × 10 specimens).

Despot et al. (2007) and Hasan et al. (2006b) found linear dependence between RIM and G for both, non-leached and leached specimens. The elapsed time after gamma radiation did not influence the linear correlation between RIM and G (Table 3).

| Leaching procedure | Time after gamma treatment [days] | Fitting curve equation | Regression coefficient, $R^2$ |
|---|---|---|---|
| non-leached | 5 | RIM = -0.0332×G + 81.989 | 0.9933 |
| | 10 | RIM = -0.0246×G + 82.006 | 0.9626 |
| | 30 | RIM = -0.048×G + 81.597 | 0.9206 |
| | 150 | RIM = -0.0462×G + 81.951 | 0.9769 |
| leached | 10 | RIM = -0.0706×G + 81.478 | 0.9613 |
| | 150 | RIM = -0.0484×G + 81.641 | 0.9763 |

Table 3. Fitting curve equations for the relationship between RIM and G for non-leached and leached specimens at different periods after gamma treatment.

Despot *et al.* (2007) and Hasan *et al.* (2006b) calculated RIM relative to the controls ($RIM_r$) for non-leached and leached specimens as the ratio between mean RIM of gamma treated specimens and mean RIM of non-leached controls. $RIM_r$ linearly depends on the radiation dose and decreases stronger for leached wood than for non-leached wood. Maximum reduction in RIM was 12 % at the highest gamma dose (Figure 12).

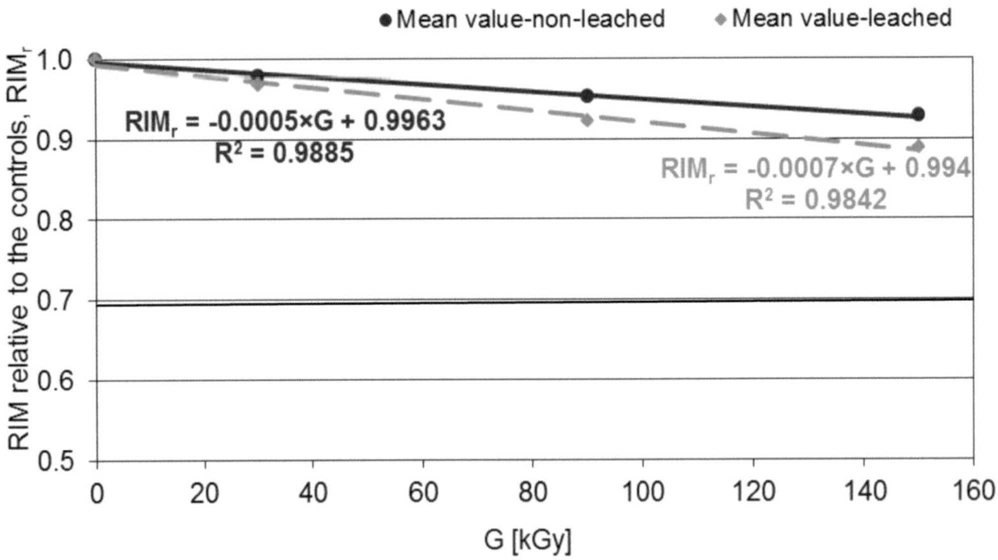


Fig. 12. Correlation between RIM relative to the controls ($RIM_r$) of non-leached and leached specimens and gamma radiation dose (Despot *et al.* 2007).

Brischke *et al.* (2006a), Rapp *et al.* (2006), and Welzbacher *et al.* (2006) described the sensitive HEMI test, which is able to reveal fine changes in "dynamic strength properties" (increased brittleness). This has been confirmed for thermally and chemically modified timber (Welzbacher *et al.* 2007, 2011; Brischke *et al.* 2012) as well as for differently degraded timber, e.g. by different basidiomycetes, soft rot fungi and bacteria (Rapp *et al.*, 2008; Brischke *et al.*, 2009; Huckfeldt *et al.*, 2010). Also the findings in this research confirmed the suitability of the HEMI test for detecting subtle differences in the mechanical properties of wood. In particular, it is the structural integrity, which affects the Resistance to impact milling RIM, wherefore the results of the HEMI test characterise structural changes on a cellular micro level. In contrast, macroscopic defects, such as cracks and splitting, will not affect the results, because they will be masked through finer fragmentation of the wood samples (Welzbacher *et al.*, 2011). As a result of various different dynamic loads (impact bending, end grain bending, compression, cleaving, shearing, and buckling) the reduced structural integrity of the wood is reflected by the HEMI test (Welzbacher *et al.*, 2011). The significant decrease in RIM through gamma radiation is therefore considered to be also attributed to the break-up of cellulose chains (Hasan *et al.*, 2006b; Despot *et al.*, 2007; 2008). This again coincides with the higher sensitivity of the HEMI-test to cellulose break down by brown rot fungi compared to white rot decay (Brischke *et al.*, 2006b; 2008).

### 4.4 Biological durability of gamma irradiated wood

Despot *et al.* (2006) and Hasan *et al.* (2006a, 2008) used pine sapwood specimens, one group steam sterilised (at 123 °C for 30 min) and second group of specimens gamma-irradiated in the range of doses of 30, 90 and 150 kGy. Specimens were subject to laboratory decay resistance tests. According to EN 113 (1996) the specimens were incubated in testing flasks with pure cultures of test fungi. They used brown- and white-rot fungi for the study. Mass loss by fungal decay (ML) was determined by weighing the oven-dry specimens before ($m_1$) and after ($m_3$) incubation to the nearest 1 mg and according to equation (3):

$$ML = \frac{m_1 - m_3}{m_1} \times 100\,[\%] \qquad (3)$$

#### 4.4.1 Resistance against decay fungi

As gamma radiation causes break-up of cellulose to shorter chains, which are water-soluble, and that leads to an "opening of additional microcracks", in which water molecules can easily penetrate. Consequently, gamma irradiated wood is also more accessible to enzymes of wood decaying fungi.

Only ten days after incubation, it was clearly visible, that irradiated specimens were more overgrown, than control steam sterilised controls, which indicates a higher susceptibility to biodegradation of gamma irradiated wood compared to non-irradiated wood (Hasan, 2006a; Hasan *et al.*, 2008; Figure 13).

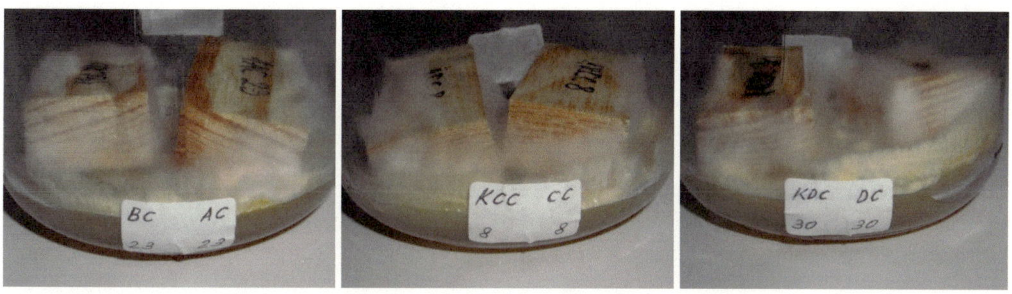


Fig. 13. Specimen pairs in testing flasks 10 days after exposure to brown-rot fungus *Gloeophyllum trabeum* (AC, KCC, KDC - non-irradiated controls, BC - specimen irradiated with 30 kGy, CC - 90 kGy, DC - specimen irradiated with dose of 150 kGy).

## Resistance against brown-rot fungi

Visible difference in appearance (irregular shape and darker colour) between irradiated and autoclaved specimen after 16 weeks of exposure to the fungus *G. trabeum* indicated a significantly higher decay of the irradiated specimens (Hasan, 2006; Hasan *et al.*, 2008). After 12 and 16 weeks of exposure to the fungus *G. trabeum*, average ML of irradiated and control specimens was more than 25 %, which proved that the fungus was virulent (EN 113: 1996). The difference in ML between irradiated and autoclaved specimens was slight and not significant after 4 weeks of incubation, except specimens irradiated with 150 kGy, whose ML was greatest. The difference in ML between controls and specimens irradiated with 30 kGy reached maximum after 8 weeks of incubation, while other two irradiated groups had no significant difference in ML comparing to control. During further incubation time the differences in ML between irradiated and control specimens decreased and after 16 weeks became insignificant (Figure 14).

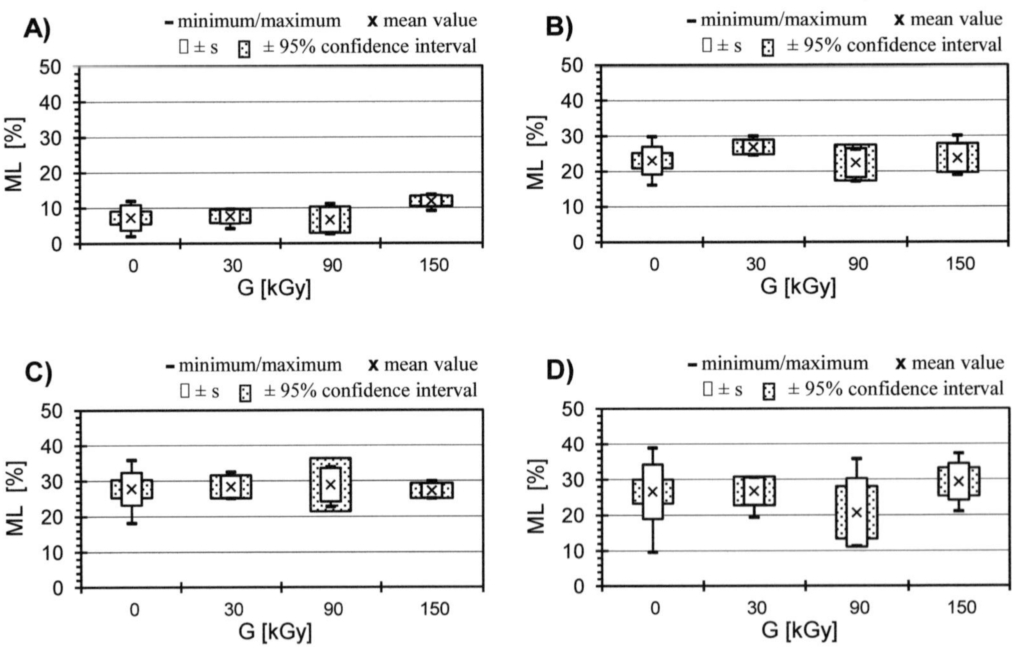


Fig. 14. Mass loss (ML) of autoclaved and irradiated specimens during exposure to fungus *Gloeophyllum trabeum*: A) 4 weeks incubation; B) 8 weeks incubation; C) 12 weeks incubation; D) 16 weeks incubation; (irradiated specimens n=7; G=0 kGy - autoclaved controls n=21).

Fungus *P. placenta* causes brown rot with broad and deep cracks. Clearly visible difference in appearance (cracks and irregular shape) between irradiated and autoclaved specimens after 16 weeks of incubation has been shown in Figure 15.

Fig. 15. Appearance of irradiated and autoclaved specimens after 16 weeks of exposure to fungus *Poria placenta*.

The difference in ML between specimens irradiated with 30 and 90 kGy and autoclaved ones is slight and not significant after 4 weeks of incubation, while specimens irradiated with 150 kGy had significantly greater ML. During further incubation differences in ML between autoclaved and irradiated specimens increased but also between irradiated groups. During the all incubation period ML increased with radiation dose. Since *P. placenta* was more virulent and more aggressive fungus compared to *G. trabeum*, the obtained results are logical (Figure 16).

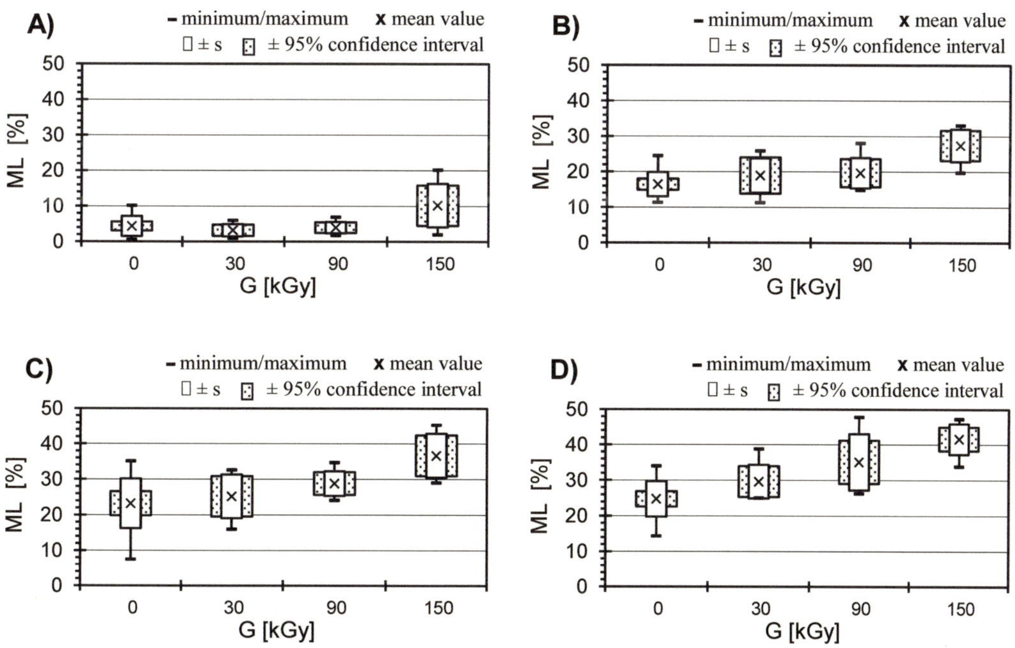


Fig. 16. Mass loss (ML) of autoclaved and irradiated specimens during exposure to fungus *Poria placenta*: A) 4 weeks incubation; B) 8 weeks incubation; C) 12 weeks incubation; D) 16 weeks incubation; (irradiated specimens n=7; G=0 kGy - autoclaved controls n=21).

**Resistance against white-rot fungi**

Considering the patterns and mechanisms of decay of white-rot fungi, in the beginning of decay they utilise simple carbohydrates - the ones incurred by gamma radiation. During further incubation, white rot fungi mainly utilise lignin, and they use radical mechanisms to degrade lignin. Broken (somehow modified) lignin accompanied with radicals incurred in wood by gamma radiation could negatively interfere with these decay patterns. Therefore no significant difference in ML caused by white-rot fungi between autoclaved and irradiated specimens was determined (Despot *et al.*, 2006; Hasan *et al.* 2006a).

In contrary to white rot, brown rot decay mechanisms, particularly decay mechanism of *P. placenta,* are less radical dependent, higher radiation doses do not influence the decay patterns of this brown rot fungus. In contrast, higher irradiation resulted in higher degree of depolymerisation, what makes wood significantly more susceptible to dacay (Despot *et al.* 2006, 2007; Figure 17).

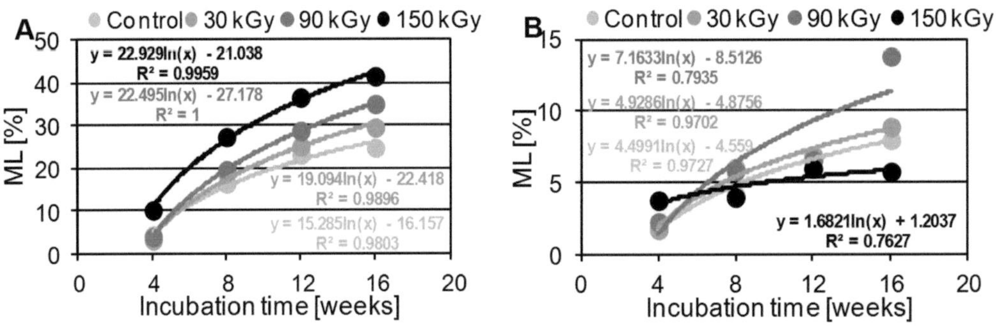


Fig. 17. Correlation between Mass loss (ML) of autoclaved and irradiated specimens and incubation time to fungus: A) *Poria placenta;* B) *Trametes versicolor.*

However, it has to be taken into consideration that depolymerisation of wood components by gamma irradiation is one aspect that could explain differences in changed "natural" durability of pine wood and needs to be considered for wood durability testing.

## 5. Summary

Gamma radiation at a level of 30 to 150 kGy causes irreversible and permanent changes in chemical and mechanical properties of wood. Using the HEMI test method it was possible to detect small but significant changes in structural integrity and brittleness of wood caused by gamma radiation. With increasing radiation dose the total amount of water-soluble sugars increased linearly, while the maximum swelling seems to be unchanged.

The method of sterilisation has a considerable influence on the natural durability of pine wood. However, wood sterilisation by gamma radiation has different influence on white rot and brown rot causing fungi. Significantly greater mass loss of gamma irradiated wood than autoclaved wood has been established in the beginning of incubation to white-rot fungi. During further incubation, the differences diminished. In contrast, higher irradiation resulted in higher degree of cellulose depolymerisation, what makes wood significantly more susceptible to biodegradation by brown-rot fungi. Degradation of gamma-irradiated wood is greater and faster due to easier accessibility of simpler carbohydrates to fungi.

## 6. Acknowledgements

Part of the research on the influence of gamma radiation onto wood properties was supported by the COST Action E37 in the frame of a Short-Term Scientific Mission (STSM) awarded to Marin Hasan, PhD. Parts of the experimental work were carried out at the Federal Research Centre for Forestry and Forest Products (BFH), Hamburg, Germany, which was the STSM host institution, and whose hosting is gratefully acknowledged. The authors extend their particular thanks to BFH and to Ruđer Bošković Institute in Zagreb, Croatia for gamma radiation provision and to its scientists for sharing their wide-ranging scientific expertise on gamma radiation.

## 7. References

Ardica, S., Calderaro, E., Cappadona, C. (1984) Radiation pretreatments of cellulose materials for the enhancement of enzymatic hydrolysis - II. Wood chips, paper, grain straw, hay, kapok. Instituto di Applicazioni e Impianti Nucleari, University of Palermo, Italy.

Bakraji, E. H, Salman, N, Al-kassiri, H. (2001) Gamma-radiation-induced wood-plastic composites from Syrian tree species. Radiation Physics and Chemistry. 61(3): 137-141.

Bakraji, E.H., Salman, N., Othman, I. (2002) Radiation-induced polymerization of acrylamide within Okoume (Aucoumea klaineana pierre). Radiation Physics and Chemistry. 64(4): 277-281.

Bogner, A. (1993) Modifikacija površine bukovine radi poboljšanja lijepljenja. Doktorska disertacija. Šumarski fakultet, Zagreb. pp. 141. (in Croatian language).

Bogner, A., Ljuljka, B., Grbac, I. (1997) Improving the glued joint strength by modifying the beech wood (*Fagus sylvatica* L.) with gamma rays. Drvna industrija 47(2): 68-73.

Brischke, C., Rapp, A.O., Welzbacher, C.R. (2006a) High-energy multiple impact (HEMI) - test - Part 1: A new tool for quality control of thermally modified timber. Document No. IRG/WP 06-20346, International Research Group on Wood Protection, Tromsø, Norway.

Brischke, C., Welzbacher, C.R., Rapp, A.O. (2006b) Detection of fungal decay by high-energy multiple impact (HEMI) testing. Holzforschung 60 (2): 217-222.

Brischke, C., Welzbacher, C.R., Huckfeldt, T. (2008) Influence of fungal decay by different basidiomycetes on the structural integrity of Norway spruce wood. Holz als Roh- und Werkstoff 66: 433-438.

Brischke, C., Welzbacher, Huckfeldt, T., Schuh, F. (2009) Impact of decay and blue stain causing fungi on the structural integrity of wood. Document No. IRG/WP/09-10699. International Research Group on Wood Protection, Beijing, China.

Brischke, C., Zimmer, K., Ulvcrona, T., Bollmus, S., Welzbacher, C.R., Larsson-Brelid, P.. Thomsen, O. (2012) Structural integrity of differently modified wood. 6th European Conference on Wood Modification, Ljubljana, Slovenia, 17-18 September 2012. In print.

Chawla, J. S. (1985) Degradation of Ligno-Cellulose Biomass, Holzforschung und Holzverwertung 37(2): 101-105.

Csupor, K., Divos, F., Gonczol, E. (2000) Radiation induced effects on wood material and fungi, In: Proceedings of 12th Int. symposium on nondestructive testing of wood, Sopron, 13. - 15. September, 2000.

Curling, S., Winandy, J.E. (2008) Comparison of the effects of gamma irradiation and steam sterilization on southern pine sapwood. Forest Products Journal 58(1/2): 87-90.

Cutter, B.E., McGinnes Jr, E.A., Schmidt, P.W. (1980) X-ray scattering and X-ray diffraction techniques in studies of gamma-irradiated wood. Wood and Fiber Science 11(4): 228-232.

Davis, J. R., Ilic, J., Wells, P. (1993) Moisture content in drying wood using direct scanning gamma-ray densitometry. Wood and Fiber Science 25(2): 153-162.

Despot, R., Hasan, M., Glavaš, M., Rep, G. (2006) On the changes of natural durability of wood sterilised by gamma radiation. Document No. IRG/WP 06-10571, International Research Group on Wood Protection, Tromsø, Norway.

Despot, R., Hasan, M., Brischke, C., Welzbacher, C.R., Rapp, A.O. (2007) Changes in physical, mechanical and chemical properties of wood during sterilisation by gamma radiation. Holzforschung 61(3): 267-271.

Despot, R., Hasan, M., Rapp, A.O., Brishke, C., Welzbacher, C.R. (2008) The Effect of Gamma Radiation on Selected Wood Properties. Document No. IRG/WP 08-40394, International Research Group on Wood Protection, Istanbul, Turkey.

Divos, F., Bejo, L. (2005) The effect of gamma irradiation on the MOE of various species. Wood Science and Technology, 40(2): 87-93.

El-Osta, M.L.M., El-Miligy, A.A., Kandeel, S.E., El-Lakany, M.H., El-Morshedy, M.M. (1985) Mathematical description of the change in properties of *Causaria* wood upon exposure to gamma radiation. 1. Changes in the compression and tensile strength. Wood and Fiber Science 17(1): 2-10.

European Standard (1996) EN 113. Wood preservatives - Test method for determining the protective effectiveness against wood destroying basidiomycetes - Determination of the toxic values.

Fadel, M. A., Kasim, S. A. (1977) A devised light absorption method for measuring fast neutron fluences and gamma doses in mixed radiation fields using a cellulose acetate detector. Nuclear Instruments & Methods. 146(3): 513-516.

Fairand, B., Ražem, D. (2010) Radiation sterilization. Chapter in the book:Pharmaceutical Dosage Forms: Parenteral Medications. Third Edition, Vol.2: Facility Design,

Sterilization and Processing. Ed. Nema, S., Ludwig, J.D. Informa Healthcare USA, Inc. New York, N.Y. 268-296.

Fengel, D., Wegener, G. (1989) Wood, Chemistry, Ultrastructure, Reactions. Reprint - Kessel Verlag, 2003. Germany. pp. 613.

Freitag, C.M., Morrell, J.J. (1998) Use of gamma radiation to eliminate fungi from wood. Oregon State University (OSU). 76-78.

Hasan, M. (2006): New knowledge on the wood sterilisation by gamma radiation. Master Thesis. Faculty of Forestry, University of Zagreb. Zagreb. pp. 120.

Hasan, M., Despot, R., Humar, M., Pohleven, F., Rep, G. (2006a): Contribution to Understanding the Biodegradation Mechanism of Wood Sterilised by Gamma Radiation. International Journal of Medicinal Mushrooms. 9(3-4): 308-309.

Hasan, M., Despot, R., Rapp, A.O., Brischke, C., Welzbacher, C.R. (2006b) Some Physical and Mechanical Properties of Wood Sterilised by Gamma Radiation. International Conference Wood Structure and Properties '06. Zvolen, Slovakia. Arbora Publishers. pp. 239-244.

Hasan, M., Despot, R., Sinković, T., Jambreković, V., Bogner, A., Humar, M. (2008) The influence of sterilisation by gamma radiation on natural durability of wood. Wood Research. 53(4): 23-34.

Huckfeldt, T.; Eichhorn, M.; Koch, G.; Welzbacher, C.R.; Brischke, C. (2010) Bewertung von Schäden an dauerhaften Hölzern am Beispiel von Bongossi (*Lophira alata*). Europäischer Sanierungskalender 2010: 95-108.

Ifju, G. (1964) Tensile strength behaviour as a function of cellulose in wood. Forest Products Journal 14(8): 366-372.

Karsulovic, J. T., Dinator, M. I., Morales, R. (2002) Nondestructive gamma radiation methods for detection of central rot in loges of lenga (*Nothogagus pumilo*), Forest Products Journal 52(11/12): 87-93.

Katušin-Ražem, B., Ražem, D., Braun, M. (2009) Irradiation treatment for the protection and conservation of cultural heritage artefacts in Croatia. Radiation Physics and Chemistry. 78(7/8): 729-731.

Kenaga, D.L., Cowling, E.B. (1959) Effect of gamma radiation on ponderosa pine: hygroscopicity, swelling and decay susceptibility. Forest Products Journal 9(3): 112-116.

Klimentov, S.A., Bysotskaia, I.F. (1979) Investigation of radiation-destroyed wood. 1. Constants of hydrolysis rate of hard hydrolysable polysaccharides of gamma-irradiated lingocellulose. Khimiya Drevesiny, Riga 5: 30-32.

Klimentov, S.A., Skvortsov, S.V., Ershov, B.G. (1981) Influence of gamma radiation on the chemical composition of wood hydrolysates. Journal of Applied Chemistry of the USSR. 53(2/7): 1254-1257.

Kunstadt, P. (1998) Radiation disinfestations of wood products. Radiation Physics and Chemistry 52(1-6):617-623.

Lester. P.J., Rogers, D.J., Petry, R.J., Connolly, P.G., Roberts, P.B. (2000) The lethal effects of gamma irradiation on larvae of the Huhu beetle, *Prionoplus reticularis*: a potential quarantine treatment for New Zealand export pine trees. Entomologia Experimentalis et Applicata. 94(3): 237-242.

Lhoneux, B. de, Antonie, R., Cote, W.A. (1984) Ultrastructural implications of gamma-irradiation of wood [*Pseudotsuga mensziessi, Liridendron tulipifera*], Wood Science and Tecnology. 18(3): 161-176.

Loos, W.E. (1962) Effect of Gamma Radiation on the Toughness of Wood. Forest Products Journal 12(5): 261-264.

Lu, C., Lam, F. (1999) Study on the X-ray calibration and overlap measurements in robot formed flakeboard mats, Wood Science and Technology 33(2): 85-95.

Magaudda, G., Adamo, M., Rocchetti, F. (2001) Damage caused by destructive insects to cellulose previously subjected to gamma-ray irradiation and artificial ageing. Restaurator - International Journal for the Preservation of Library and Archival Material 22(4): 242-250.

Oldham, S.C., Nolan, P.F., Maclenan, M.P. (1990) A gamma-radiographic study of wood and polymer combustion and the effects of flame retardants. Nuclear Instruments and Methods in Physics Research Section A: Accelerators, Spectrometers, Detectors and Associated Equipment. 299(1-3): 661-665.

Panshin, A. J., deZeeuw, C. (1980) Textbook of Wood Technology, Structure, Identification, Properties, and Uses of the Commercial Woods of the United States and Canada, Fourth Edition. McGraw-Hill, Inc., pp: 722.

Pratt, L.H., Smith, D.G., Thornton, R.H., Simmons, J.B., Depta, B.B., Johnson, R.B. (1999) The effectiveness of two sterilisation methods when different precleaning techniques are employed. Journal of Dentistry. 27(3): 247-248.

Rapp, A.O., Sailer, M., Brand, K. (2003) Umweltfreundliche Konservierung von Massivholz und Massivholzprodukten für den Außenbereich. Schlussbericht für das BMBF-Projekt 0339862. Bundesforschungsanstalt für Forst- und Holzwirtschaft (BFH), Hamburg.

Rapp, A.O., Brischke, C., Welzbacher, C.R. (2006) Interrelationship between the severity of heat treatments and sieve fractions after impact ball milling: a mechanical test for quality control of thermally modified wood. Holzforschung. 60(1): 64-70.

Rapp, A.O., Brischke, C., Welzbacher, C.R., Nilsson, T., Björdal, C. (2008) Mechanical strength of wood from the Vasa shipwreck. Stockholm: The International Research Group on Wood Protection, IRG/WP/08-20381.

Seaman, J.F., Millett, M.A., Lawton, E.J. (1952) Effect of high energy cathode rays on cellulose. Industrial & Engineering Chemistry Research 44(12): 2848-2852.

Seifert, K. (1964) Zur Chemie gammabestrahlten Holzes. Holz als Roh- und Werkstoff. 22(7): 267-275.

Sheikh, N., Afshar Taromi, F. (1993) Radiation induced polymerization of vinyl monomers and their application for preparation of wood-polymer composites. Radiation Physics and Chemistry. 42(1-3): 179-182.

Sharman, C.V., Smith, R.S. (1970) Gamma radiation sterilization of ponderosa pine and birch sapwood. Wood and Fiber Science 2(2): 134-143.

Shuler, C.E. (1971) Gamma Irradiation Effects On Modulus of Elasticity Of Engelman Spruce. Forest Products Journal 21(1):49-51.

Shuler, C.E., Shottafer, J.E., Campana, R.J. (1975) Effect of Gamma Irradiation *In Vivo* On the Flexural Properties of American Elm. Wood Science and Technology 7(3): 209-212.

Šimkovic, I., Mlynár, J., Alföldy, J., Lübke, H., Micko, M.M. (1991) Increased extractability of irradiated wood meal. Holzforschung 45(3): 229-232.

Severiano, L.C., Lahr, FA.R., Bardi, M.A.G., Santos, A.C., Machado, L.D.B. (2010) Influence of gamma radiation on properties of common Brazilian wood species used in artwork. Progress in Nuclear Energy. 52(8): 730-734.

Struszczyk, H., Ciechańska, D., Wawro, D., Niekraszewicz, A., Strobin, G. (2004) Review of alternative methods applying to cellulose and chitosan structure modification. Radiation processing of polysaccharides, International Atomic Energy Agency - IAEA. November, 2004. pp. 55-65.

Subrahmanyam, V.S., Das, S.K., Ganguly, B.N., Bhattacharya, A.DeA. (1998) Positron anihilation study on gamma-irradiated cellulose acetate matrix. Polymer. 39(6-7): 1507-1508.

Tabirih, P.K., McGinnes Jr, E.A., Kay, M.A., Harlow. C.A. (1977) A note on anatomical changes of white oak wood upon exposure to gamma radiation. Wood and Fiber Science 9(3): 211-215.

Tišler, V., Medved, S. (1997) Changes in wood structure due to irradiation by gamma rays. Les-*Wood* 49(4): 85-89. (in Slovenian language).

Tsutomu, A., Misato, N., Tadashi Y. (1977) Some Physical Properties of Wood and Cellulose Irradiated with Gamma Rays. Wood Research. 62(1): 19-28.

Unger, A., Schiewind, A. P., Unger, W. (2001) Conservation of Wood Artifacts, A Handbook. Berkeley and Eberswalde, Berlin. pp. 346-348 and 497-498.

Welzbacher, C.R., Brischke, C., Rapp, A.O. (2006) High-energy multiple impact (HEMI) -test - Part 2: A mechanical test for the detection of fungal decay. Document No. IRG/WP 06-20339, International Research Group on Wood Protection, Tromsø, Norway.

Welzbacher, C.R., Brischke, C., Rapp, A.O. (2007) Influence of treatment temperature and duration on selected biological, mechanical, physical, and optical properties of thermally modified timber (TMT). Wood Material Science and Engineering 2: 66-76.

Welzbacher, C.R., Brischke, C., Rapp, A.O. (2009) Estimating the heat treatment intensity through various properties of thermally modified timber (TMT). Document No. IRG/WP/09-40459. International Research Group on Wood Protection, Beijing, China.

Welzbacher, C.R., Rapp, A.O., Haller, P., Wehsener, J. (2004) Thermisch vergütete und verdichtete Fichte für tragende Anwendungen im Ingenieurholzbau. AiF Forschungsbericht 63 ZBR ½.

Welzbacher, C.R., Rassam, G., Talaei, A., Brischke, C. (2011) Microstructure, strength and structural integrity of heat-treated beech and spruce wood. Wood Material Science and Engineering 6: 219-227.

Zamani, M., Savides, E., Charalambous, S. (1981) The response of cellulose nitrate to gamma radiation. Nuclear Tracks & Radiation Measurements-International Journal of Radiation Applications & Instrumentation, Part D. 4(3): 171-176.

# 15

# The Effects of Gamma Radiation in Nectar of Kiwifruit (*Actinidia deliciosa*)

Marcia N. C. Harder[1] and Valter Arthur[2]
[1]*Institute of Energetic and Nuclear Research – IPEN/CNEN/USP*
*Technology College of Piracicaba – FATEC Piracicaba*
[2]*Center of Nuclear Energy in Agriculture, University of Sao Paulo – CENA/USP*
*Brazil*

## 1. Introduction

### 1.1 The kiwifruit

The kiwifruit (*Actinidia deliciosa*) is an exotic fruit to Brazil belonging to the family Actinidiaceae, the genus *Actinidia*, which has some edible species such as *A. chinensis, A. arguta, A. kolomikta, A. polygama* and *A. eriantha* and contrary what of imagines whose origin center are the China mountainous regions which have from 800 to 2000m altitude, where it grows in the forest shade and where it was called the Chinese gooseberry. Also found were records of the presence of kiwi to the northeast of India and Japan. New Zealand has always taken this credit but the first productions of the *A. delicious* species appeared only in mid 1910 and definitely where it was named in honor of the bird symbol of this country due to his likeness and that receives the same name: Kiwi.

The kiwifruit is a plant belonging to the order Theales, *Actinidiaceae* family and genus *Actinidia* with more than 53 species including the *Actinidia deliciosa* is the most important (Ferri, Kersten, Machado, 1996).

The genus *Actinidia* in 1959 has been classified by Wallich. Hutichon that divided the family into genera actinideácea only climbers: *Actinidia, Clamatochetra* and *Sladenita* (Almeida, 1996).

They are typical plants of locations with temperate or subtropical mountain. The varieties of fruit most widely marketed varieties are produced by several species of *Actinidia deliciosa* and to a much lesser extent, by some varieties of *Actinidia chinensis.* The fruit pulp has coloration ranging from bright green to yellow, with several small seeds and black in the middle and characterized by the oval form and to be flattened. The bark has a light brown color covered with small fells. Being a low-calorie fruit is widely used in energy-restricted diets because about 90% of its weight consists of water (Demczuk Junior, 2007; Oliveira, 2011).

Another feature of the kiwifruit plant is to be clambering and its production be conducted preferably in tutoring systems (Figure 1) as the vines in order that the fruits do not touch the ground preserving them of the early deterioration.

Fig. 1. Kiwifruit tutored system plantation (SEMAPI, 2011)

The species of *Actinidia deliciosa* (A. Chev.) several delicious varieties are known to be preferred by the consumer and therefore more widespread the variety 'Hayward' (Figure 2) fruit with better taste and aroma (flavor) and a greater capacity for storage.

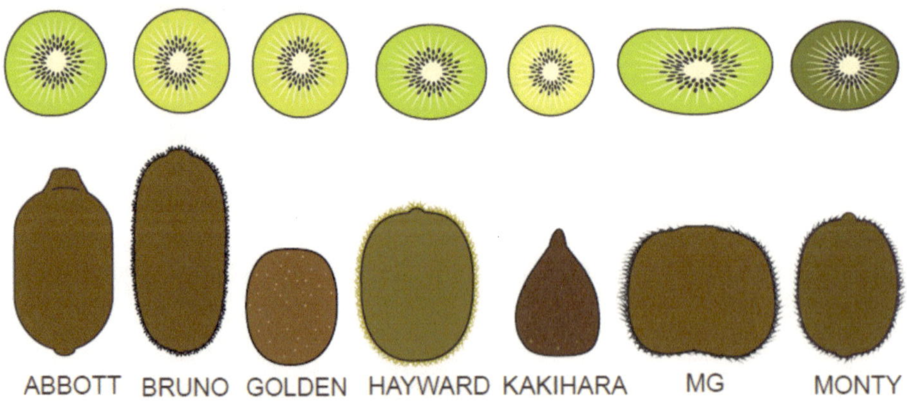


Fig. 2. Kiwifruit varieties (CEAGESP, 2011)

### 1.1.2 Kiwifruit cultivars

Also like all other fruits the kiwifruit has many cultivars, which differ in pollinator (male) and producing (female) and the adaptation to the environmental conditions; harvest time; chances of post-harvest conservation; productivity; format; size; taste;, resistance to pests and diseases; among others (Souza; Marodin; Barradas, 1996).

Oliveira (2011) shows which varieties are part of the division between male and female cultivars:

#### 1.1.2.1 Female cultivars

a. Abbot

It is a vigorous plant; very productive; early flowering; with flowers grouped about two or three. The flowers are creamy white and the petals have rounded elliptical, with the edges slightly wrinkled. The leaves are usually short and round with pointed apex (ALMEIDA, 1996).

The fruit has an oblong shape; weighing 65 to 70 grams; with a period of maturation medium; good storage under refrigeration and successfully supporting the handling and transport. The pulp is bright green color; fragrant; mildly acidic; considered as has a good quality and very popular in New Zealand for family consumption. It features low demand hours in the cold and high bud fertility and therefore a high productive power can be 400 kilograms of fruit per foot (Souza; Marodin; Barradas, 1996).

b. Bruno

It is a vigorous and productive plant. Have solitary flowers (rarely combined) with cream-white elliptical petals; more or less rounded and flat. The leaves have circular shape with the bottom rope. The fruits are medium in size - 60 to 70 grams - very uniform; cylindrical in shape and elongated; bark brown and covered with dense fell, short and spiky (Almeida, 1996).

The pulp is sweet and a little tart, very goodly, with the highest concentration of vitamin C than other varieties. The maturity is very early with the possible preservation in a refrigerator however had a reduced resistant to handling and transportation. In southern Brazil, the harvest usually occurs in April opening. It is not very demanding in cold times requiring a half hours of cold necessary to cultivate Hayward. In the conditions of the Brazil Southern the Bruno cultivar has shown the best yield among the varieties grown kiwifruit tree (Souza; Marodin; Barradas, 1996).

c. Hayward

Presents late-flowering with slightly lower productivity for other varieties of flowering coincide. Its flowers are solitary and rarely in pairs. It is very sensitive to chlorosis; drought and nematodes. Has multiplication by cutting easier than the others (Souza; Marodin; Barradas, 1996).

It is a plant with medium vigor. It has the harvest beginning around the first half of November. At harvest the fruit must reach a level of maturity as measured by the method of soluble solids of at least 6.2 percent. The fruits are heavy (90 to 150g); oval with the elliptical cross section; with brown bark greenish; covered with fine fells (Almeida, 1996).

The leaves are rounded and the base strings with overlapping. The pulp is green with hints of straw-colored; moderately juicy; slightly tart and mildly fragrant and aromatic with good taste qualities (Almeida, 1996). According to Souza, Marodin and Barradas (1996), the perfect balance between acids and sugars that make the variety is preferred by markets around the world.

Have good resistance to transportation; handling and excellent cold storage. Although not determined it is estimated that this growing need 700 to 1,000 hours of temperatures below 7.2°C during winter (Souza; Marodin; Barradas, 1996).

This cultivar is now widely cultivated in the world due to its excellent features and the fact that the packaging and the alveoli are thought to their fruits.

d. Monty

It is a very vigorous plant and induces major productions. The flowers are cream-white with variable petals form and grouped into two or three branches. The leaves are usually round-based trunk (Almeida, 1996).

The fruits are elliptical medium size and uneven with average weight of 65g but in some places is around 40g. Maturation takes place in mid-season and the fruits are difficult to preserve. They are little resistant to transport and handling. Its flesh is bright green, sweet-tart and fragrant (Souza; Marodin; Barradas, 1996).

#### 1.1.2.2 Male cultivars

a. Matua

It is an early flowering cultivar by abundant; prolific and persistent form for a long time. Its flowers are grouped in variable number from 1 to 5. In some situations presents an early flowering in relation to 'Hayward'. Features of a high-pollen than the 'Tomur' variety (Souza; Marodin; Barradas, 1996).

b. Tomur

It is a plant variety which starts blooming in mid-season and lasts a long time. The flowers are grouped in varying levels from 1 to 7 usually 5. It blooms a little later than 'Matua' and seems especially suited to pollinate 'Hayward'. Its pollen has a low germination (Souza; Marodin; Barradas, 1996).

### 1.1.3 World production

There are few data about the world production which is known is that today in the worldwide the main producer of kiwifruits are Italy; China; New Zealand and Chile (Figure 3).

Kiwifruits can be found throughout the year because there is a mismatch between the different harvest times for each producing country: since mid-May until late November in New Zealand; and the rest of the time in the Mediterranean countries; Chile; California and Australia. It is a product that was developed commercially in recent years through to a marketing job done by New Zealand which is an example of how to disclose a 'new product'.

### 1.1.4 The fruit characteristics

Kiwi fruits display tolerance to low temperatures for conservation, allowing storage for up to 8 months under these conditions (Schuck, 1992).

Kiwi is one of the only fruits which does not have a specific regulation from the Brazilian Ministry of Agriculture about standardization and classification. Therefore, there is no official rules for commercialization, classification, selection and packing for kiwi in Brazil (Harder et al., 2009).

**Kiwi: U.S. import-eligible countries; world production and exports**

**Total production, exports and export value (2007) for countries eligible to ship kiwi to the United States**

| Country 1/ | Production | Total exports | Export value |
|---|---|---|---|
| | *1,000 metric tons* | | *1,000 US$* |
| Argentina | nd | 0,41 | 213,00 |
| **Australia** | 5,35 | 0,50 | 842,00 |
| **Canada** | 0,08 | 0,07 | 125,00 |
| Chile | 170,00 | 160,19 | 143.053,00 |
| **France** | 70,16 | 26,84 | 50.776,00 |
| **Greece** | 70,10 | 36,79 | 34.356,00 |
| **Italy** | 417,00 | 333,97 | 389.450,00 |
| **Japan** | 32,80 | 0,01 | 37,00 |
| **Korea, Republic of** | 10,50 | 0.00 | 0.00 |
| **New Zealand** | 365,00 | 347,92 | 573.490,00 |
| **Spain** | 13,00 | 9,71 | 13.753,00 |
| **Percent eligible** | 94,87 | 82,02 | 77,30 |
| **World median** | 11,40 | 0,01 | 8,00 |
| **World average** | 56,38 | 11,54 | 16.000,94 |

Note: nd = no data; Percent eligible = Percentage of total world production, export, or export value for this commodity for countries eligible to export this commodity to the United States according to fruit and vegetable import regulations from USDA's Animal and Plant Health Inspection Service (APHIS). Median = Median global production quantity, export quantity, or value. Median is the value for which half of all values in a series are greater and half are smaller. Average = Average global production quantity, export quantity, or value. Average is the sum of all numbers in a series divided by the total number of entries in that series.

1/ Countries eligible to export this commodity to the United States according to APHIS regulations as of June 2010. See Documentation for more information. Countries in bold are high-income nations, all others are middle- and low-income nations according to the 2010 country classification developed by World Bank.

* Only certain region(s) within this country is (are) eligible to export this commodity to the United States or is (are) regulated differently than the rest of the country according to APHIS regulation as of June 2010. See Documentation for more information.

**Top world producers and exporters of kiwi (2007) 1/**

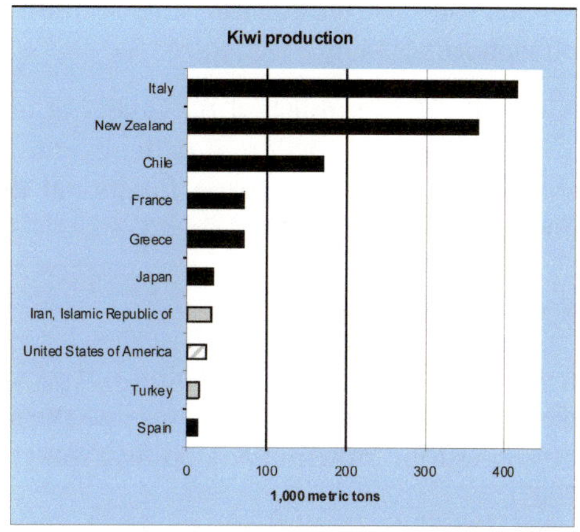


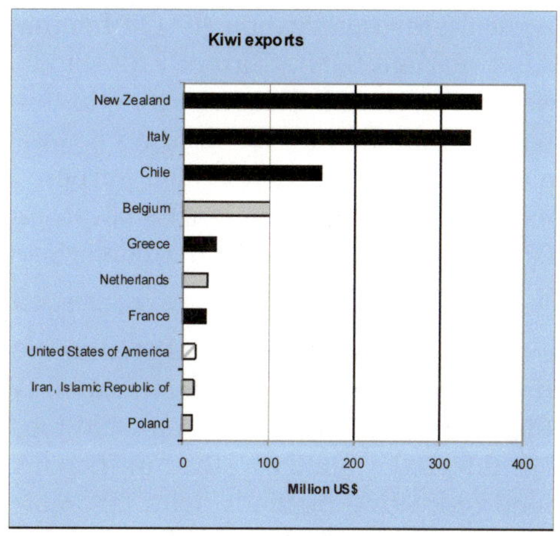


Fig. 3. Kiwi world production and exports (USDA, 2011)

The kiwi fruit presents a high nutritional value, rich mainly in vitamin C and fibers, calcium, iron and phosphorus, which turns it an excellent nutritional option, with an important association between quality attributes and flavor, with great acceptance in consuming markets, mainly among children. Kiwi contains an enzyme actinine with meat softening properties by the reason it can be used to soften the meat before cooking and reduce the cooking time and addition flavor with the fruit pulp. Furthermore the actinidina prevents the coagulation of gelatin and is responsible for the bitter taste that dairy products are consumed when presented with kiwi fruit. (Carvalho, 2000).

One of the fibers is found in kiwi fruit pectin that helps lower blood cholesterol levels. Other bioactive compounds found in kiwifruit are phenolic flavonoids, amino acids such as arginine and glutamate and chlorophyll. All compounds of the kiwi provide a potential anticancer and anti-inflammatory that helps in increasing the immunity of individuals who have the habit on the frequency of fruit intake.

The nutritional composition of kiwifruit per 100 grams of fruit is the following: 67 calories; carbohydrates 15g; 1g of protein; 0.5g of lipid; 1g of fiber; 17g of Vitamin A; Vitamin C 80-300mg; 0.4g of iron; calcium 26g; sodium 5.2g; 30g of magnesium; potassium 280-340g and phosphorus 20-40g.

A recent study published by Chang et al. (2010) reveals that the kiwifruit is excellent for improves bowel function in patients with irritable bowel syndrome with constipation (IBS), characterized by preexisting functional diarrhea, ranging from normal functioning of the intestine and constipation.

### 1.2 Food irradiation

Irradiation is an excellent method of food conservation, as well as reinforces the action of other applied processes for the same purpose. Irradiation satisfies completely the objectives of giving food nutritional stability, sanitary conditions and a long shelf life (EMBRARAD, 2007).

The Brazilian law follows the international recommendations suggested by the Food and Agriculture Organization (FAO); International Atomic Energy Agency (IAEA) and Codex Alimentarium (UN). Currently all standards for the use of this technology are described in Resolution No. 21. That according to this any food can be irradiated subject to compliance with the minimum and maximum dosage applied where the minimum dose should be enough to achieve the intended purpose and the maximum must to be less than that which would compromise the functional properties and/or sensory attributes of food (Oliveira, 2011; Oliveira et. al 2011 and Modolo; Silva; Arthur, 2011).

According to the same author, food irradiation has been the subject of intense research for over forty years. International organizations such as the Food and Agriculture Organization (FAO) and World Health Organization (WHO) has been reviewed all researches about this theme and concluded that irradiated food is safe and beneficial (Ornellas; Gonçalves; Silva; Martins, 2006). Similarly, the nutritional value of irradiated foods was compared with other food processing methods, with favorable results showing that the irradiation uses are the same with common treatments or better.

Treatment of fresh fruit by irradiation is performed with the primary purpose of delaying the process of ripening and decay resulting in a significant increase in shelf life of irradiated fruit. On the existence of many studies are still needed further research on the appropriate doses and the effects of radiation on the qualities of the fruit. Therefore, the application of radiation associated with cold storage can keep the quality attributes of kiwifruit adding value to cultivate (Levite; Santos; Foes, 2008).

Investigations demonstrated that macronutrients, such as proteins and carbohydrates are relatively stable at doses of up to 10 kGy, and that micronutrients, mainly vitamins can be sensitive to any method of food conservation. The sensitivity of various types of vitamins to irradiation and other methods for food conservation is variable; vitamin C and B1 (thiamin) are the most sensitive to irradiation. In general, the process of irradiation with acceptable dose cause little chemical changes in foods, whereas the food nutritional quality is no more affected than when it is treated with other conventional methods of preservation (Villavicencio et. al., 1998; Wiendl 1997).

In the contemporary environment people have requested more time to work and less time to make a good meal so read-to-eat food became a practical way for a more balanced meal. The minimally processed foods are a reality and a perfectly viable option.

The use of minimally processed products in Brazil began in the 90s by some companies attracted by the new market trend. The success of this undertaking depends however by the use of raw materials of high quality; handled and processed with high hygienic condition. It is necessary to use appropriate packaging and temperature control in the processing; distribution and marketing since they are critical to the reduction of physiological deterioration and/or microbiological (Leite; Gêa; Arthur, 2006).

The use of radiation to control microbial contamination in ready-to-eat foods becomes perfectly satisfactory from the point of view of health ensuring food safety for consumption as the technological point of view because several studies have shown that the radiation does not interfere or interfere not significantly in the quality of some foods that was submitted of this process of conservation.

Furthermore of this kind of technology is considered "cold" and has almost no influence on sensory parameters, including the item texture.

### 1.3 Food irradiation in kiwifruit

Oliveira (2011) in a study with Kiwifruits minimally processed and irradiated at doses of 0 (control), 1 and 2 kGy and stored under refrigeration at 6°C found the following considerations:

- No significant difference between the doses for the parameters pH, chlorophyll and acidity;
- Mass loss decreased with increasing dose and storage period;
- The soluble solids content showed a significant difference between treatments, the results of the samples during analysis showed variation probably due to lack of homogeneity of the samples on the period of maturation;
- A reduction in ascorbic acid content in the treatments and during the storage period;
- The humidity and acidity did not differ between treatments and periods of analysis;
- Therefore gamma radiation did not induce harmful changes in physical-chemical properties of the kiwifruit which can be used for the preservation of minimally processed kiwifruit.

### 1.4 Nectar

Still about the context of ready-to-eat food a good option is the use of technological concepts for the production of juices. It is an alternative fast; easy and enjoyable for take fruit.

Nectar is a sweet drink non-alcoholic and non-fermented base of fruit juice in which the concentration of the pulp is at least 20 to 30% depending of the fruit.

The fruit is valued for the health benefits and contribution to improving the quality of life. The need for convenient products for the consumer market brought about the emergence of fast food chains leading consumer foods low in fiber; vitamins and minerals; rich in salt value; fat and sugar. A reflection of these eating habits was apparent by the increasing incidence of obesity and cardiovascular disease. Today consumers already have access to food at the same time convenient and healthy. The light processing adds to the fruit (products healthy by nature) the value of convenience. With the advancement of technology it is possible to find in the market peeled and sliced fruit or juice form with cool features ready for consumption (Demczuk Junior, 2007; Oliveira, 2011; Oliveira et. al., 2011).

The principal processed products obtained from fruits are juices and nectars (sweetened drink). In this work, a no-alcoholic sweetened drink was prepared from Kiwi, containing 50% pure juice and sugar, and ready to consume (Tocchini, 1995), which was treated with irradiation at 0.5; 1.0 and 2.0 kGy doses.

The aim of this work was to formulate a no-alcoholic sweetened drink, starting from kiwi fruits and to submit the drink to gamma radiation derived from a source of Cobalt-60, and evaluates changes in physical and chemical quality attributes cause by irradiation.

## 2. Material and methods

Kiwi fruits (Actinidia deliciosa) in natura were locally purchased. The fruits were washed in a bleach solution, peeled, and cut in half. The juice was extracted by home centrifuge Walita, pre-filtered, centrifuged at 5000 rpm, and filtered. Mineral water was added to 50% volume, and sucrose to reach 16° Brix. It was made an ultra filtration with ceramic membrane of 0.10μm at the pressure of 3bar at 45°C for clarification, according to Lopes et al. (2005). The drink was kept in 500 mL plastic bottles (PET), and exposed to the following treatments:

1. Control;
2. Irradiation with 0.5 kGy;
3. Irradiation with1.0 kGy
4. Irradiation with 2.0 kGy

Irradiation was conducted in a Cobalt-60 source; model Gammabeam 650, of Nordion Canada, installed in the Center of Nuclear Energy in Agriculture of the University of Sao Paulo - CENA/USP., Piracicaba city, State of São Paulo, Brazil. The dose rate was 0.712 kGy/hour. The analyses were conduced in the Laboratory of Radiobiology and Environment of CENA/USP.

Dosimetry was performed using 5 mm diameter alanine dosimeters (Bruker Instruments, Rheinstetten, Germany), and the free radical signal was measured with Bruker EMS 104 EPR Analyzer. The actual dose was within 0.02 of the target dose. Samples was turned 360 continuously during the irradiation process to achieve uniform target doses and the non-irradiated control was placed outside the irradiation chamber to have the same environmental temperature effect with the irradiating sample.

### 2.1 Physico-chemical analyses

### 2.2 Soluble solids content

Estimated in refractometer RT-30ATC and expressed n Brix, according to AOAC (1995).

### 2.3 Title acidity

Determined and estimated as the volume in mL of NaOH require to titrate 100mL of drink to reach pH 8,2, expressed in percent of citric acid, with the drink diluted 1:10, according to AOAC (1995)

### 2.4 pH

Determined using pHmeter MB-10, according to AOAC (1995).

### 2.5 Total ascorbic acid

Determined according to Jacobs (1985) using oxalic acid and titrating with a dichloro-benzeneindophenol solution. Estimated based on volume in mL of the the solution of 2,6 dichlophenolindophenol used to titrate 50 mL of drink, until becoming light pink, expressed in mg ascorbic acid/100mL kiwi drink.

### 2.6 Color

The colorimeter Minolta CR-200 b was used, previously calibrated in White according to pre-determined standards, according to (Bible and Singha, 1993).

Three values of chroma were evaluated: a*, b* and L (Figure 4). The value a* characterizes the color from the red (+a*) to the green (-a*); the value b* indicates the color from the

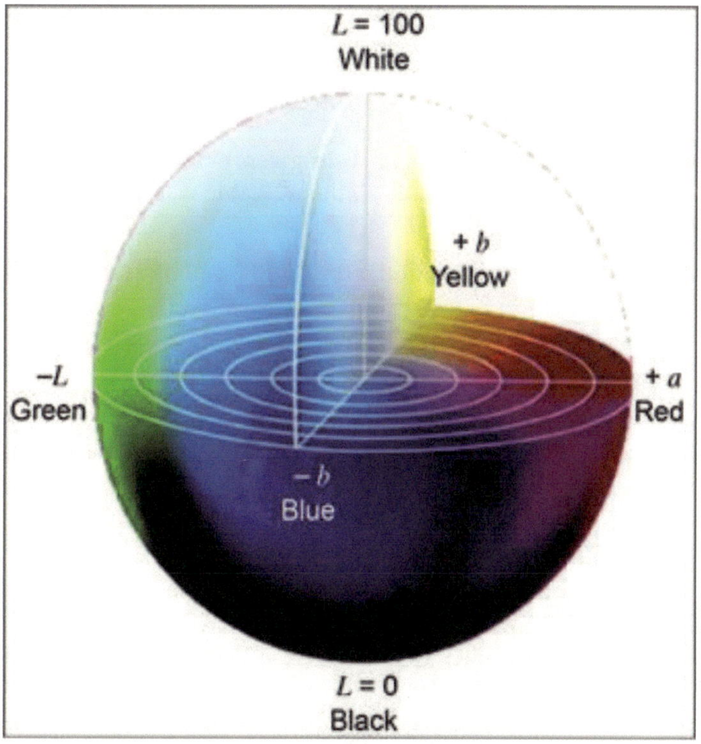


Fig. 4. Representation of a solid color in the space color L* a* b* (Konica Minolta, 2011)

yellow (+b*) to the blue (-b*). The value L determine the light ranging from white (L=100) to black (L=0). The chroma is the ratio between a* and b*, where the real color can be obtained. Hue-Angle is the angle between a* and b*, indicating the color saturation of the analyzed object.

To estimate chroma value, the following formula was adopted (1) and to estimate the Hue-Angle, formula (2) (Estevez & Cava, 2004).

$$C=\sqrt(a2+b2) \tag{1}$$

$$H^{o}=arcpg\ b^*/a^* \tag{2}$$

### 2.7 Sensorial analyses

Sensory acceptances of the irradiated and non-irradiated kiwi nectar were done with 41 non-expert panelists.

It was requested the participation of healthy adults (from 18 to 50 years old) of both sexes who declared kiwi's consumers and, that did not present any reaction concerning the consumption of the same and the same ones manifested its consent. They were selected by aleatoric base. The sensorial analysis was constituted by color; smell; flavor; texture and overall acceptance.

Panelists were instructed to evaluate each attribute using a nine-point hedonic scale ranging from "extremely dislike" to "extremely like". Test of acceptability of hedonic scale was used, since it is necessary to know the consumers "affectionate status" regarding the product inferring the preference, in other words, the most favorite samples are the more accepted and vice versa. The scales were balanced once they present equal number of positive and negative categories (1 to 9) (Ferreira, 2000).

About 50 ml of each kiwi nectar treatment were given individually to the panelists and a three digit code was used for the sample. Mineral water was provided to wash the oral cavity after tasting each treatment

To red light was used to mask possible induction by the color in others attributes (Ferreira, 2000).

### 2.8 Statistical analyses

The experimental design was complete randomized with three replications. Results were analyzed (ANOVA) using the F test, and mean comparisons were tested based on Tukey ($p<0,05$) using SAS (Statistical Analysis System, 1996).

## 3. Results and discussion

### 3.1 Tenor of soluble solids

The obtained variations of Brix (Brix degrees) of the kiwi nectar irradiated at doses of: 0 (control); 0.5; 1.0; 2.0 kGy are in Table 1.

Is possible to observe in agreement with the Table 1, that the treatment which received radiation dose of 2.0 kGy presented a larger amount of soluble solids, following by dose of

0.5 kGy. For the control the nectar irradiated at 1.0 kGy was found the same amount of soluble solids, in other words, the same value for °Brix.

| Dose | °Brix | Acidity titled (%) | pH | Ascorbic acid |
|---|---|---|---|---|
| Control | 16.70±0.01[1b2] | 8.8±0.11[b] | 3.5±0.1[b] | 108.0±0.01[a] |
| 0.5 | 16.60±0.01[c] | 9.2±0.1[a] | 3.2±0.1[a] | 108.0±0.01[a] |
| 1.0 | 16.70±0.01[b] | 8.2±0.1[c] | 3.0±0.1[c] | 54.0±0.01[b] |
| 2.0 | 16.80±0.01[a] | 8.8±0.1[b] | 3.0±0.1[d] | 54.0±0.01[b] |

[1] Media ± Standard Deviation
[2] medias with different word(s) in the vertical they differ significantly at the level of 5%.

Table 1. Variation of Brix (degrees Brix), acidity titled, pH and ascorbic acid of the kiwi nectar irradiated to 0.5; 1.0; 2.0kGy and, in the control

Statistical differences were verified among the treatments, indicating that the irradiation influenced this parameter, confirmed in a study accomplished by Spoto and Veruna-Bernardi (2002) in orange juice irradiated in several doses an increase was observed in the tenor of soluble solids.

In the research they found values above the literature for fruit the kiwi nectar needs sucrose addition to reach to 15 °Brix (ideal to nectar) what allowed an ideal palate for the consumption in terms of sugar percentage

### 3.2 Titled acidity

The obtained variations of the titled acidity (% of citric acid) of the kiwi nectar irradiated at doses of: 0 (control); 0.5; 1.0; 2.0 kGy, are in Table 1.

Is possible to observed through of the Table 1, that the dose of (0.5 kGy) which received the smallest radiation presented an increase in the acidity tenor indicating that the treatment influenced for the increase following by the control that did not differ statistically of the doses of 2.0 kGy and 1.0 kGy presented a smaller acidity among the analyzed samples.

The tenors of total acids vary from 1.0-1.5% being the citric acid the principal. In this work were found values above the one of the literature for kiwi.

Study accomplished by Leite; Gêa; Arthur; 2006, showed tenors of total acids for the sample non- irradiated 16.04%, and for the sample irradiated 16.58 and 15.85% for the doses of 1.0 and 2.0 kGy, respectively.

These values differ to the values found in this work; what could be explained by the degree of fruit maturation.

### 3.3 pH

The obtained variations of the pH of the kiwi nectar irradiated at doses of: 0 (control); 0.5; 1.0; 2.0 kGy are in Table 1.

Is possible to observe that with the irradiation there was a decrease in the value of the pH, what resulted in more acid nectar, as increased with the dose of gamma radiation.

In agreement with Matsumoto et al. (1983), the pH value for the kiwi nectar should be around of 3.3. However, the sample that more approached this research was submitted to the irradiation with dose of 0.5 kGy.

Study accomplished by Leite; Gêa; Arthur; 2006 showed values of pH 3.38 for the samples irradiated in the doses of 1.0 and 2.0 kGy, and 3.41 for the sample no- irradiated. These values are same to the values found for in this research.

### 3.4 Total ascorbic acid

The results of total tenor of ascorbic acid (nectar mg/100mL) of the kiwi nectar irradiated at doses of: 0 (control); 0.5; 1.0; 2.0 kGy are in Table 1.

With relation the total tenor of ascorbic acid presents (Table 1), is noticed that the irradiation in the doses of 1.0 and 2.0 kGy promoted a reduction of 50% regarding the control and the samples irradiated with dose of 0.5 kGy.

### 3.5 Color analysis

According to Table 2, the treatment does not promote significant alterations in the color parameters.

| Dose | Parameters | | | | |
|---|---|---|---|---|---|
| | L | a* | b* | Croma | Hue-Angle |
| Control | 22.03±0,00[1] [a2] | -2.23±0.00[a] | 5.33±0.00[a] | 5.78 | -1.17 |
| 0.5 | 23.20±0,00 [a] | -2.05±0.00[a] | 4.95±0.00[a] | 5.36 | -1.18 |
| 1.0 | 23.68±0,00 [a] | -2.19±0.00[a] | 5.40±0.00[a] | 4.91 | -1.11 |
| 2.0 | 22.23±0,00 [a] | -1.76±0.00[a] | 4.47±0.00[a] | 4.80 | -1.20 |

[1]Mean ± Standard Deviation
[2]Means with different word(s) in the vertical differ significantly at the level of 5%.

Table 2. Mediun Values of L, a*, b*, Croma e Hue-Angle of kiwi nectar

There was not a lineal decrease in the treatments, but the treatments did not promote statistical differences between the samples. (Lee et al., 2008), in coherence with (Jo et al., 2003), present in their study a decrease of color values in their irradiation treatments for tamarind juice and fresh green tea, respectively. But (Kim et al., 2006) study, Curcuma aromatica extracts got an improvement in color by gamma radiation.

### 3.6 Sensorial analysis

For the sensorial analysis was not found difference in significant statistics among the aspects analyzed in the present study (color, aroma, flavor, texture and overall acceptance) for doses used, according to Table 3.

Those results are similar to (Lee at al., 2008) and they were in coherence with the previous studies of (Song et al., 2007), that did not had significant change observed in the irradiated tamarind juice and fresh vegetable juices, when compared with juices non-irradiated. Thus also maintains the sensory qualities of the kiwi nectar.

| Samples | Parameters | | | | |
|---|---|---|---|---|---|
| | Color | Smell | Flavour | Texture | Acceptance |
| Control | 6.19±1.92[1] [a2] | 6.14±1.84[a] | 6.32±1.79[a] | 6.26±1.82[a] | 6.41±1.98[a] |
| 0.5 | 5.68±1.93[a] | 5.44±1.93[a] | 5.76±1.82[a] | 5.78±1.71[a] | 5.75±1.92[a] |
| 1.0 | 5.48±2.10[a] | 5.88±2.33[a] | 6.29±1.69[a] | 6.22±1.95[a] | 5.95±1.80[a] |
| 2.0 | 5.24±1.92[a] | 5.60±2.18[a] | 5.43±1.56[a] | 5.95±1.95[a] | 5.51±1.78[a] |

[1]Mean ± Standard Deviation
[2]Means with different capital word(s) in the vertical differ significantly at the level of 5%.

Table 3. Sensorial Analisys about color, smell, flavor, texture and acceptance.

All results are according to Harder et al. (2009) that found the same data for these parameters.

## 4. Conclusion

The results show that gamma radiation when used as a process of treatment for kiwi nectar did not promote significant alterations in the physiochemical and sensorial characteristics of the kiwi nectar, with exception for total ascorbic acid in the doses of 1.0 and 2.0 kGy. Then the process of irradiation can used to conserved kiwi nectar, because the gamma radiation has no negative influence in characteristics of the kiwi nectar.

## 5. References

Almeida, J. R. Kiwi - Cultura de Actinídeas: como Produzir, como Vender. Clássica Editora. Lisboa, mar. 1996.

AOAC, Official methods of analysis of AOAC international, AOAC, Gaithersburg, USA. 2005.

Bible, B.B.; Singha, S. Canopy position influences cielab coordinates of peach color. *Hortscience*, 28, p. 992-993, 1997.

Carvalho, A. V., Lima, L. C. O. Qualidade de kiwi minimamente processados e submetidos a tratamentos com ácido ascórbico, ácido cítrico e cloreto de cálcio. *Pesquisa Agropecuária Brasileira*, 37, p. 679-685, 2002.

CEAGESP. Kiwi. In: www.ceagesp.gov.br.

Chang, C.C.; Lin Y.T.; Lu, Y.T.; Liu, Y.S. Kiwi improves bowel function in patients with irritable bowel syndrome with constipation. *Asia Pacific Journal of Clinic Nutrition*, 19, p. 451-457, 2010.

Demczuk Junior, B. Influência de pré-tratamentos químicos nas características físico-químicas e sensoriais do kiwi submetido à desidratação osmótica e armazenado sob refrigeração. 2007. Dissertação (Mestrado em Tecnologia em Alimentos) - Universidade Federal do Paraná, Curitiba.

Embrarad. http://www.embrarad.com.br ., (2007).

Estevez, M.; Cava, R. Lipid and protein oxidation, release of iron from heme molecule and colour deterioration during refrigerated storage of liver pate. *Meat Science*, 68, p. 551-558, 2004.

Ferreira, L. P. Análise sensorial: testes discriminativos e afetivos, SBCTA, Campinas, Brasil,185p., 2000.

Ferri, V. C.; Kersten, E.; Machado, A. A. Efeito do ácido indolbutírico no enraizamento de estacas semilenhosas de kiwi (*Actinidia deliciosa*, A. Chev.) cultivar Hayward. *Revista Brasileira de Agrociência*, v. 2, n. 1, p.63-66, jan.-abr. 1996.

Harder, M. N. C.; Toledo, T. C. F.; Ferreira, A. C. P.; Arthur, V. Determination of changes induced by gamma radiation in nectar of kiwi fruit (Actinidia deliciosa). *Radiation Physics and Chemistry*, 78, p. 579-582, 2009.

Jacobs, M. B. The chemical analysis of foods and food products, Van Nostrand, New York, USA. 356p. 1985.

Jo, C.; Son, J.H.; Lee, H.j.; Byun, M.W. Irradiation application for color renoval and purification of green tea leaves extract. *Radiation Physics Chemistry*, 66, p. 179-184, 2003.

Kim, J.K.; Jo, C.; Hwang, H.J.; Park, H.J.; Kim, Y.J.; Byun, M.W. (2006) Color improvement by irradiation of Curcuma aromatica extract for industrial application. *Radiation Physics Chemistry*, 75, p. 449-452, 2006.

KONICA MINOLTA. Comunicação precisa da cor. Controle de qualidade da percepção à instrumentação. Japan, 2001.

Lee, J.W.; Kim, J.K.; Periasamy, S.; Jong-il, C.; Kim, J.H.; Han, S.B.; Kim, D.J.; Byun, M.W. Effect of gamma irradiation on microbial analysis, antioxidant activity, sugar content and color of ready-to-use tamarind juice during storage. *Food Science and Tecnology*, 42, p. 101-105, 2008.

Leite, D.T.S.; Gêa, A.S.; Arthur, V. Efeito de diferentes doses de radiação nas características físico-químicas de kiwi minimamente processado. In: http://www.cena.usp.br/ecpg/trabalhos/16.PDF. , 2006.

Levit, V.; Santos, A. da S. dos; Foes, A. D. R., Preservação de pêssegos por irradiação. In: REUNIÃO ANUAL DA SOCIEDADE BRASILEIRA DE QUÍMICA, 31., 2008, Águas de Lindóia, *Anais eletrônicos da Sociedade Brasileira de Química*. Águas de Lindóia, 2008. Disponível em: <http://www.sbq.org.br/ranteriores/23/resumos/1174-2/index.html>. Acesso em: 18 abr. 2011.

Lopes, M.S; Lopes, N.E.C.; Gomes, E.R.S; Pereira, N.C. . Análise de minerais no suco de acerola ultrafiltrado e concentrado por osmose inversa. *IV Congresso Brasileiro de Engenharia Química e Iniciação Científica*. 6 p., 2005.

Matsumoto, S. T., Obara, B., Luh, S. Changes in chemical constituents of kiwifruit during postharvest ripening., *Journal of Food Science*, 48, p. 607-611, 1983.

Modolo, D. M.; Silva, L. C. A. S.; Arthur, V. Irradiação de frutos de kiwi (*Actinidia deliciosa*). *Simpósio Internacional de Iniciação Cientifica da Universidade de São Paulo*. Piracicaba, Resumo, pg. 189., 2011.

Oliveira, A. C. S. Avaliação dos efeitos da radiação gama nas características físico-químicas de kiwi (Actinídia deliciosa, A. Chev. ) cv. Hayward minimamente processado. 2011. Dissertação (Mestrado em Tecnologia Aplicada) - Instituto de Pesquisas Energéticas e Nucleares - IPEN, São Paulo.

Oliveira, A. C. S.; Silva, L. C. A.; Oliveira, M.; Modolo, D.; Arthur, V. Evaluation of gamma radiation on minimally processed kiwifruit. *International Meeting on Radiation Processing*. Montreal, Abstract, p.189, 2011.

Ornellas, C. B. D.; Gonçalves, M. P. J.; Silva, P. R., Martins, R. T. Atitude do consumidor frente à irradiação de alimentos. *Ciência e Tecnologia de Alimentos*, v.26, n.1, p. 211-213, jan.-mar., 2006.

Schuck, E., (1992). Cultivares de quiwi., *Agropecuária Catarinense*, 5, p. 9-12. Ano

SEMAPI. Sistema de produção de kiwi tutorado. In: http://sustentabilidadesemapi.blogspot.com/2009_01_01_archive.html.

Song, H.P.; Byun, M.W.; Jô, C.; Lee, C.H.; Kim, K.S.; Kim, D.H. (2007) Effects of gamma irradiation on the microbiological, nutritional, and sensory properties of fresh vegetable juice. *Food Control*, 18, p. 5-10. Ano

Souza, P. V. D.; Marodin, G. A. B.; Barradas, C. I. N. Cultura do quivi. Porto Alegre, RS : Cinco Continentes, 1996.

Spoto, M. H. F., Verruma-Benardi, M. R.. Estudo microbiológico e físico-químico do suco de laranja fresco irradiado., *Higiene Alimentar*, 16, p. 76-80, 2002

Statistical Analysis System Institute,). Sas/Qc. Software: usage and reference, Statpoint, Inc., Herndon, USA, 1996.

Tocchini, R. P.). Industrialização de polpas, sucos e néctar de frutas, *Ital*, Campinas, Brasil, 167p., 1995.

USDA. Kiwi world production and exports. In: www.ers.usda.gov/data/fruitvegphyto/Data/fr-kiwi.xls.

Villavicencio, A. L. C. H., Manzini-Filho, J., Delincée, H. Application of different techniques to identify the effect of irradiation on Brazilian beans after six months storage., *Radiation Physics and Chemistry*, 52, p. 161-166, 1998.

Wiendl, F. M. Irradiação de alimentos. *Biológico*, v 59, n.1, p. 75-76, 1997.